Understanding Battlefield Coalitions

This book improves our understanding of battlefield coalitions, providing novel theoretical and empirical insight into their nature and capabilities, as well as the military and political consequences of their combat operations.

The volume provides the first dataset of battlefield coalitions, uses primary sources to understand how non-state actors of varying types form such groupings, reports interviews with policymakers illuminating North Atlantic Treaty Organization operations, and uses cases studies of various wars waged throughout the nineteenth, twentieth, and twenty-first centuries to understand how other such collectives have operated. Part I introduces battlefield coalitions as an object of study, demonstrating how they are distinct from other wartime collectives. Using a novel dataset of actors fighting in 492 battles during interstate wars waged between 1900 and 2003, it provides, for the first time, a comprehensive portrait of the universe of battlefield coalitions. Part II explores processes and dynamics involved in the formation of battlefield coalitions, addressing how potential coalition members prepare for future battles in peacetime (as well as the consequences of such preparations) and the dynamics of mission design. Part III focuses on how battlefield coalitions are organised and fight when combat ensues, notably their decision-making rules and practices, command structures, and learning capacities. Part IV addresses three curious tendencies observed in the operations of battlefield coalitions: partners under-providing effort in combat, rebels and terrorist networks persisting in cooperation even when their interests diverge, and members defecting from the collective. Part V concludes with a chapter outlining for future researchers what we know about battlefield coalitions and what remains to be understood.

This book will be of much interest to students of military and strategic studies, defence studies and International Relations.

Rosella Cappella Zielinski is an Associate Professor of Political Science, Boston University, USA. She is the author of *How States Pay for Wars* (2016).

Ryan Grauer is an Associate Professor of International Affairs, Graduate School of Public and International Affairs, University of Pittsburgh, USA. He is the author of *Commanding Military Power* (2016).

Cass Military Studies

Counterinsurgency Warfare and Brutalisation
The Second Russian-Chechen War
Roberto Colombo and Emil Aslan Souleimanov

Managing Security
Concepts and Challenges
Edited by Laura R. Cleary and Roger Darby

Understanding the Impact of Social Research on the Military
Reflections and Critiques
Edited by Eyal Ben-Ari, Helena Carreiras, and Celso Castro

Civil-Military Cooperation in International Interventions
The Role of Soldiers
Agata Mazurkiewicz

Contemporary Military Reserves
Between the Civilian and Military Worlds
Edited by Eyal Ben-Ari and Vincent Connelly

Military Strategies of the New European Allies
A Comparative Study
Håkan Edström and Jacob Westberg

Proxy War in Yemen
Bernd Kaussler and Keith A. Grant

Understanding Battlefield Coalitions
Edited by Rosella Cappella Zielinski and Ryan Grauer

For more information about this series, please visit: https://www.routledge.com/Cass-Military-Studies/book-series/CMS

Understanding Battlefield Coalitions

Edited by Rosella Cappella Zielinski
and Ryan Grauer

R Routledge
Taylor & Francis Group

LONDON AND NEW YORK

First published 2024
by Routledge
4 Park Square, Milton Park, Abingdon, Oxon OX14 4RN

and by Routledge
605 Third Avenue, New York, NY 10158

Routledge is an imprint of the Taylor & Francis Group, an informa business

© 2024 selection and editorial matter, Rosella Cappella Zielinski and Ryan Grauer; individual chapters, the contributors

The right of Rosella Cappella Zielinski and Ryan Grauer to be identified as the authors of the editorial material, and of the authors for their individual chapters, has been asserted in accordance with sections 77 and 78 of the Copyright, Designs and Patents Act 1988.

British Library Cataloguing-in-Publication Data
A catalogue record for this book is available from the British Library

ISBN: 978-1-032-50837-5 (hbk)
ISBN: 978-1-032-50838-2 (pbk)
ISBN: 978-1-003-39989-6 (ebk)

DOI: 10.4324/9781003399896

Typeset in Sabon
by MPS Limited, Dehradun

For John, Lewis, Lauren, and Nigel

Contents

Data Availability Statement

The data that support the findings of Chapter 2, 'A Century of Coalitions in Battle: Incidence, Composition, and Performance, 1900–2003,' are openly available at the Harvard Dataverse at https://doi.org/10.7910/DVN/YAKVTB.

The data that support the findings of Chapter 10, 'Speaking with One Voice: Coalitions and Wartime Diplomacy,' are openly available at the Harvard Dataverse at https://doi.org/10.7910/DVN/KLQFAP.

Acknowledgements

The following chapters were previously published in a *Journal of Strategic Studies* special issue, entitled *Battlefield Coalitions* (volume 45, number 2). When citing this material, please use the original publication information, as noted here:

Chapter 1: 'Understanding Battlefield Coalitions'
Rosella Cappella Zielinski and Ryan Grauer
Journal of Strategic Studies, volume 45, number 2 (2022): pp. 177–85
https://doi.org/10.1080/01402390.2021.2011231

Chapter 2: 'A Century of Coalitions in Battle: Incidence, Composition, and Performance, 1900–2003'
Rosella Cappella Zielinski and Ryan Grauer
Journal of Strategic Studies, volume 45, number 2 (2022): pp. 186–210
https://doi.org/10.1080/01402390.2021.2011233

Chapter 4: 'When the Coalition Determines the Mission: NATO's Detour in Libya'
Stéfanie von Hlatky and Thomas Juneau
Journal of Strategic Studies, volume 45, number 2 (2022): pp. 258–79
https://doi.org/10.1080/01402390.2021.2011234

Chapter 6: 'Command and Military Effectiveness in Rebel and Hybrid Battlefield Coalitions'
Dan Reiter
Journal of Strategic Studies, volume 45, number 2 (2022): pp. 211–33
https://doi.org/10.1080/01402390.2021.2002692

Chapter 7: 'Learning from Losing: How Defeat Shapes Coalition Dynamics in Wartime'
Sara Bjerg Moller
Journal of Strategic Studies, volume 45, number 2 (2022): pp. 280–302
https://doi.org/10.1080/01402390.2022.2030716

Chapter 9: 'Why Rebels Rely on Terrorists: The Persistence of the Taliban-al-Qaeda Battlefield Coalition in Afghanistan'
Barbara Elias
Journal of Strategic Studies, volume 45, number 2 (2022): pp. 234–57
https://doi.org/10.1080/01402390.2021.2002691

Chapter 10: 'Speaking with One Voice: Coalitions and Wartime Diplomacy'
Eric Min
Journal of Strategic Studies, volume 45, number 2 (2022): pp. 303–27
https://doi.org/10.1080/01402390.2021.2011232

These essays are reprinted with the permission of the publisher, Taylor & Francis, Ltd. For permission information, please visit http://www.tandfonline.com.

Contributors

Kathryn M.G. Boehlefeld is an Assistant Professor at Air University's Air Command and Staff College and is a faculty member for the School of Advanced Nuclear Deterrence Studies (SANDS), USA. Her research centres on national security policy, civil-military relations, and nuclear deterrence. She holds a PhD in Political Science from the University of Notre Dame.

Rosella Cappella Zielinski is an Associate Professor of Political Science at Boston University and non-resident fellow at the Brute Krulak Center for Innovation and Creativity at Marine Corps University, USA. She studies the political economy of security and is the author of *How States Pay for Wars* (Cornell University Press, 2016), among other pieces.

Barbara Elias is the Sarah and James Bowdoin Associate Professor of Government at Bowdoin College, USA. Her research focuses on international relations, insurgency warfare, national security, U.S. foreign policy and Islam and politics. She received her PhD from the University of Pennsylvania in political science, and was the Director of the Afghanistan/Pakistan/Taliban Documentation Project at The National Security Archive in Washington D.C.

Ryan Grauer is an Associate Professor of International Affairs in the Graduate School of Public and International Affairs at the University of Pittsburgh, USA. His research investigates the sources and use of military power in the international system, and he is the author *Commanding Military Power* (Cambridge University Press, 2016), among other pieces.

Kelly A. Grieco is a Senior Fellow with the Reimagining US Grand Strategy Program at the Stimson Center and an Adjunct Associate Professor in the Center for Security Studies at Georgetown University, USA. Her research addresses US foreign policy, military alliances, airpower, and the future of war. She holds a PhD in Political Science from the Massachusetts Institute of Technology.

Stéfanie von Hlatky is the Canada Research Chair in Gender, Security, and the Armed Force and Associate Dean (research) in the Faculty of Arts and Science at Queen's University, Canada. Her research focuses on NATO, military cooperation and the Women, Peace and Security agenda. She recently published the book *Deploying Feminism: The Role of Gender in NATO Military Operations* with Oxford University Press.

Thomas Juneau is an Associate Professor at the University of Ottawa's Graduate School of Public and International Affairs, Canada. His research focuses on the Middle East, the role of intelligence in national security and foreign policy-making, and Canadian foreign and defence policy. From 2003 until 2014, he worked as a policy analyst with Canada's Department of National Defence.

Casey Mahoney is a PhD Candidate in the Department of Political Science at the University of Pennsylvania, USA. He is completing a predoctoral fellowship at the Clements Center for National Security at the University of Texas at Austin and is an adjunct researcher at the RAND Corporation.

Eric Min is an Assistant Professor of Political Science at the University of California, Los Angeles, USA. He received his PhD in Political Science from Stanford University. His research focuses on the application of machine learning, text, and statistical methods to the analysis of interstate war, diplomacy, and conflict management. His research has been published in *American Political Science Review*, *International Organization*, and *Journal of Peace Research*.

Sara Bjerg Moller is an Associate Teaching Professor in the Security Studies Program at Georgetown University, USA. Her current research agenda is centred on organisational adaptation in peacetime alliances and wartime coalitions. She has published articles in *The Journal of Strategic Studies, Asian Security, International Politics, The Washington Quarterly,* and elsewhere.

Dan Reiter is Samuel Candler Dobbs Professor of Political Science at Emory University, USA. He is the awarded-winning author, coauthor, or editor of dozens and articles and books on international relations, including *Democracies at War* (with Allan C. Stam; Princeton, 2002), *How Wars End* (Princeton, 2009), and *The Sword's Other Edge: Tradeoffs in the Pursuit of Military Effectiveness* (Cambridge, 2017).

Alastair Smith is the Bernhardt Denmark Professor of International Relations in the Wilf Family Department of Politics at New York University, USA. He uses advanced game theoretic and statistical techniques to research the interface between international relations and comparative politics, including how leaders best survive in office under different institutional contexts. He has published five books and approximately 50 articles with leading presses and journals.

Alex Weisiger is an Associate Professor of Political Science at the University of Pennsylvania, USA. His work examines topics such as war termination, reputation in international politics, and the democratic peace. His book, *Logics of War: Explanations for Limited and Unlimited Conflicts*, was published by Cornell University Press in 2013.

Part I

Introducing Battlefield Coalitions

1 Battlefield Coalitions

Preparation, Organisation, Execution

Rosella Cappella Zielinski and Ryan Grauer

War is an inherently collective activity. At minimum, two belligerents bring together large numbers of soldiers and officers to fight in coordination against a similar collection of individuals as the sides violently contest some political objective. Often, there is an additional layer of collective effort as, on one or both sides of a battle, multiple belligerents band together to combat their common foe. Collectives featuring forces of multiple belligerents fighting side-by-side in combat recur throughout history. Alexander confronted, and defeated, many different battlefield coalitions on his march through the Balkans, Asia Minor, Persia, and India between 336 and 323 BCE. Various Gallic tribes often combined in their attempts to stem Julius Caesar's drive into their territory between 58 and 50 BCE. During the First Crusade, waged 1096–1099, loose confederations of forces totalling approximately 100,000 men and headed by various European nobles marched on the Holy Land and eventually took Jerusalem. During the Imjin War, fought 1592–1598, a coalition of Chinese and Korean forces repulsed repeated Japanese attempts to invade and conquer territory on the Asian mainland. Between 1600 and 1783, myriad Native American tribes of the Iroquois nations formed coalitions to fight combat actions with and against European explorers and settlers. Both world wars featured coalitions of belligerent states on both sides of many battlefields. More recently, the United States brought together coalitions of forces to fight combat actions in the Balkans, Iraq, and Afghanistan, often against other coalitions comprised of both state and non-state actors.

There is little reason to think that these battlefield coalitions – amalgams of officers, troops, and materiel brought together by multiple distinct actors for the purpose of jointly waging combat in the same operational battlespace – are likely to disappear from the international arena anytime soon. In the wake of Russia's intensification of its war against Ukraine in 2022, states around the world increased their efforts to prepare for a potential conflict of the same sort, and many of them bolstered their plans to fight in battle alongside others. In Europe, for example, Sweden and Finland applied to join the North Atlantic Treaty Organisation (NATO) in order to facilitate their abilities to deploy their forces alongside those of other members of the alliance in the

DOI: 10.4324/9781003399896-2

event of a Russian strike west. In the Indo-Pacific, the United States military continued its efforts to train with and prepare for conflict in the Taiwan Strait alongside forces of partners in the region. China, for its part, sent forces to participate in Russia's quadrennial multilateral Vostok military exercises in September 2022.

Though battlefield coalitions are common phenomena in the historical record and actors around the world today continue to plan to fight as part of them, they have been largely overlooked in the study of collective warfighting. Instead, scholars have focused on the creation, organisation, operation, and termination of alliances and wartime coalitions. By the terms of one prominent formulation, alliances are 'written agreements, signed by official representatives of at least two independent states, that include promises to aid a partner in the event of military conflict, to remain neutral in the event of conflict, to refrain from military conflict with one another, or to consult/cooperate in the event of international crises that create a potential for military conflict.'[1] Wartime coalitions, per common usage, are 'group[s] of states that coordinate military activity during a war, regardless of the nature of the pre-war relationship.'[2]

Battlefield coalitions are distinct from these well-studied warfighting collectives in two important ways. First, they may be formed by a wide variety of actors. Battlefield coalitions may be created from the militaries of states that are formal allies, the militaries of states that have no written agreement to cooperate during war, or both. Crucially, they may also include – or be solely comprised of – forces fielded by non-state actors. Second, in contrast to broader collectives united around political or military agreements to wage war against a common foe, battlefield coalitions are the groups that actually apply combined force in combat. That is, they operate at the operational and tactical levels of warfighting. Battlefield coalitions thus occupy a distinct and vital place in the causal chain connecting the application of military means with political ends. Understanding more about them – their creation, composition, organisation, performance, and associated military and political consequences – will illuminate important relationships between military force, political objectives, and patterns of outcomes in the international arena.

Existing scholarship that sheds light directly on battlefield coalitions has overwhelming focused on combined forces drawn from groups of states that are part of formal alliances like NATO,[3] led by the United States,[4] or fighting during the world wars and Korean War.[5] While these studies provide important insights, battlefield coalitions more often than not are ad hoc, drawing together forces from states that do not have prior agreements to fight alongside one another in war; do not involve the United States; and fight in conflicts other than the world wars and Korean War.[6] It is unclear whether and to what degree existing understandings of battlefield coalitions can be generalised to the broader universe of such collectives.

An additional consequence of the focus in existing scholarship is the under-examination of non-state actors as part of such groups. There is a small literature on battlefield coalitions that include forces fielded by non-state actors, but it tends to focus on the formation and disintegration of such groups rather than how they operate in combat.[7] A substantial minority of battlefield coalitions fighting in the twentieth century featured a wide array of non-state actors, ranging from hyperlocal partisans like the Boxers in China to transnational millenarian terrorists like al-Qaeda, fighting both alongside states and in combination with other non-state actors; under-standing how such groups function in combat requires further investigation.[8]

The universe of battlefield coalitions is far more vast, and the questions to be answered about them are much more numerous, than is reflected in existing scholarship. Especially because such groupings are increasingly common in conflicts around the world, a renewed, more focused effort at understanding them and their role in the international arena is needed.[9]

1.1 Chapters in This Volume

The chapters in this volume begin the necessary work of invigorating schol-arship on battlefield coalitions, providing insight into such groups' creation, composition, organisation, and performance as well as the military and political consequences of their operations in combat. They do so by drawing on ideas, theories, and findings from various literatures in international rela-tions, strategic studies, and military operations – including, but not limited to, the literatures on formal alliances, organisational theory, bargaining, and military adaption – to develop new frameworks for analysis. They illustrate the value and potential of a scholarly focus on battlefield coalitions with rich empirical details, generated from newly collected data, robust case studies, close readings of primary sources, and participant interviews.

1.1.1 A Portrait of Battlefield Coalitions

To begin to understand battlefield coalitions, it is necessary to first know something about them as a phenomenon in the international arena. Accordingly, the volume begins by introducing such collectives with what is to date the only systematic overview of their presence, composition, and performance in interstate wars. Using newly collected data, Chapter 2, 'A Century of Coalitions in Battle: Incidence, Composition, and Performance, 1900–2003,' reports that battlefield coalitions are common and powerful actors in interstate wars, and such groups are becoming more common as well as more likely to include non-state actors over time. Crucially, however, there are important variations across battlefield coalitions in terms of their com-position and performance.

With this newfound understanding of what the universe of battlefield coalitions looks like, the other chapters in this volume investigate how such groups prepare to fight together, organise to conduct combined operations, and execute their collective efforts. The final chapter, a critical read for scholars, considers the paths forward in studying battlefield coalitions.

1.1.2 Preparation

Part II, 'Preparation,' addresses battlefield coalitions prior to combat, taking up the issues of who fights in a battlefield coalition, how potential members prepare to fight alongside each other and the ramifications of that preparation, and how members design their missions. In Chapter 3, 'Exercising Escalation: Do Multinational Military Exercises Provoke Interstate Security Crises?,' Kathryn M.G. Boehlefeld and Kelly A. Grieco consider the intended and unintended consequences of preparing to fight together. Future battlefield coalition members often engage in multinational military exercises (MMEs) to both increase effectiveness on the battlefield and deter common adversaries from engaging in unwanted behaviour. Yet, such exercises may provoke adversaries or carry increased risk of escalation. This chapter offers a framework for classifying MMEs along two dimensions to understand, assess, and forecast their impact on strategic stability.

In Chapter 4, 'When the Coalition Determines the Mission: NATO's Detour in Libya,' Stéfanie von Hlatky and Thomas Juneau explore how coalitions, once formed, shape the design of the mission. They argue that, in the face of unclear mandates, a variety of domestic and international political factors drive coalitions' definition of their combined mission. Complex considerations of partners' relative power, freedom of action, and role within the group all shape intra-coalitional negotiations and, ultimately, the content of the mission they pursue together in combat.

1.1.3 Organisation

The chapters in Part III, 'Organisation,' investigate how battlefield coalitions organise themselves to execute combined operations. Command and control, or the arrangement and exercise of authority, is particularly salient in this regard, with differing structures and patterns sometimes facilitating and sometimes inhibiting collective action in combat. In Chapter 5, 'Battlefield Coalitions as International Institutions: A Conceptual Framework,' Casey Mahoney argues that battlefield coalitions should be understood as bundles of international institutions that govern the decision-making rules of the collective. As such, assessing battlefield coalitions along the lines that other international institutions are studied – including in terms of formalisation, membership, scope, control, centralisation, and flexibility – promises to shed important light on the variety, performance, and achievements of such groups.

In Chapter 6, 'Command and Military Effectiveness in Rebel and Hybrid Battlefield Coalitions,' Dan Reiter considers the way in which the inclusion of non-state actors in battlefield coalitions alters the incentives for and dynamics of collective action in combat. In particular, he argues that state and non-state actor incentives to invest in contextually appropriate command structures – the authority relations that condition groups' chances of success in combat – diverge along predictable dimensions and have implications for the effectiveness of such coalitions.

In Chapter 7, 'Learning from Losing: How Defeat Shapes Coalition Dynamics in Wartime,' Sara Bjerg Moller takes up the question of how battlefield coalitions evolve over time and argues that battlefield performance is likely to drive organisational change. In particular, she contends that, when combined losses begin to mount, battlefield coalitions are likely to sacrifice their autonomy in combat for additional coordination capabilities and, ideally, improved chances for victory.

1.1.4 Execution

Part IV, 'Execution,' considers the performance and achievements of battlefield coalitions as they perform their intended function: fighting. In Chapter 8, 'Regime Type, War Aims, and Coalition Member Effort in Combat,' Rosella Cappella Zielinski, Ryan Grauer, and Alastair Smith investigate how the composition of a battlefield coalition shapes individual member forces' effort. They note that under provision of effort by battlefield coalitions members, or the phenomena of belligerents fighting less hard when in partnership with others than they would on their own, is common and, given the stakes of combat, puzzling. Using selectorate theory insights, they argue that different battlefield coalition members under-provide effort for different reasons: democracies work to free-ride on the efforts of others as they pursue common goods while non-democracies shirk and pursue private goods when fighting as part of a collective.

In Chapter 9, 'Why Rebels Rely on Terrorists: The Persistence of the Taliban-al-Qaeda Battlefield Coalition in Afghanistan,' Barbara Elias investigates the performance and persistence of a particular type of battlefield coalition: one comprised of rebels and terrorists. She contends that, when they are not rivals for authority, the differing objectives of the partners incentivise them to continue fighting together, battle after battle, even in the face of significant pressure from external actors to desist. Institutionalisation of cooperation ensures that the fighting itself continues to serve the interests of the battlefield coalition members.

In Chapter 10, 'Coalitions and Wartime Diplomacy: Speaking with One Voice,' Eric Min considers the conditions under which battlefield coalition partners are likely to defect from the collective effort to combat a common foe and individually initiate negotiations that could end their own participation in the war, or perhaps the conflict more generally. In this, he contends, the relative balance of power within a battlefield coalition and the group's performance on the battlefield interact, with both victories and defeats motivating loyalty and defection under different circumstances.

1.1.5 Looking Ahead

In Part V, 'Looking Ahead,' the final chapter in this volume, 'Next Steps in the Study of Battlefield Coalitions,' provides a path for future researchers, albeit one paved with challenges to be overcome. Alex Weisiger argues that, though the treatment of battlefield coalitions presented in this volume covers a vast array of both collective types and activities, there are still more varieties of such groups to consider as well as a pressing need for clarity on how generalisable the theories advanced in these chapters may be. Building on this foundation, he contends, future scholars would do well to consider the role that factors like geography, technology, and logistics play in the formation, organisation, operation, and performance of battlefield coalitions. It is only through considering such additional factors that we might arrive as a more general theory of battlefield coalitions.

1.2   Battlefield Coalitions in International Relations

The chapters in this volume thus offer important new insights into frequency, composition, performance, and political achievements of battlefield coalitions. Their arguments and findings have important implications for both scholars and policymakers. For the former, they represent new views into the causal chain connecting the use of force to patterns of outcomes in the international arena as well as the nature and challenges of collective action in perhaps the most inhospitable environment imaginable. For the latter, they point towards the need to think more carefully about how domestic and international factors may shape the potential for collaboration in combat as well as the need to take seriously the challenges and best practices of working well with non-state actors. These chapters, however, represent only the beginning of the work that remains to be done.

Even a short list of questions about battlefield coalitions that still require investigation is daunting. For example, sovereignty concerns loom large in creating and organising battlefield coalitions – are certain types of states more or less likely to relinquish some sovereign rights than others? Does the identity of coalition partners matter for such decisions? Do the same factors driving states' actions with regard to their sovereignty drive non-state actors and their concerns about autonomy in the context of battlefield coalition operations? Once battlefield coalitions are created and organised, how can common barriers to collective action like shirking and free-riding be minimised and exploitation of the partners' comparative advantages be maximised? How well do such barriers and advantages need to be managed to facilitate effective combined performance in combat? What role do time and experience play in conditioning how battlefield coalition partners work together? Thinking more broadly, are actors that come together to cooperate as battlefield coalition partners likely to build on that experience and cooperate in other realms? Or perhaps do such cooperative spill-over effects depend on performance?

The research agenda on battlefield coalitions and their impact in international relations is thus quite robust; increasing our collective understanding of these groups and their impact on broader military and political dynamics will require the efforts of many scholars. We regard this volume as the start of a vigorous conversation, and hope that it inspires others to take up the many questions that remain to be answered.

Notes

1 Brett Ashely Leeds et al., 'Alliance Treaty Obligations and Provisions, 1815–1944,' *International Interactions* 28, no. 3 (2002): 238, https://doi.org/10.1080/03050620213653. For recent reviews of the copious literature on alliances, see Patricia A. Weitsman, 'Alliances and War,' in *Oxford Research Encyclopedia of International Studies* (Oxford: Oxford University Press, 2017); Sten Rynning and Olivier Schmitt, 'Alliances,' in *The Oxford Handbook of International Security*, ed. Alexandra Gheciu and William C. Wohlforth (Oxford: Oxford University Press, 2018).

2 Daniel S. Morey, 'Military Coalitions and the Outcome of Interstate Wars,' *Foreign Policy Analysis* 12, no. 4 (October 2016): 535, https://doi.org/10.1111/fpa.12083. On wartime coalitions, see, for example, Patricia Weitsman, 'Wartime Alliances versus Coalition Warfare: How Institutional Structure Matters in the Multilateral Prosecution of Wars,' *Strategic Studies Quarterly* 4, no. 2 (Summer 2010): 113–36, https://www.jstor.org/stable/26269800; Scott Wolford, *The Politics of Military Coalitions* (New York: Cambridge University Press, 2015); Patrick A. Mello and Stephen M. Saideman, 'The Politics of Multinational Military Operations,' *Contemporary Security Policy* 40, no. 1 (January 2019): 30–7, https://doi.org/10.1080/13523260.2018.1522737; Marina E. Henke, *Constructing Allied Cooperation: Diplomacy, Payments, and Power in Multilateral Military Coalitions* (Ithaca, NY: Cornell University Press, 2019).

3 See, for example, Ivo H. Daalder and Michael E. O'Hanlon, *Winning Ugly: NATO's War to Save Kosovo* (Washington, DC: Brookings Institution Press, 2004); Patricia Weitsman, *Waging War: Alliances, Coalitions, and Institutions of Interstate Violence* (Stanford, CA: Stanford University Press, 2013); David P. Auerswald and Stephen M. Saideman, *NATO in Afghanistan: Fighting Together, Fighting Alone* (Princeton: Princeton University Press, 2014); Olivier Schmitt, 'International Organization at War: NATO Practices in the Afghan Campaign,' *Cooperation and Conflict* 52, no. 4 (December 2017): 502–18, https://doi.org/10.1177/0010836717701969.

4 See, for example, Weitsman, 'Wartime Alliances versus Coalition Warfare'; Sarah E. Kreps, *Coalitions of Convenience: United States Military Interventions after the Cold War* (New York: Oxford University Press, 2011); Weitsman, *Waging War*; Olivier Schmitt, *Allies That Count: Junior Partners in Coalition Warfare* (Washington, DC: Georgetown University Press, 2018).

5 See, for example, Patricia Weitsman, *Dangerous Alliances: Proponents of Peace, Weapons of War* (Stanford, CA: Stanford University Press, 2003); Kelly Ann Grieco, 'War by Coalition: The Effects of Coalition Military Institutionalization on Coalition Battlefield Effectiveness' (PhD thesis, Cambridge, MA, Massachusetts Institute of Technology, 2016); Sara Bjerg Moller, 'Fighting Friends: Institutional Cooperation and Military Effectiveness in Multinational War' (PhD thesis, New York, Columbia University, 2016); Alex Weisiger, 'Exiting the Coalition: When Do States Abandon Coalition Partners during War?,' *International Studies Quarterly* 60, no. 4 (December 2016): 753–65, https://doi.org/10.1093/isq/sqw029; Rosella Cappella Zielinski and Ryan Grauer, 'Organizing for Performance: Coalition Effectiveness on the Battlefield,' *European Journal of International Relations* 26, no. 4 (December 2020): 953–78, https://doi.org/10.1177/1354066120903369.

6 Rosella Cappella Zielinski and Ryan Grauer, 'A Century of Coalitions in Battle: Incidence, Composition, and Performance, 1900–2003,' Chapter 2, this volume.
7 See, for example, Fotini Christia, *Alliance Formation in Civil Wars* (New York: Cambridge University Press, 2012); Seden Akcinaroglu, 'Rebel Interdependencies and Civil War Outcomes,' *Journal of Conflict Resolution* 56, no. 5 (October 2012): 879–903, https://doi.org/10.1177/0022002712445741; Michael Woldemariam, 'Battlefield Outcomes and Rebel Cohesion: Lessons From the Eritrean Independence War,' *Terrorism and Political Violence* 28, no. 1 (January 2016): 135–56, https://doi.org/10.1080/09546553.2014.886575; Peter Rudloff and Michael G Findley, 'The Downstream Effects of Combatant Fragmentation on Civil War Recurrence,' *Journal of Peace Research* 53, no. 1 (January 2016): 19–32, https://doi.org/10.1177/0022343315617067; Emily Kalah Gade, Mohammed M Hafez, and Michael Gabbay, 'Fratricide in Rebel Movements: A Network Analysis of Syrian Militant Infighting,' *Journal of Peace Research* 56, no. 3 (May 2019): 321–35, https://doi.org/10.1177/0022343318806940.
8 Cappella Zielinski and Grauer, 'A Century of Coalitions in Battle,' Chapter 2, this volume.
9 Cappella Zielinski and Grauer, 'A Century of Coalitions in Battle,' Chapter 2, this volume.

Bibliography

Akcinaroglu, Seden. 'Rebel Interdependencies and Civil War Outcomes.' *Journal of Conflict Resolution* 56, no. 5 (October 2012): 879–903. 10.1177/0022002 712445741.

Auerswald, David P., and Stephen M. Saideman. *NATO in Afghanistan: Fighting Together, Fighting Alone*. Princeton: Princeton University Press, 2014.

Cappella Zielinski, Rosella, and Ryan Grauer. 'Organizing for Performance: Coalition Effectiveness on the Battlefield.' *European Journal of International Relations* 26, no. 4 (December 2020): 953–78. 10.1177/1354066120903369.

Christia, Fotini. *Alliance Formation in Civil Wars*. New York: Cambridge University Press, 2012.

Daalder, Ivo H., and Michael E. O'Hanlon. *Winning Ugly: NATO's War to Save Kosovo*. Washington, DC: Brookings Institution Press, 2004.

Gade, Emily Kalah, Mohammed M. Hafez, and Michael Gabbay. 'Fratricide in Rebel Movements: A Network Analysis of Syrian Militant Infighting.' *Journal of Peace Research* 56, no. 3 (May 2019): 321–35. 10.1177/0022343318806940.

Grieco, Kelly Ann. 'War by Coalition: The Effects of Coalition Military Institutionalization on Coalition Battlefield Effectiveness.' PhD thesis, Massachusetts Institute of Technology, 2016.

Henke, Marina E. *Constructing Allied Cooperation: Diplomacy, Payments, and Power in Multilateral Military Coalitions*. Ithaca, NY: Cornell University Press, 2019.

Kreps, Sarah E. *Coalitions of Convenience: United States Military Interventions after the Cold War*. New York: Oxford University Press, 2011.

Leeds, Brett Ashely, Jeffrey Ritter, Sara Mitchell, and Andrew Long. 'Alliance Treaty Obligations and Provisions, 1815–1944.' *International Interactions* 28, no. 3 (2002): 237–60. 10.1080/03050620213653.

Mello, Patrick A., and Stephen M. Saideman. 'The Politics of Multinational Military Operations.' *Contemporary Security Policy* 40, no. 1 (January 2019): 30–7. 10.1080/13523260.2018.1522737.

Moller, Sara Bjerg. 'Fighting Friends: Institutional Cooperation and Military Effectiveness in Multinational War.' PhD thesis, Columbia University, 2016.

Morey, Daniel S. 'Military Coalitions and the Outcome of Interstate Wars.' *Foreign Policy Analysis* 12, no. 4 (October 2016): 533–51. 10.1111/fpa.12083.

Rudloff, Peter, and Michael G. Findley. 'The Downstream Effects of Combatant Fragmentation on Civil War Recurrence.' *Journal of Peace Research* 53, no. 1 (January 2016): 19–32. 10.1177/0022343315617067.

Rynning, Sten, and Olivier Schmitt. 'Alliances.' In *The Oxford Handbook of International Security*, edited by Alexandra Gheciu and William C. Wohlforth, 653–67. Oxford: Oxford University Press, 2018.

Schmitt, Olivier. *Allies That Count: Junior Partners in Coalition Warfare*. Washington, DC: Georgetown University Press, 2018.

Schmitt, Olivier. 'International Organization at War: NATO Practices in the Afghan Campaign.' *Cooperation and Conflict* 52, no. 4 (December 1, 2017): 502–18. 10.1177/0010836717701969.

Weisiger, Alex. 'Exiting the Coalition: When Do States Abandon Coalition Partners during War?' *International Studies Quarterly* 60, no. 4 (December 2016): 753–65. 10.1093/isq/sqw029.

Weitsman, Patricia. 'Alliances and War.' In *Oxford Research Encyclopedia of International Studies*. Oxford: Oxford University Press, 2017. 10.1093/acrefore/9780190846626.013.118.

Weitsman, Patricia. *Dangerous Alliances: Proponents of Peace, Weapons of War*. Stanford, CA: Stanford University Press, 2003.

Weitsman, Patricia. *Waging War: Alliances, Coalitions, and Institutions of Interstate Violence*. Stanford, CA: Stanford University Press, 2013.

Weitsman, Patricia. 'Wartime Alliances versus Coalition Warfare: How Institutional Structure Matters in the Multilateral Prosecution of Wars.' *Strategic Studies Quarterly* 4, no. 2 (Summer 2010): 113–36. https://www.jstor.org/stable/26269800.

Woldemariam, Michael. 'Battlefield Outcomes and Rebel Cohesion: Lessons From the Eritrean Independence War.' *Terrorism and Political Violence* 28, no. 1 (January 2016): 135–56. 10.1080/09546553.2014.886575.

Wolford, Scott. *The Politics of Military Coalitions*. New York: Cambridge University Press, 2015.

2 A Century of Coalitions in Battle

Incidence, Composition, and Performance, 1900–2003

Rosella Cappella Zielinski and Ryan Grauer

Battlefield coalitions – amalgams of officers, troops, and materiel brought together by multiple distinct political communities for the purpose of jointly waging combat in the same operational battlespace – abound. From the earliest recorded conflicts in Mesopotamia to the present day, forces fielded by separate political communities have fought alongside one another against common adversaries. Such coalitions have been recorded on all continents, save Antarctica. In the current era, battlefield coalitions are especially common; as we report for the first time in this chapter, since the end of the Cold War, more than half of the belligerent sides fighting major land battles in interstate wars waged around the world were coalitions.

Given their presence across time and space, battlefield coalitions are remarkably diverse in terms of their composition and performance. The belligerents fighting the war in Afghanistan exemplify this diversity. When the United States invaded the country after the 11 September 2001 terror attacks, it began operations against the Taliban regime by forming a battlefield coalition with the Northern Alliance, itself a coalition of various non-state actor militias. The Americans and Northern Alliance often fought against another battlefield coalition: the Taliban, Al-Qaeda, and elements of the Pakistani Taliban. That latter battlefield coalition, which was initially an alignment of state and non-state actors, comprised only non-state actors after the Taliban's fall from power in November 2001.[1] Over time, the United States and its Afghan partners were joined by forces fielded by other states. Some actors joining subsequent battlefield coalitions were members of the North Atlantic Treaty Organization (NATO), which invoked Article 5 of the agreement and committed the organisation to assist in the defence of the United States on 4 October 2001; others were not. Diversity in battlefield coalition performance in Afghanistan is as great as the variability in the composition of such groups; there were complete victories and collapses, as in the early fights prior to the fall of the Taliban, and murkier outcomes, as in Operation Anaconda (2002) and Operation Moshtarak (2010).

The increasingly common incidence of remarkably diverse battlefield coalitions underscores the pressing need for scholars and policymakers concerned with the generation, use, and consequences of military power in

DOI: 10.4324/9781003399896-3

the international arena to understand under what conditions such groups fight as greater or less than the sum of their parts. In this chapter, we argue that understanding battlefield coalition performance requires first understanding their composition. Only by investigating and assessing the relationship between different types of battlefield coalitions and how well they have fought can we begin to address the impact of regime type, treaty commitments, and non-state actors on the frequency, course, and results of combat in the past and moving forward.

At present, we know relatively little about the composition and, in turn, performance of battlefield coalitions. Despite the many forms such groups may take, scholars have tended to concentrate their energies on US-led and NATO collectives. While such studies are important in their own right, generalising from them to the larger population of battlefield coalitions necessitates caution, as such groupings are distinct in many ways from most battlefield coalitions. For example, US-led battlefield coalitions are often characterised by significant asymmetry in partner resource contributions while battlefield coalitions drawn solely from NATO members benefit from their members having engaged in decades of pre-conflict cooperation on defence matters. Understanding whether these and other differences between US-led or NATO-created groupings and other battlefield coalitions matter – and whether and to what extent insights from the study of the former can be generalised to the latter – requires a more comprehensive portrait of the composition and performance of battlefield coalitions across time and space.

This chapter provides the first such view of battlefield coalitions. Using our new *Belligerents in Battle* dataset, which contains information on belligerents fighting 492 major land battles waged during 62 interstate wars in the twentieth century, we explore the frequency of battlefield coalitions, their effectiveness and efficiency relative to actors fighting alone, and what kinds of groups perform best on the battlefield. We show that, between 1900 and 2003, battlefield coalitions comprised almost one quarter of all belligerent sides and were more than half of such actors after the end of the Cold War. Battlefield coalitions proved more effective and efficient in combat than solo belligerents, winning more often and suffering fewer casualties than forces fighting alone. Some battlefield coalitions performed better than others; however, groups including forces fielded by the United States or states with pre-existing treaty agreements were particularly effective and efficient. Similarly, battlefield coalitions comprised of forces fielded by democracies were quite powerful. By contrast, battlefield coalitions that included non-state actors were much more ineffective, losing the majority of their fights.

Our data and findings, beyond shedding new light on crucial debates among academics, speak to significant areas of concern in contemporary warfare. As the United States seeks to line democracies up against non-democratic threats like Russia and China, a variety of voices have raised questions about the value of maintaining formal military alliances such as

NATO; understanding the actual combat consequences of such relationships is essential for advancing the debate. Additionally, regardless of the future of American counter insurgency operations, non-state actors will continue to take on more prominent roles in the international security arena, making improved understandings of how states and non-state actors might better work together to combat shared threats crucial moving forward.

This chapter proceeds in four parts. First, we summarise existing scholarship as it relates to the investigation of battlefield coalitions. Second, we briefly describe our *Belligerents in Battle* data and its unique advantages. Third, we use the *Belligerents in Battle* data to paint a portrait of the composition and performance of battlefield coalitions in the twentieth century. We conclude with a discussion of the many avenues for research and implications for policy suggested by our findings.

2.1 Assessing Battlefield Coalitions

The majority of existing scholarship on military collectives considers such groupings in the context of wars rather than battles – that is, it focuses on alliances and wartime coalitions rather than battlefield coalitions.[2] This body of work has generated a number of important findings regarding the formation of wartime collectives,[3] the cohesion of such groups during wars,[4] the incidence and practice of co-belligerents abandoning each other during conflicts,[5] the expansion of wartime collectives and conflict,[6] and the relationship between the existence of such groups and war outcomes.[7] Useful as these insights are, there is no clear reason to think the factors driving alliance and wartime coalition creation, cohesion, performance, and such collectives' impact on international relations are the same as those that drive dynamics in groups that combine, coordinate, and operate in battle, under fire at the sharp end of the spear.

Scholarship that does seek to understand battlefield coalitions tends to narrowly focus on a few sets of cases. Here, scholars privilege the study battlefield coalitions in which the United States is a co-belligerent,[8] World War and Korean War coalitions,[9] or NATO.[10] While there are a few exceptions to these trends,[11] it is difficult to draw generalisations from this body work. Relying too heavily on the insights of studies of NATO or coalitions in which the United States is the dominant member likely skews understandings of how extensive pre-war institutionalisation and the presence of a major power alters aspects of coalition warfare like individual states' incentives to join battle,[12] the efforts required of and made by junior partners,[13] and battle outcomes. Over-generalising from the experience of World War coalitions, by contrast, risks implicitly biasing assessments of non-European collectives as well as those multilateral groups that are not engaged in large-scale, existential conflicts. To fully appreciate both the contributions these studies make to the corpus of scholarly knowledge on battlefield coalitions and their limitations, it is necessary to have a systemic understanding of multilateral forces in combat that extends beyond this narrow range of cases.

At present, there are few tools to help gain such a comprehensive understanding. As noted, there are no suitable qualitative treatments, and most datasets created to explore belligerents and their performance in conflicts aggregate information over entire wars.[14] The few datasets that contain information on belligerents and their performance in battle are problematic for surveying battlefield coalitions. For example, the Historical Evaluation and Research Organization's CDB-90 collection[15] is opaque with respect to criteria defining specific variable values, biased in its inclusion of individual combat actions, and does not account for the presence or facilitate the examination of coalition forces at the battle level.[16] Other datasets, like those compiled by McNabb Cochran and Long, Lehmann and Zhukov, and Min, do not include the detail necessary to account for the wide variation observed across space and time in both the composition of battlefield coalitions and their performance in battle.[17] Relatedly, many existing datasets rely exclusively on military encyclopaedias – specifically those authored by Clodfelter, Eggenberger, Jaques, and Showalter – to populate their observations.[18] Dependence on such encyclopaedias raises questions about the quality of the information included; as Reiter, Stam, and Horowitz note, 'military encyclopaedias often lack the accuracy of careful historical scholarship or primary sources.'[19] Even when their information is of high quality, however, encyclopaedias are often missing relevant values for key phenomena.[20]

While there are thus many existing data resources offering useful information on coalitions at the war level, and a few resources offering some information on coalitions at the battle level, they all have important shortcomings that limit their utility for studying such groups.

2.2 Belligerents in Battle Dataset

To improve our understanding of battlefield coalitions, we created the *Belligerents in Battle* dataset. Drawing upon official government histories, intelligence documents, conflict-specific texts, and military history encyclopaedia, *Belligerents in Battle* contains information on 984 belligerent sides engaged in 492 major land battles fought during the 62 interstate wars waged between 1900 and 2003.[21] For every battle, we provide information on onset and termination dates; the identity, regime type, and state/non-state status of all belligerents involved; the troops deployed and casualties suffered by each belligerent side; the initiator of combat and location of fighting; and the outcome. In battles where at least one belligerent side was a battlefield coalition, we provide information on whether at least two of the co-belligerents had a pre-existing defence treaty; whether at least two of the co-belligerents had prior experience of fighting together in a major battle; and whether any member of the coalition limited its contribution to only airstrikes during the engagement. We provide a brief discussion of the creation of the *Belligerents in Battle* dataset here; for more detail on our battle identification process, coding rules, and case summaries for all 492 battles, consult the *Belligerents in Battle Codebook*.[22]

The most significant decisions made in the creation of any battle-level dataset are the determination of what constitutes a battle and what engagements count as battles.[23] We define a battle as a major discrete combat action within a larger war that involves significant ground activity by actors on both sides. This definition deliberately avoids thresholds for size, length, and lethality, as what counts as significant for any of such measures necessarily depends on the larger context in which the battle was fought. For example, actions fought by relatively small numbers of forces during the Third Central American War (1906) might barely register among the cataclysmic clashes of the First World War (1914–1918), but may nevertheless be quite significant in context. Our conception in this regard aligns with the conventional disciplinary approach to studying battles and facilitated the creation of a list of engagements that are commonly recognised as discrete combat actions in the historical record.

We identify battlefield coalitions when multiple belligerents fought in the same operational space during a given battle – that is, when the forces of multiple distinct actors were required by the needs of proximity to coordinate their combat actions in some manner.[24] For such actors, we account for coalition size, the presence of non-state actors, and members' regime type. Coalition size is the number of co-belligerents that fought together. The smallest battlefield coalition is two (e.g., Nicaragua and Honduras at the Battle of San Marcos, waged during the Fourth Central American War) and the largest is 11 (i.e., the group that attacked the 'Saddam Line' on 24 February 1991 at the outset of the Persian Gulf War). Non-state actor presence codes whether at least one non-state actor was part of the battlefield coalition. Regime type codes battlefield coalitions on the basis of the governance structures of the members: democratic battlefield coalitions are comprised of forces contributed solely by states that scored 6 or higher on the Polity IV scale at the time of the battle, non-democratic battlefield coalitions include forces contributed solely by actors that scored less than 6, and mixed battlefield coalitions include forces contributed by actors that score both above and below 6.[25] Non-state actors were uniformly coded as non-democratic.

We also account for characteristics that pertain to both solo belligerents and battlefield coalitions. Initiator designates which belligerent side struck first in the battle. Home Turf indicates whether the battle was fought on territory claimed by the attacker, the defender, or a third party. Battle Outcome is a dichotomous measure coding which belligerent side won or lost the engagement. We judge a belligerent side to have won if it achieved its operational and tactical goals in the engagement.[26] In cases where scholarly consensus is that the result was a draw, in which by definition neither side achieved its operational and tactical goals, both sides are coded as having lost and an additional variable, Battle Draw, is used to indicate that the engagement ended in a stalemate. Twenty-seven of the 492 battles included in *Belligerents in Battle* are coded as ending in draws.

We also account for the number of troops employed and casualties suffered by all actors. For troops figures, we count each belligerent side's ground forces involved in combat.[27] In some cases, official and historical sources concur on numbers of ground forces engaged. For example, in large, set-piece battles initiated by Western, and particularly American, militaries, available numbers on forces employed are often precise and consistent. In many cases, however, reporting is inconsistent. Inconsistencies take three forms. First, sources sometimes offer a range of troops engaged. In such cases, we attempted to identify a scholarly consensus on the likely number. If no such consensus could be identified, we used the average of the range. Second, sources sometimes only offer the number of units that were engaged (e.g., three divisions). In such cases, we used order of battle information on the size of units in the belligerent's military and calculated the approximate number of troops engaged.[28] Finally, sources sometimes offer coverage too scant to identify probable troop numbers. In such cases, like in battles in the 1934 war between Saudi Arabia and Yemen, we code the variable as missing. These coding decisions necessarily inject noise and uncertainty into our data on troop counts. For that reason, we discourage the use of individual battles' troop counts for estimation purposes. We rely instead on Troop Share, which is each belligerent side's proportion of the total number of friendly and adversary forces engaged in each battle. By focusing on the relative number of troops engaged in battle rather than the absolute number, any particular inaccuracy in an individual belligerent side's specific troop count is less likely to impact estimation of overall patterns and trends. Additionally, Troop Share normalises the relative presence of belligerents in their battles, permitting more meaningful comparisons between engagements of different sizes.

For casualties, we adopt a broad definition of the term and count all reported killed, wounded, captured, and missing. We do so because there is inconsistency in reporting, with some sources disaggregating killed, wounded, captured, and missing; some counting only one or two classes of casualty; and some not distinguishing between the various types. Given such inconsistencies, we again caution against the use of individual casualty numbers in *Belligerents in Battle*. Instead, we employ Casualty Ratio, which reports casualties suffered as a percentage of troops fielded. Casualty Ratio, like Troop Share, helps in buffering against any specific inaccuracies in casualty counts having a disproportionate effect on assessments of overall patterns and trends and in making battles of various sizes more comparable.

The final type of information contained in *Belligerents in Battle* concerns pre-existing defence relations between battlefield coalitions. We measure such relations in two ways. First, Prior Defence Obligations codes whether two or more members of the battlefield coalition had concluded a formal treaty obligating them to provide military assistance in the defence of one another prior to the onset of the battle. To code this variable, we rely on the information contained in the Alliance Treaty Obligations and Provisions (ATOP) Project.[29] Second, Prior Fighting Experience codes whether two or more of the battlefield coalition members had fought alongside one another in a previous battle. This variable is graduated, indicating whether the co-belligerents had fought a battle together in a recent interstate war, if they had fought one battle together in the current war, or if they had fought three or more battles together in the current war.

2.3 Coalitions in Combat, 1900–2003

Belligerents in Battle provides an unprecedented view of coalitions in battle. We exploit this view and assess the conditions under which battlefield coalitions fight as greater or less than the sum of their parts in two steps in this section. First, to establish a baseline for thinking about the relative capabilities of battlefield coalitions, we compare their operations and performance with that of solo belligerents since 1900, both in the aggregate and over time. Second, we conduct a provisional investigation of the capabilities of different types of battlefield coalitions by comparing the operations and performance of variously composed groups in the aggregate and over time. While we limit our discussion here to descriptive statistics, our figures indicate that battlefield coalitions warrant special attention from scholars, that many of the insights present in the current literature may be biased in important ways, and that the necessity of understanding more about these phenomena is only likely to grow moving forward.

2.3.1 Fighting with Others vs. Fighting Alone

To begin thinking about the performance of battlefield coalitions, simply distinguishing between such groups and belligerents that fought on their own is revealing. Table 2.1 reports the frequency with which actors have fought major interstate land battles with others and alone since 1900, as well as several differences in how both types of belligerents have both operated and performed in combat.

Notably, battlefield coalitions were a common occurrence – approximately 23% of the 984 belligerent sides that fought major land battles in interstate wars waged between 1900 and 2003 were comprised of multiple actors. They are also systematically different from solo belligerents. First, battlefield coalitions were more effective in combat than solo belligerents, winning their battles more than 53% of the time while those fighting alone won less than 46% of the time. This difference is statistically and substantively significant. In practice, shifting the battlefield coalitions' win rate from 53.5% to 45.3% would mean that nearly one-sixth of all victorious such collectives would have lost; shifting solo belligerents' win rates from 45.3%

Table 2.1 Belligerents in Battle, 1900–2003

Belligerent Type	Belligerent Sides	Battle Victory	Average Casualties	Average Troop Share	Battle Initiator	Battle on Home Turf
All belligerent sides	984	47.2%	24.2%	50.0%	50.0%	42.3%
Solo belligerents	756	45.3%	25.2%	49.0%	48.3%	40.1%
Battlefield coalitions	228	53.5%**	20.9%*	53.2%**	55.7%**	49.6%**

Notes
* Difference with solo belligerents significant at the $p < 0.1$ level.
** Difference with solo belligerents significant at the $p < 0.05$ level.

to 53.5% would mean that over 60 additional such actors would have won. Second, battlefield coalitions were also more efficient than solo belligerents. While battlefield coalitions tended to be larger relative to their opponents than solo belligerents – fielding on average more than 53% of all troops engaged in combat during any given battle – they do not seem to have won simply due to the weight of numbers: multilateral forces suffered, on average, more than 4% fewer casualties per battle (measured as a proportion of troops fielded) than did forces fighting alone. Again, this difference is statistically and substantively significant. Consider a hypothetical case in which a belligerent side deploys 40,000 troops in battle (the median figure in our data). If it suffered casualties at the average rate of solo belligerents, it could expect losses slightly greater than 10,000; if it suffered casualties at the average rate of battlefield coalitions, it could expect more than 1,500 fewer losses, or approximately 8,350. Additional differences between battlefield coalitions and solo belligerents suggest possible reasons why the former were more effective and efficient than the latter: they were more likely to initiate combat and fight on territory controlled by a member of the group. Choosing when to fight and engaging the adversary on terrain with which troops are familiar have long been thought to afford significant advantages to the side that can do so.[30]

Examining the performance of battlefield coalitions and solo belligerents over time reveals further evidence in support of the notion that the former were consistently more effective and efficient than the latter. To assess temporal variation in the performance of battlefield coalitions and solo belligerents between 1900 and 2003, we divided the time covered by our data into eight periods of 13 years; beyond offering equal time periods for comparison, the breaks between eras fall at historically convenient points, including just before both World Wars and at the end of the Cold War.

Figure 2.1 depicts the presence of battlefield coalitions over time, where presence is measured as a percentage of all belligerent sides fighting during

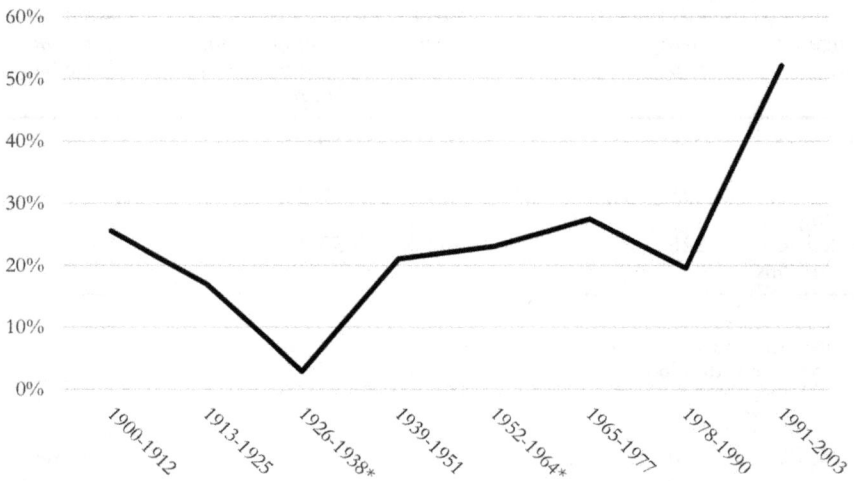

Figure 2.1 Battlefield Coalitions as Belligerent Sides, 1900–2003.

the period. Battlefield coalitions consistently accounted for between 15% and 30% of all belligerent sides for the majority of the twentieth century. Notably, the world wars do not stand out in this regard; battlefield coalitions were relatively more common in the periods before the First World War and after the Second World War than they were during either of those conflicts. An exception to this stability was the 1926–1938 interwar period, when only two of the 70 belligerent sides engaged in battle were coalitions – both involving China and non-state partners fighting the Soviet Union during the 1929 Manchurian War. The relatively stable presence of battlefield coalitions evaporated after the Cold War, however; between 1991 and 2003, 52% of the belligerent sides fighting major battles (49/94) were coalitions.

Though there was a steep rise in relative frequency of battlefield coalitions after the Cold War, there is little reason to think that such groups' impressive combat performance in the aggregate is driven by those collectives. First, the 49 battlefield coalitions that fought together after the end of the Cold War represent only 21% of the total number of such groups that engaged in combat between 1900 and 2003. Second, the post-Cold War battlefield coalitions were roughly as effective and efficient as those that fought prior to 1991.[31]

Figure 2.2 depicts the relative effectiveness of battlefield coalitions and solo belligerents over time. There is a remarkable degree of consistency in the performance of both types of actors during much of the twentieth century. Except for two periods, battlefield coalitions won between 40% and 60% of their fights before the end of the Cold War. The exceptions prove the rule, however: in the 1926–1938 and 1952–1964 periods, there were only eight

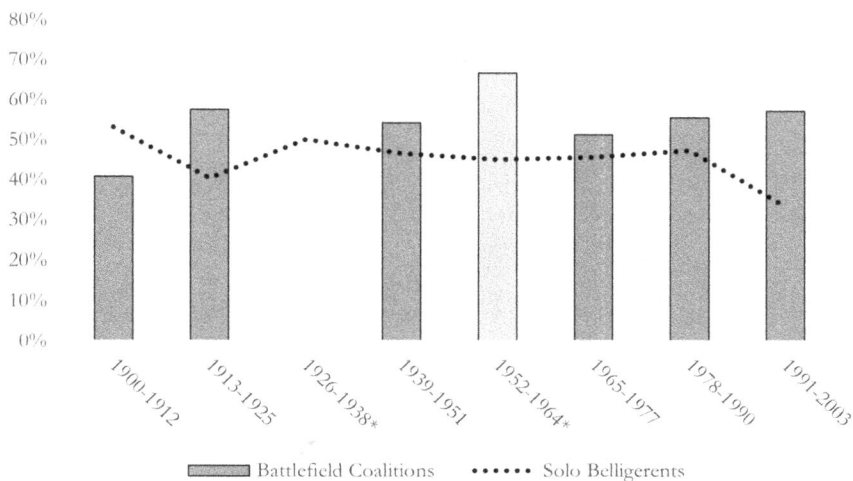

Figure 2.2 Belligerent Victories, 1900–2003.

Note: *Average of 2–6 battlefield coalitions fighting during the period.

total battlefield coalitions that fought major engagements. The Chinese and their partners lost both battles they fought together during the interwar period, then four of the six battlefield coalitions that fought between 1952 and 1964 were engaged in combat against other such groups. For their part, solo belligerents were similarly consistent, winning between 40% and 53% of their battles in each period until 1991. After the end of the Cold War, the relative effectiveness of the two types of forces diverged: battlefield coalitions won 57% of their battles while solo belligerents won only one-third of their fights. This split – its origins and possible implications for thinking about belligerents' likely effectiveness moving forward – merits further investigation, but it does not call into question the consistency of battlefield coalitions' manifest effectiveness over time. Such groups' effectiveness since the end of the Cold War is on par with their performance during the First World War period and squarely within the band in which battlefield coalitions' success has ranged since 1900.

Figure 2.3 depicts the relative efficiency of battlefield coalitions and solo belligerents over time. Here, too, there is remarkable consistency in the performance of both battlefield coalitions and solo belligerents. Excepting the two periods during which there were very few battlefield coalitions, such groups' casualty rates ranged between 9% and 28% during the twentieth century. The interwar period and the post-Korean War period were extreme outliers. Casualty data is available for only one of the two battlefield coalition fights during the Manchurian War in the interwar period: the Soviet offensive that routed and resulted in the killing, wounding, or capturing of

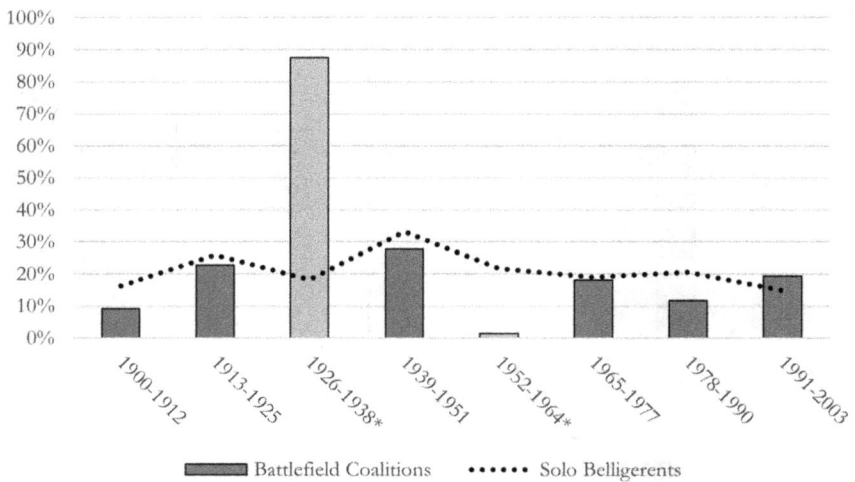

Figure 2.3 Belligerent Casualty Rates, 1900–2003.

Note: *Average of 2–6 battlefield coalitions fighting during the period.

88% of the Chinese and White Russian forces, ending the war. Casualty data is available for five of the six battlefield coalition fights in the years following the Korean War, but the combined fights were especially low-cost: the British and French suffered less than 1% casualties during Operation Musketeer in the 1956 Suez War and neither side lost many soldiers during the major engagements of the 1957–1958 Ifni War in the Spanish Sahara. Solo belligerents' casualty rates ranged between 15% and 33% during the eight periods spanning 1900 and 2003. Notably, excepting the outlier interwar period, the only time span during which solo belligerents were, on average, more efficient than battlefield coalitions was the post-Cold War era. This is further evidence that the performance of the 49 battlefield coalitions that came together after 1990 is almost certainly not driving the impressive aggregate efficiency of such groups.

2.3.2 Composition of Battlefield Coalitions and Performance

While battlefield coalitions as a whole were relatively consistent in their effectiveness and efficiency over time, there are some marked differences in performance across types of groups. Table 2.2 reports the performance and a variety of characteristics of several kinds of battlefield coalitions: those that include the United States, those that fought during the world wars, those comprised of forces drawn from varying mixes of regime types, those comprised of forces from states with pre-existing defence obligations, and those that include non-state actors. In the table, the referent category for

Table 2.2 Coalitions in Comparison, 1900–2003[32]

Battlefield Coalition Type	Battlefield Coalitions	Battle Victory	Average Casualties	Coalition Size	Average Troop Share	Prior Fighting Experience	Battle Initiator	Battle on Home Turf
United States as member	78	←***	→*	←***	←***	←***	←***	←***
Fought during world wars	60	←	←**	→	←	←**	←	→
Democratic	45	←***	←	←***	←**	←***	←***	←***
Mixed	63	←→***	→→	←→***	←→***	←→	←→***	←→***
Non-democratic	120	→***	→	→***	→***	→**	→***	→**
Pre-existing defence obligations	85	←***	→	←***	←	←***	←**	→**
Non-state actors as member(s)	90	→***	→	→	→*	→***	→*	→

Notes
* Difference significant at the p < 0.1 level.
** Difference significant at the p < 0.05 level.
*** Difference significant at the p < 0.01 level.

each row is the set of battlefield coalitions that do not possess the noted characteristic. For example, for the battlefield coalitions including the United States noted in the first row, the comparison group is all battlefield coalitions that did not include the United States as a participant and, for the democratic battlefield coalitions noted in the third row, the comparison group is all battlefield coalitions with forces contributed by at least one non-democracy.

Considering first the battlefield coalitions that included the United States, such groups performed especially well between 1900 and 2003. American forces fought as part of approximately 34% (78/228) of the battlefield coalitions since 1900 and those groups were significantly more effective and efficient than collectives without them. Battlefield coalitions with United States forces won 85.9% of their battles while suffering, on average, 16.6% casualties; battlefield coalitions without US forces won only 36.7% of their battles while suffering, on average, 23.6% casualties. These differences may be driven by the composition of the different battlefield coalitions, how they fought on the battlefield, or both. Battlefield coalitions, including US forces, were larger than those without in terms of both the number of members and the relative proportion of all troops deployed in discrete battles.[33] Groups including US forces also tended to be more familiar with one another; members of such coalitions had prior experience fighting together in 84.6% of their battles while members of coalitions without US participation had such experience in 68.0% of their battles. When fighting, battlefield coalitions that included US forces initiated combat 73.1% of the time and fought on the turf of a coalition member in 51.3% of their battles; the equivalent rates for battlefield coalitions without US forces are 46.7% and 48.7%.[34] Crucially, all of these differences, save the location of combat, are statistically significant, meaning that battlefield coalitions that include American forces are fundamentally different from other battlefield coalitions without US participation. This suggests scholarship that focuses on the composition, internal dynamics, and performance of battlefield coalitions that include the United States is important, but almost certainly cannot be generalised to the population of such groups more broadly.

Significantly, it does not seem that the 60 battlefield coalitions that fought during the world wars are substantially different from the 165 that did not. On average, world war battlefield coalitions were more familiar with one another and suffered more casualties than those that fought in other conflicts.[35] These differences are understandable, given that the length of the wars afforded co-belligerents more time to familiarise themselves and fight with one another as well as the tendency for coalition forces to be routed at the beginning of both conflicts.[36] In all other respects, however, the world war battlefield coalitions are statistically indistinguishable from the collective belligerents in other wars. This finding lends some extra weight to existing research on the composition and operation of battlefield coalitions in the world wars; while caution is needed, findings from such cases may well be informative about the general trends in and behaviours of such groupings across time and space.

Examining the performance of battlefield coalitions categorised according to the regime type of the actors contributing forces yields a host of intriguing differences across groupings. Non-democratic battlefield coalitions stand apart from democratic and mixed battlefield coalitions: while the average casualty rates of the different types were statistically indistinguishable, both democratic and mixed battlefield coalitions won approximately twice as many of their battles as non-democratic battlefield coalitions. Democratic and mixed battlefield coalitions emerged victorious 71.1% and 73.0% of the time, respectively, while non-democratic collectives achieved their combat objectives only 36.7% of the time. This discrepancy in battlefield effectiveness could be the result of groups' composition, including that non-democratic battlefield coalitions tend to be smaller both in terms of number of members and relative to their adversaries than democratic and mixed battlefield coalitions: democratic, mixed, and non-democratic battlefield coalitions averaged 2.6, 3.3, and 2.2 members and enjoyed average troop shares of 60.1%, 55.1%, and 49.5%, respectively.[37] Additionally, members of non-democratic battlefield coalitions tended to be less familiar with one another than their counterparts in democratic and mixed battlefield coalitions, having had prior experience fighting alongside one another in 67.5% of their battles while members of democratic and mixed coalitions had such experience in 91.1% and 73.0% of their fights. It is perhaps because of their relative lack of familiarity that non-democratic battlefield coalitions are less aggressive, initiating combat only 46.7% of the time, compared to 77.8% of the time for democratic battlefield coalitions and 57.1% of the time for mixed battlefield coalitions. Mixed battlefield coalitions were especially likely to fight on the territory of a member, doing so 66.7% of the time, compared to 46.7% for democratic and 41.7% for non-democratic groupings. Of the three types, non-democratic battlefield coalitions proved the least likely to maximise the benefits of fighting alongside partners.

Like battlefield coalitions that include the United States, groups comprised of forces fielded by actors that have pre-existing treaty commitments to defend one another appear to be qualitatively different from those that do not. Such groups tend to win more often – 64.7% compared to 46.9% for groups without such commitments – and suffer fewer casualties, though the difference in efficiency is not statistically significant.[38] Battlefield coalitions with pre-existing defence agreements may be more effective than those without because they tend to be larger in terms of membership (3.2 members, compared to 2.3) and have more prior fighting experience (83.5%, compared to 67.8%). They also tend to field a slightly larger percentage of all engaged troops in their battles (55.0%, compared to 52.0%), though that difference is not statistically significant. Perhaps surprisingly, battlefield coalitions with pre-existing defence commitments are significantly more aggressive than those without: they initiate battle more often (65.9%, compared to 49.7%) and fight on the territory of a coalition member less often (38.8%, compared to 55.9%).[39] These systematic differences suggest that scholarship focused on the composition and operation of battlefield coalitions grounded in pre-existing defence commitments, like those fielded by NATO, offers important insights into the composition and operations of many belligerents – approximately 37% of all battlefield coalitions had such agreements. However, they also suggest that existing work on such groups, like that on groups that include the United States, likely is not generalisable to the population of battlefield coalitions more broadly.

The final group of battlefield coalitions described in Table 2.2 is the set that includes at least one non-state actor co-belligerent. These battlefield coalitions are significantly less effective than those comprised solely of states, with the former winning 42.2% of the time and the latter emerging victorious 60.9% of the time. This may be due to the tendency of battlefield coalitions to field a smaller overall share of troops in their fights (49.7%, compared to 55.0%). At the same time, such groups may suffer fewer casualties than those without non-state actors – 17.9%, compared to 22.1% – though the difference is not statistically significant. Perhaps surprisingly, there is no statistically meaningful difference between the two types of battlefield coalition in terms of the number of members. As might be expected, battlefield coalitions with non-state actors tended to have less prior experience fighting with one another than those without (62.2%, compared to 81.2%). When fighting, groups with non-state actors also tended to be less aggressive, initiating combat in only 48.9% of their battles, compared to state-only collectives, which initiated combat 60.1% of the time. There is no statistically meaningful difference between battlefield coalitions with non-state actors and those without in terms of where their combat actions took place.

Belligerents in Battle thus reveals that battlefield coalitions tend to perform better than solo belligerents, winning more often and suffering fewer casualties in the process. The advantage of fighting battles with partners held throughout the twentieth century, though certain types of battlefield coalition were particularly effective and most likely to fight as greater than the sum of their parts: those including US forces, those including troops fielded by democracies, and those created by partners with pre-existing defence obligations. Battlefield coalitions that performed particularly poorly, as less than the sum of their parts, were those comprised solely of forces fielded by non-democracies and those that include non-state actors. While much more research is needed to determine the robustness of and mechanisms underpinning these patterns, the portrait of battlefield coalitions that emerges from a systematic examination of their twentieth century experience suggests that, generally, there are distinct advantages to going into battle alongside partners, but choices about who to fight with must be made carefully.

2.4 Studying and Creating Battlefield Coalitions

Belligerents often fight their wars alongside others and frequently take the additional step of deploying their forces to fight side-by-side in the same operational battlespace. While, on average, fighting battles with partners pays, some battlefield coalitions appear to be better bets than others. During the twentieth century, forming battlefield coalitions with forces fielded by the United States, treaty partners, and democracies proved more efficacious than other types of groupings, particularly those raised by non-state actors.

This portrait of battlefield coalitions provided by *Belligerents in Battle* suggests a wide range of new avenues of research for scholars investigating collective warfighting. First, while existing scholarship on US-led and NATO battlefield coalitions is valuable, the insights derived from such work likely do not travel cleanly to the study of other battlefield coalitions. Defining the scope conditions and reach of such claims would shed important new light on the incidence, process, and achievements of collective action on the battlefield. A potentially useful starting point for that research is scholarship on battlefield coalitions formed during the world wars; they appear to be similar to the larger population of such collectives along several dimensions.

Second, battlefield coalitions fighting between 1900 and 2003 were distinct from solo belligerents in several ways, reflecting and contradicting various elements of conventional wisdom about fighting with partners. Why do battlefield coalitions tend to outnumber and defeat their adversaries more often than solo belligerents while suffering relatively fewer casualties? Are these trends related to battlefield coalitions' tendencies to initiate combat and fight on their own territory more often than solo belligerents? Napoleon, Churchill, and Eisenhower all famously lamented the pitfalls of fighting with partners; the apparent advantages require further investigation.

Third, given the friction inherent to the process of creating, organising, and operating a battlefield coalition, how is it that multiple actors have overcome the centrifugal forces of battlefield cooperation to not only fight together, but do so more efficiently and effectively than belligerents fighting alone? The presence of a powerful actor like the United States or pre-existing defence obligations that have institutionalised military cooperation to some degree cannot explain the achievements of the majority of battlefield coalitions.

Fourth, our findings regarding the composition and performance of democratic, mixed, and non-democratic battlefield coalitions raise important questions about the reasons why we observe such significant variation. Is there something specific about the forces fielded by democracies that allow them to fight well both together and in combination with forces drawn from non-democratic actors? Alternatively, are characteristics inherent to non-democracies that make them less likely to build large, aggressive, effective battlefield coalitions?[40]

Fifth, the increasing frequency of non-state actor participation in battlefield coalitions requires study. What types of states are likely to partner with non-state actors? Under what circumstances? Given that there seem to be few benefits to sharing operational battlespace with non-state actors, why are groups including such actors increasingly common? Further, are there meaningful differences across different types of non-state actors that are not captured by our data but do matter for the performance of battlefield coalitions?

Finally, thinking about battlefield coalitions over time reveals additional questions worthy of investigation. Why have such groupings grown more common in recent years? Given the remarkable variation observed in the composition of battlefield coalitions, what explains the relative consistency in the effectiveness of such groups in combat over time?

Our findings also have implications for policymakers. The remarkable performance of battlefield coalitions comprised of militaries fielded by states with pre-existing defence obligations suggests there is operational and tactical merit to sustaining, and even reinvigorating, existing alliance obligations. This finding also indicates that policymakers thinking about emergent military challenges in regions where alliances are in short supply have reason to be concerned about the operational and tactical capabilities of any potential ad hoc battlefield coalitions of which they may be a part. Also significant is the relatively poor performance of battlefield coalitions that include non-state actors; as such forces are increasingly common in the international arena, devising ways to build and sustain more effective cooperative relations with them is likely to pay dividends both within and beyond counterinsurgency wars.

Battlefield coalitions are extremely diverse in terms of their composition and performance, and scholars have much to learn about such how such groups come together, organise, operate, and perform. That they appear to be increasingly common belligerents in interstate wars means that the need to improve such understandings is urgent.

Notes

1 Barbara Elias, 'Why Rebels Rely on Terrorists: The Persistence of the Taliban-al-Qaeda Battlefield Coalition in Afghanistan,' Chapter 9, this volume.
2 Rosella Cappella Zielinski and Ryan Grauer, 'Battlefield Coalitions: Preparation, Organisation, Execution,' Chapter 1, this volume.
3 Scott Wolford, *The Politics of Military Coalitions* (New York: Cambridge University Press, 2015); Scott Wolford and Emily Hencken Ritter, 'National Leaders, Political Security, and the Formation of Military Coalitions,' *International Studies Quarterly* 60, no. 3 (September 2016): 540–51, https://doi.org/10.1093/isq/sqv023; Marina E. Henke, *Constructing Allied Cooperation: Diplomacy, Payments, and Power in Multilateral Military Coalitions* (Ithaca, NY: Cornell University Press, 2019).
4 Thomas Stow Wilkins, 'Analysing Coalition Warfare from an Intra-Alliance Politics Perspective: The Normandy Campaign 1944,' *Journal of Strategic Studies* 29, no. 6 (December 2006): 1121–50, https://doi.org/10.1080/01402390601016592; Ulrich Pilster, 'Are Democracies the Better Allies? The Impact of Regime Type on Military Coalition Operations,' *International Interactions* 37, no. 1

(March 2011): 55–85, https://doi.org/10.1080/03050629.2011.546259; Ajin Choi, 'Fighting to the Finish: Democracy and Commitment in Coalition War,' *Security Studies* 21, no. 4 (October 2012): 624–53, https://doi.org/10.1080/0963 6412.2012.734232.

5 Ulrich Pilster, Tobias Böhmelt, and Atsushi Tago, 'Political Leadership Changes and the Withdrawal from Military Coalition Operations, 1946–2001,' *International Studies Perspectives* 16, no. 4 (November 2015): 463–83, https:// doi.org/10.1111/insp.12058; Alex Weisiger, 'Exiting the Coalition: When Do States Abandon Coalition Partners during War?,' *International Studies Quarterly* 60, no. 4 (December 2016): 753–65, https://doi.org/10.1093/isq/sqw029; Kathleen J. McInnis, *How and Why States Defect from Contemporary Military Coalitions* (New York: Palgrave Macmillan, 2019).

6 Scott Wolford, 'Power, Preferences, and Balancing: The Durability of Coalitions and the Expansion of Conflict,' *International Studies Quarterly* 58, no. 1 (March 2014): 146–57, https://doi.org/10.1111/isqu.12036.

7 Pilster, 'Are Democracies the Better Allies?'; Daniel S. Morey, 'Military Coalitions and the Outcome of Interstate Wars,' *Foreign Policy Analysis* 12, no. 4 (October 2016): 533–51, https://doi.org/10.1111/fpa.12083; Benjamin A. T. Graham, Erik Gartzke, and Christopher J. Fariss, 'The Bar Fight Theory of International Conflict: Regime Type, Coalition Size, and Victory,' *Political Science Research and Methods* 5, no. 4 (October 2017): 613–39, https://doi.org/1 0.1017/psrm.2015.52. For an analysis of coalitions and crisis outcomes, see Skyler J. Cranmer and Elizabeth J. Menninga, 'Coalition Quality and Multinational Dispute Outcomes,' *International Interactions* 44, no. 2 (March 2018): 217–43, https://doi.org/10.1080/03050629.2017.1369410.

8 Sarah E. Kreps, *Coalitions of Convenience: United States Military Interventions after the Cold War* (New York: Oxford University Press, 2011); Patricia Weitsman, *Waging War: Alliances, Coalitions, and Institutions of Interstate Violence* (Stanford, CA: Stanford University Press, 2013); Olivier Schmitt, *Allies That Count: Junior Partners in Coalition Warfare* (Washington, DC: Georgetown University Press, 2018).

9 Patricia Weitsman, *Dangerous Alliances: Proponents of Peace, Weapons of War* (Stanford, CA: Stanford University Press, 2003); Kelly Ann Grieco, 'War by Coalition: The Effects of Coalition Military Institutionalization on Coalition Battlefield Effectiveness' (Thesis, Massachusetts Institute of Technology, 2016), https://dspace.mit.edu/handle/1721.1/104572; Sara Bjerg Moller, 'Fighting Friends: Institutional Cooperation and Military Effectiveness in Multinational War' (Columbia University, 2016), https://doi.org/10.7916/D8125SNZ; Weisiger, 'Exiting the Coalition'; Rosella Cappella Zielinski and Ryan Grauer, 'Organizing for Performance: Coalition Effectiveness on the Battlefield,' *European Journal of International Relations* 26, no. 4 (December 2020): 953–78, https://doi.org/10.1177/1354066120903369; Rosella Cappella Zielinski, Ryan Grauer, and Alastair Smith, 'Regime Type, War Aims, and Coalition Member Effort in Combat,' Chapter 8, this volume.

10 Ivo H. Daalder and Michael E. O'Hanlon, *Winning Ugly: NATO's War to Save Kosovo* (Washington, DC: Brookings Institution Press, 2004); Weitsman, *Waging War*; David P. Auerswald and Stephen M. Saideman, *NATO in Afghanistan: Fighting Together, Fighting Alone* (Princeton: Princeton University Press, 2014); Olivier Schmitt, 'International Organization at War: NATO Practices in the Afghan Campaign,' *Cooperation and Conflict* 52, no. 4 (December 1, 2017): 502–18, https://doi.org/10.1177/0010836717701969.

11 See, for example, Kenneth M. Pollack, 'The Influence of Arab Culture on Arab Military Effectiveness' (PhD thesis, Cambridge, MA, Massachusetts Institute of

Technology, 1996). The historical literature on other coalitions is also quite large, though coalitions in major wars remain overrepresented and there is no effort to systematically generalise insights from individual cases; see, for example, Richard L. DiNardo, *Germany and the Axis Powers: From Coalition to Collapse* (Lawrence: University Press of Kansas, 2005); J. Lee Ready, *Forgotten Allies: The Military Contribution of the Colonies, Exiled Governments, and Lesser Powers to the Allied Victory in World War II* (Jefferson, NC: McFarland & Company, 2012); Barbara W. Tuchman, *Stilwell and the American Experience in China: 1911–1945* (New York: Random House, 2017).

12 Atsushi Tago, 'Why Do States Join US-Led Military Coalitions? The Compulsion of the Coalition's Missions and Legitimacy,' *International Relations of the Asia-Pacific* 7, no. 2 (May 2007): 179–202, https://doi.org/10.1093/irap/lcl001; Randall Newnham, '"Coalition of the Bribed and Bullied?" U.S. Economic Linkage and the Iraq War Coalition,' *International Studies Perspectives* 9, no. 2 (May 2008): 183–200, https://doi.org/10.1111/j.1528-3585.2008.00326.x; Srdjan Vucetic, 'Bound to Follow? The Anglosphere and US-Led Coalitions of the Willing, 1950–2001,' *European Journal of International Relations* 17, no. 1 (March 2011): 27–49, https://doi.org/10.1177/1354066109350052; Henke, *Constructing Allied Cooperation*.

13 Cranmer and Menninga, 'Coalition Quality and Multinational Dispute Outcomes'; Schmitt, *Allies That Count*; Olivier Schmitt, 'More Allies, Weaker Missions? How Junior Partners Contribute to Multinational Military Operations,' *Contemporary Security Policy* 40, no. 1 (January 2019): 70–84, https://doi.org/10.1080/13523260.2018.1501999.

14 Bruce M. Russett, 'An Empirical Typology of International Military Alliances,' *Midwest Journal of Political Science* 15, no. 2 (1971): 262–89, https://doi.org/1 0.2307/2110272; Brett Ashely Leeds et al., 'Alliance Treaty Obligations and Provisions, 1815–1944,' *International Interactions* 28, no. 3 (2002): 237–60, https://doi.org/10.1080/03050620213653; Meredith Reid Sarkees and Frank W. Wayman, *Resort to War: 1816–2007* (Thousand Oaks, CA: CQ Press, 2010); Morey, 'Military Coalitions and the Outcome of Interstate Wars'; Dan Reiter, Allan C. Stam, and Michael C. Horowitz, 'A Deeper Look at Interstate War Data: Interstate War Data Version 1.1,' *Research & Politics* 3, no. 4 (2016), https://doi.org/10.1177/2053168016683840; Jason Lyall, *Divided Armies: Inequality and Battlefield Performance in Modern War* (Princeton, NJ: Princeton University Press, 2020).

15 Trevor N. Dupuy, *Understanding War* (New York: Paragon House Publishers, 1987).

16 Michael C. Desch, *Power and Military Effectiveness: The Fallacy of Democratic Triumphalism* (Baltimore: Johns Hopkins University Press, 2008), 58–9; Ryan Grauer and Michael C. Horowitz, 'What Determines Military Victory? Testing the Modern System,' *Security Studies* 21, no. 1 (March 2012): 11n34, https://doi.org/10.1080/09636412.2012.650594; Kathryn McNabb Cochran and Stephen B. Long, 'Measuring Military Effectiveness: Calculating Casualty Loss-Exchange Ratios for Multilateral Wars, 1816–1990,' *International Interactions* 43, no. 6 (November 2017): 1020, https://doi.org/10.1080/03050629.2017.1273 914.

17 Cochran and Long, 'Measuring Military Effectiveness'; Todd C. Lehmann and Yuri M. Zhukov, 'Until the Bitter End? The Diffusion of Surrender across Battles,' *International Organization* 73, no. 1 (Winter 2019): 133–69, https://doi.org/10.1017/S0020818318000358; Eric Min, 'Interstate War Battle Dataset (1823–2003),' *Journal of Peace Research* 58, no. 2 (March 2021): 294–303, https://doi.org/10.1177/0022343320913305.

18 Michael Clodfelter, *Warfare and Armed Conflicts*, 3rd ed. (Jefferson, NC: McFarland & Company, 2007); David Eggenberger, *An Encyclopedia of Battles: Accounts of Over 1,560 Battles from 1479 B.C. to the Present* (New York: Dover Publications, 1985); Tony Jaques, *Dictionary of Battles and Sieges: A Guide to 8,500 Battles from Antiquity through the Twenty-First Century* (Westport, CT: Greenwood, 2006); Dennis Showalter, Stephen Hart, and Ralph Ashby, *The Encyclopedia of Warfare* (London: Sterling Publishing, 2013).

19 Reiter, Stam, and Horowitz, 'A Deeper Look at Interstate War Data,' 13.

20 For example, relying exclusively on Clodfelter, the Lehmann and Zhukov dataset excludes multiple wars, including the Franco-Thai War of 1940–1941, Offshore Islands War of 1954, Ifni War of 1957–1959, Taiwan Straits War of 1958, War of Attrition of 1969–1970, Sino-Vietnamese Border War of 1987, and Kargil War of 1999, and fails to incorporate values on surrender – their phenomenon of interest – in vast numbers of cases.

21 Interstate wars are drawn from COW 4.0. Sarkees and Wayman, *Resort to War: 1816–2007.*

22 Available at http://www.ryangrauer.com.

23 For different perspectives on the utility of battles as a unit of analysis and the feasibility of identifying what counts as a battle, see Alex Weisiger, 'Learning from the Battlefield: Information, Domestic Politics, and Interstate War Duration,' *International Organization* 70, no. 2 (Spring 2016): 347–75, https://doi.org/10.1017/S0020818316000059; Min, 'Interstate War Battle Dataset (1823–2003).'

24 When colonial forces were under command of their empire state, they were counted as part of the latter. When such forces exercised independent command at the operational and tactical levels of warfighting activity, we treat them as separate belligerents. Free-state forces are not included as belligerents.

25 Monty G. Marshall and Keith Jaggers, 'Polity IV Annual Time Series 1800–2011' (College Park, MD: Center for Systemic Peace, 2012), http://systemicpeace.org/polity/polity4.htm. We use Polity rather than alternative measures of democracy like those contained in the V-Dem Project or Freedom House assessments because it is the most liberal metric; that is, it counts more states as democracies than do other measures. As such, *Belligerents in Battle* inherently biases against finding a statistically significant relationship between regime type and belligerent behaviour. There is nothing in the dataset that would preclude future scholars from adopting alternative measures of democracy. On the merits of different measures of regime type, see Vanessa A Boese, 'How (Not) to Measure Democracy,' *International Area Studies Review* 22, no. 2 (June 2019): 95–127, https://doi.org/10.1177/2233865918815571.

26 Grauer and Horowitz, 'What Determines Military Victory?,' 96.

27 While we exclude air and naval forces from these counts, given the difficulty of accurately capturing the number of such forces involved in the use of aerial and seaborne platforms, we designate when a co-belligerent in a battlefield coalition contributed only air sorties to the fight.

28 Often, order of battle data often reports maximum unit size. As a result, by converting counts of units into counts of soldiers, we undoubtedly overcount the number of troops engaged; units inevitably suffer attrition during war and almost no unit ever enters battle at full strength. Any bias this introduces into our data will trend towards overestimation in all cases. If order of battle information could not be found, we code the variable as missing.

29 Leeds et al., 'Alliance Treaty Obligations and Provisions, 1815–1944.'

30 Sun Tzu, *The Art of War*, trans. Samuel Griffith (New York: Oxford University Press, 1971); Carl von Clausewitz, *On War*, ed. Michael Howard and Peter Paret (Princeton: Princeton University Press, 1976).

31 For a replication of Table 2.1 without the post-Cold War period, see the Appendix; the average figures are very similar to those for the entire period, though some differences between battlefield coalitions and solo belligerents lose statistical significance. The Appendix is available at http://www.ryangrauer.com.

32 For a fully reported table with all statistics, see the Appendix, available at http://www.ryangrauer.com.

33 Battlefield coalitions with US forces averaged 3.2 members while those without averaged 2.3 members; the former fielded, on average, 58.3% of all troops in their battles while the latter fielded only 50.5%.

34 There is some reason to think that groups including US forces are driving the relative performance of battlefield coalitions and solo belligerents in the aggregate. Without the United States, battlefield coalitions won less often than solo belligerents, though the difference is not statistically significant. Battlefield coalitions without US forces enjoyed a greater manpower advantage, suffered fewer casualties, and both initiated combat and fought on their own territory more often than solo belligerents, though the differences in these statistics are not statistically significant. In short, when battles involving the US are dropped from the data, battlefield coalitions and solo belligerents are statistically indistinguishable.

35 Battlefield coalitions fighting during the World Wars had members with prior experience fighting with one another 85.0% of the time and an average casualty rate of 27.0%; such groups fighting in all other wars had members with prior fighting experience 69.6% of the time and suffered an average casualty rate of 17.6%.

36 Sara Bjerg Moller, 'Learning from Losing: How Defeat Shapes Coalition Command Arrangements in Wartime,' Chapter 7, this volume.

37 Non-democratic battlefield coalitions were also more likely to include non-state actors: 65.8% (79/120) of such groups included at least one non-state actor, compared to 17.5% (11/63) of mixed battlefield coalitions and no democratic battlefield coalitions.

38 Battlefield coalitions with members that concluded pre-war defence treaties with one another suffered an average casualty rate of 19.6%; those without such members suffered an average casualty rate of 21.8%.

39 As with US participation, there is some reason to think that groups with pre-war defence agreements are driving the relative performance of battlefield coalitions and solo belligerents in the aggregate. Excluding such cases, battlefield coalitions win slightly more often than solo belligerents (46.9%, compared to 45.3%), though the difference is statistically insignificant. They also field a slightly higher percentage of troops, suffer lower levels of casualties, and initiate combat more often than solo belligerents, though none of the differences are statistically significant. The only statistically significant difference between battlefield coalitions without prior defence obligations and solo belligerents is that the former fight on their home turf more frequently than the latter (55.9%, compared to 40.9%; p = 0.000). In short, ad hoc battlefield coalitions perform in ways that are largely indistinguishable from solo belligerents.

40 See, for example, Dan Reiter and Allan C. Stam, *Democracies at War* (Princeton: Princeton University Press, 2002); Cappella Zielinski, Grauer, and Smith, 'Regime Type, War Aims, and Coalition Member Effort in Combat,' Chapter 8, this volume.

Bibliography

Auerswald, David P., and Stephen M. Saideman. *NATO in Afghanistan: Fighting Together, Fighting Alone*. Princeton: Princeton University Press, 2014.

Boese, Vanessa A. 'How (Not) to Measure Democracy.' *International Area Studies Review* 22, no. 2 (June 2019): 95–127. 10.1177/2233865918815571

Cappella Zielinski, Rosella, and Ryan Grauer. 'Organizing for Performance: Coalition Effectiveness on the Battlefield.' *European Journal of International Relations* 26, no. 4 (December 2020): 953–78. 10.1177/1354066120903369

Choi, Ajin. 'Fighting to the Finish: Democracy and Commitment in Coalition War.' *Security Studies* 21, no. 4 (October 2012): 624–53. 10.1080/09636412.2012. 734232

Clausewitz, Carl von. *On War*. Edited by Michael Howard and Peter Paret. Princeton: Princeton University Press, 1976.

Clodfelter, Michael. *Warfare and Armed Conflicts*. 3rd ed. Jefferson, NC: McFarland & Company, 2007.

Cochran, Kathryn McNabb, and Stephen B. Long. 'Measuring Military Effectiveness: Calculating Casualty Loss-Exchange Ratios for Multilateral Wars, 1816–1990.' *International Interactions* 43, no. 6 (November 2017): 1019–40. 10.1080/0305 0629.2017.1273914

Cranmer, Skyler J., and Elizabeth J. Menninga. 'Coalition Quality and Multinational Dispute Outcomes.' *International Interactions* 44, no. 2 (March 2018): 217–43. 10.1080/03050629.2017.1369410

Daalder, Ivo H., and Michael E. O'Hanlon. *Winning Ugly: NATO's War to Save Kosovo*. Washington, DC: Brookings Institution Press, 2004.

Desch, Michael C. *Power and Military Effectiveness: The Fallacy of Democratic Triumphalism*. Baltimore: Johns Hopkins University Press, 2008.

DiNardo, Richard L. *Germany and the Axis Powers: From Coalition to Collapse*. Lawrence: University Press of Kansas, 2005.

Dupuy, Trevor N. *Understanding War*. New York: Paragon House Publishers, 1987.

Eggenberger, David. *An Encyclopedia of Battles: Accounts of Over 1,560 Battles from 1479 B.C. to the Present*. New York: Dover Publications, 1985.

Graham, Benjamin A. T., Erik Gartzke, and Christopher J. Fariss. 'The Bar Fight Theory of International Conflict: Regime Type, Coalition Size, and Victory.' *Political Science Research and Methods* 5, no. 4 (October 2017): 613–39. 10.1017/ psrm.2015.52

Grauer, Ryan, and Michael C. Horowitz. 'What Determines Military Victory? Testing the Modern System.' *Security Studies* 21, no. 1 (March 2012): 83–112. 10.1080/09636412.2012.650594

Grieco, Kelly Ann. 'War by Coalition: The Effects of Coalition Military Institutionalization on Coalition Battlefield Effectiveness.' Thesis, Massachusetts Institute of Technology, 2016. https://dspace.mit.edu/handle/1721.1/104572

Henke, Marina E. *Constructing Allied Cooperation: Diplomacy, Payments, and Power in Multilateral Military Coalitions*. Ithaca, NY: Cornell University Press, 2019.

Jaques, Tony. *Dictionary of Battles and Sieges: A Guide to 8,500 Battles from Antiquity through the Twenty-First Century*. Westport, CT: Greenwood, 2006.

Kreps, Sarah E. *Coalitions of Convenience: United States Military Interventions after the Cold War*. New York: Oxford University Press, 2011.

Leeds, Brett Ashely, Jeffrey Ritter, Sara Mitchell, and Andrew Long. 'Alliance Treaty Obligations and Provisions, 1815–1944.' *International Interactions* 28, no. 3 (2002): 237–60. 10.1080/03050620213653

Lehmann, Todd C., and Yuri M. Zhukov. 'Until the Bitter End? The Diffusion of Surrender across Battles.' *International Organization* 73, no. 1 (Winter 2019): 133–69. 10.1017/S0020818318000358

Lyall, Jason. *Divided Armies: Inequality and Battlefield Performance in Modern War*. Princeton, NJ: Princeton University Press, 2020.

Marshall, Monty G., and Keith Jaggers. 'Polity IV Annual Time Series 1800–2011.' College Park, MD: Center for Systemic Peace, 2012. http://systemicpeace.org/polity/polity4.htm

McInnis, Kathleen J. *How and Why States Defect from Contemporary Military Coalitions*. New York: Palgrave Macmillan, 2019.

Min, Eric. 'Interstate War Battle Dataset (1823–2003).' *Journal of Peace Research* 58, no. 2 (March 2021): 294–303. 10.1177/0022343320913305

Moller, Sara Bjerg. 'Fighting Friends: Institutional Cooperation and Military Effectiveness in Multinational War.' Columbia University, 2016. 10.7916/D8125SNZ

Morey, Daniel S. 'Military Coalitions and the Outcome of Interstate Wars.' *Foreign Policy Analysis* 12, no. 4 (October 2016): 533–51. 10.1111/fpa.12083

Newnham, Randall. '"Coalition of the Bribed and Bullied?" U.S. Economic Linkage and the Iraq War Coalition.' *International Studies Perspectives* 9, no. 2 (May 2008): 183–200. 10.1111/j.1528-3585.2008.00326.x

Pilster, Ulrich. 'Are Democracies the Better Allies? The Impact of Regime Type on Military Coalition Operations.' *International Interactions* 37, no. 1 (March 2011): 55–85. 10.1080/03050629.2011.546259

Pilster, Ulrich, Tobias Böhmelt, and Atsushi Tago. 'Political Leadership Changes and the Withdrawal from Military Coalition Operations, 1946–2001.' *International Studies Perspectives* 16, no. 4 (November 2015): 463–83. 10.1111/insp.12058

Pollack, Kenneth M. 'The Influence of Arab Culture on Arab Military Effectiveness.' PhD thesis, Massachusetts Institute of Technology, 1996.

Ready, J. Lee. *Forgotten Allies: The Military Contribution of the Colonies, Exiled Governments, and Lesser Powers to the Allied Victory in World War II*. Jefferson, NC: McFarland & Company, 2012.

Reiter, Dan, and Allan C. Stam. *Democracies at War*. Princeton: Princeton University Press, 2002.

Reiter, Dan, Allan C. Stam, and Michael C. Horowitz. 'A Deeper Look at Interstate War Data: Interstate War Data Version 1.1.' *Research & Politics* 3, no. 4 (2016). 10.1177/2053168016683840

Russett, Bruce M. 'An Empirical Typology of International Military Alliances.' *Midwest Journal of Political Science* 15, no. 2 (1971): 262–89. 10.2307/2110272

Sarkees, Meredith Reid, and Frank W. Wayman. *Resort to War: 1816–2007*. Thousand Oaks, CA: CQ Press, 2010.

Schmitt, Olivier. *Allies That Count: Junior Partners in Coalition Warfare*. Washington, DC: Georgetown University Press, 2018.

Schmitt, Olivier. 'International Organization at War: NATO Practices in the Afghan Campaign.' *Cooperation and Conflict* 52, no. 4 (December 1, 2017): 502–18. 10.1177/0010836717701969

Schmitt, Olivier. 'More Allies, Weaker Missions? How Junior Partners Contribute to Multinational Military Operations.' *Contemporary Security Policy* 40, no. 1 (January 2019): 70–84. 10.1080/13523260.2018.1501999

Showalter, Dennis, Stephen Hart, and Ralph Ashby. *The Encyclopedia of Warfare*. London: Sterling Publishing, 2013.

Sun, Tzu. *The Art of War*. Translated by Samuel Griffith. New York: Oxford University Press, 1971.

Tago, Atsushi. 'Why Do States Join US-Led Military Coalitions? The Compulsion of the Coalition's Missions and Legitimacy.' *International Relations of the Asia-Pacific* 7, no. 2 (May 2007): 179–202. 10.1093/irap/lcl001

Tuchman, Barbara W. *Stilwell and the American Experience in China: 1911–1945*. New York: Random House Trade Paperbacks, 2017.

Vucetic, Srdjan. 'Bound to Follow? The Anglosphere and US-Led Coalitions of the Willing, 1950–2001.' *European Journal of International Relations* 17, no. 1 (March 2011): 27–49. 10.1177/1354066109350052

Weisiger, Alex. 'Exiting the Coalition: When Do States Abandon Coalition Partners during War?' *International Studies Quarterly* 60, no. 4 (December 2016): 753–65. 10.1093/isq/sqw029

Weisiger, Alex. 'Learning from the Battlefield: Information, Domestic Politics, and Interstate War Duration.' *International Organization* 70, no. 2 (Spring 2016): 347–75. 10.1017/S0020818316000059

Weitsman, Patricia. *Dangerous Alliances: Proponents of Peace, Weapons of War*. Stanford, CA: Stanford University Press, 2003.

Weitsman, Patricia. *Waging War: Alliances, Coalitions, and Institutions of Interstate Violence*. Stanford, CA: Stanford University Press, 2013.

Wilkins, Thomas Stow. 'Analysing Coalition Warfare from an Intra-Alliance Politics Perspective: The Normandy Campaign 1944.' *Journal of Strategic Studies* 29, no. 6 (December 2006): 1121–50. 10.1080/01402390601016592

Wolford, Scott. 'Power, Preferences, and Balancing: The Durability of Coalitions and the Expansion of Conflict.' *International Studies Quarterly* 58, no. 1 (March 2014): 146–57. 10.1111/isqu.12036

Wolford, Scott. *The Politics of Military Coalitions*. New York: Cambridge University Press, 2015.

Wolford, Scott, and Emily Hencken Ritter. 'National Leaders, Political Security, and the Formation of Military Coalitions.' *International Studies Quarterly* 60, no. 3 (September 2016): 540–51. 10.1093/isq/sqv023

Part II
Preparation

3 Exercising Escalation

Do Multinational Military Exercises Provoke Interstate Security Crises?

Kathryn M.G. Boehlefeld and Kelly A. Grieco

Against the backdrop of the war in Ukraine, in June 2022, the North Atlantic Treaty Organization (NATO) and Russia held dueling military exercises in the Baltic Sea. Sixteen NATO allies and partners tested their 'collective readiness and adaptability,' preparing as a battlefield coalition for a possible war with Russia in order to prevent that war from ever happening in the first place.[1] Coinciding with the NATO exercise, Russia's Baltic Fleet launched its own military exercise in the same seas. If the NATO-led exercise raised anxieties in Moscow, Russia's shadowing of NATO vessels and issuing a warning that 'in the event of aggressive plans' it would 'target' NATO ships alarmed observers in the West.[2] Some Western defence analysts warn this dangerous 'action-reaction cycle in terms of military exercises' could invite conflict rather than deter it.[3]

But such exercises also have important military benefits, allowing forces to improve their collective warfighting abilities in peacetime. Battlefield coalitions are typically more effective in combat when member countries have prior experience fighting side by side, but not all battlefield coalitions have such a legacy of military cooperation.[4] In peacetime, battlefield coalitions turn to multinational military exercises to build more effective cooperative defence relations, expecting them to pay wartime dividends. Though these exercises improve battlefield performance, they may simultaneously make war itself more likely.

Do multinational exercises generate conflict spirals, or do they instead contribute meaningfully to deterrence by sending credible signals of capability and resolve? Under what conditions are multinational military exercises more likely to deter challengers and reassure allies and partners, or raise the risks of inadvertent escalation and actual conflict?

These questions are important because states routinely use multinational military exercises as visible signals of political commitments and military capabilities for achieving credible deterrence, and thus directly affecting strategic stability.[5] Though MMEs have become common tools of statecraft, remarkably little research has addressed this topic.[6] Indeed, most studies of battlefield coalitions focus on wartime cooperation and military effectiveness, even though cooperative relations, including combined military

DOI: 10.4324/9781003399896-5

trainings, often precede the war.[7] Understanding the effects of MMEs beyond the battlefield – whether exercises are more likely to deter or provoke adversaries, which exercises are more effective for deterrence or carry increased risk of escalation, and what states can do to realise the full strategic benefits of exercising alongside allies and partners – is therefore critical to international security.

To address these questions, we develop a new framework for classifying MMEs as foreign policy signals. Two dimensions of MMEs – the degree of combined military integration and the recognised status of the exercise – are particularly significant. Combined military integration – the extent to which the exercise enhances interoperability and brings together the military activities of participants across different domains, functions, and levels of command – signals the ability of allied and partner forces to fight effectively as a battlefield coalition. Status recognition – whether the exercise receives recognition of its importance – signals a strong interest in maintaining the security of allied and partner countries. Combining these two metrics produces four ideal-types of MMEs for sending distinct signals about collective military capabilities and the extent and depth of political alignments.

Based on variation of these exercise signals, we develop a theory that explains the impact of MMEs on strategic stability. We theorise high-status exercises with high levels of military integration raise the dangers of crisis instability by creating temporary, localised military advantages. High-status exercises with low levels of military integration tend to contribute to peacetime instability because they send a strong signal of potential strategic realignment, and thus a shift to the balance of power. In contrast, low-status exercises generally have positive or neutral effects on strategic stability, because they, at least individually, have only marginal effects on the balance of power.

This chapter proceeds as follows. The first section reviews arguments in the existing literature about the consequences of MMEs for strategic stability. The next section presents a new typology of MMEs. The third section describes our theory, capturing the variation in the effects of MMEs on the likelihood of conflict. The fourth section subjects our theory to a plausibility probe comparing the effects of variation in the degree of military integration and status recognition of NATO-Ukrainian exercises, 1994–2021, on strategic stability. Consistent with the theory, we find the alignment and capability concerns posited by our theory drove Russian behaviour. The chapter concludes with a discussion of policy implications and next steps for future research.

3.1 Multinational Military Exercises and Strategic Stability

Multinational exercises aim to prepare battlefield coalitions for war even as they try to avert its outbreak. Exercises create and communicate collective capabilities at the operational level, but they also send broader strategic signals about a country's level of commitment to the battlefield coalition.[8] NATO refers to them as 'excellent signaling devices' for communicating 'strong messages to several audiences: to potential foes – we can do this; to Allies – we are doing this together; and to domestic populations – we are doing this for you.'[9] Military exercises are therefore often viewed as tools to bolster deterrence, but they also create security dilemmas and carry the risks of accidental and inadvertent escalation.

On the one hand, MMEs may strengthen extended deterrence – the attempt by one or more countries to dissuade an adversary from attacking an ally or partner – by improving battlefield coalition performance and communicating information about coalition capabilities and resolve to adversaries.[10] First, these exercises bolster the collective military power of the battlefield coalition and thereby reduce the likelihood an adversary could win a quick and cheap victory.[11] Second, multinational exercises are highly visible demonstrations of both capabilities and resolve.[12] According to the bargaining theory of war, private information about capabilities and resolve and incentives to misrepresent such information promote disagreements about states' relative strength that lead to deterrence failures.[13] By showcasing capabilities, exercises can improve relative strength assessments, promoting agreement. MMEs also send credible signals of resolve by tying hands and sinking costs – they create international and domestic audience costs that leaders pay *ex post* if they renege on commitments, and sink costs, requiring states to pay the costs of these trainings *ex ante* whether or not they follow through on commitments.[14] As credible signals of capabilities and resolve, multinational exercises can thus bolster deterrence.[15]

On the other hand, such trainings can trigger security dilemma dynamics, raising the risks of inadvertent and accidental escalation.[16] Even if the exercises are defensively motivated, they involve a buildup of collective capabilities which other states may find threatening, leading them to take countermeasures leading to an action-reaction cycle of spiralling escalation.[17] Multinational exercises, like any military training, also involve a risk of accidents and mishaps.[18] Even if MMEs have avoided inadvertent or accidental wars in the past, they may still be courting disaster. As Scott Sagan argues about the dangers of nuclear accidents, 'things that have never happened before happen all the time in history' and 'the lack of earlier nuclear accidents is therefore insufficient for making such a strong statement about future possibilities.'[19]

The historical record contains too many close calls to be sanguine about the potential dangers of MMEs. In 1983, for example, a NATO command-post exercise, practicing command and staff procedures for escalating from conventional to nuclear conflict, caused a 'war scare.' Soviet leadership feared the exercise meant a surprise nuclear attack was imminent. In Poland and East Germany, Soviet aircraft, some nuclear capable, were placed on 'combat alert,' as Chief of the Soviet General Staff Nikolai Ogarkov retired to a subterranean command bunker.[20] Though scholars still debate the gravity of the situation, the episode underscores the dangers of such exercises.[21]

3.2 A Typology of Multinational Military Exercises

A better understanding of the effects of MMEs on international security requires a conceptualisation of the different forms and strategic purposes of these exercises as signalling mechanisms to affect the calculations of adversaries. There are different types of multinational exercises, and the distinctions among them matter for strategic stability – that is, the risk of a conflict breaking out (crisis stability) and the incentives to engage in intense competition (peacetime stability).[22]

We propose a novel classification of MMEs based on two dimensions: the degree of combined military integration and the level of status recognition. Combined military integration refers to the extent to which the exercise adopts 'best practices' for enhancing battlefield coalition performance, standardising planning procedures, unifying command and control structures, addressing equipment compatibility issues, and creating common doctrine and tactics. Battlefield coalitions are fundamentally tools of capability aggregation, but they vary widely in their capacities to convert national resources and capabilities into effective coalition fighting power. To create a seamless battlefield, coalitions must meld and coordinate every element of combat power across domains, functions, and levels of war.[23] MMEs build and test the ability of allies and partners to integrate their military capabilities and operate effectively, so they vary significantly in the degree to which they prepare battlefield coalitions for war.

We develop two indicators to objectively assess the degree of combined military integration: the complexity of the exercise scenario and the operational breath and/or tactical depth of the training. The exercise scenario is complex if it focuses on the planning and execution of mid- to high-intensity conventional warfare, the most difficult and complex of all battlefield coalition operations. The exercise has operational breadth if it is large and requires the integration of different types of forces and combat arms within the coalition – for example, coordinating ground movements with close air support across national military lines. An exercise has tactical depth, even if small in actual numbers, so long as coordination among participants occurs at lower levels in the military echelon – for example, testing the technical interoperability of computer systems for exchanging information across tactical units within a battlefield coalition. A 'high' level of combined military integration means the exercise scenario is complex *and* demonstrates operational breadth and/or tactical depth, whereas a 'low level' indicates the absence of these shared features. In short, the greater the degree of combined military integration the participants demonstrate in the exercise, the greater their capacity to aggregate their military power against a common enemy in wartime, to fight and win coalition wars.[24]

The second dimension – the status recognition of an exercise – captures the degree to which the exercise is recognised, esteemed, and seen to confer upon its members a sense of status or import, either as a member of a prospective battlefield coalition (club status) or on a hierarchical spectrum (positional status).[25] Exercising states cannot simply proclaim an exercise has standing, it needs to be recognised and interpreted as significant by a reference group of states, including the adversary – such as great powers, regional powers, or another grouping of states based on the situational context.[26]

To assess the recognised status of a given exercise, we examine major media coverage by the reference group and senior level political or military attendance at the exercise.[27] Major media coverage by the reference group confers club status, identifying the exercising countries as members of the same battlefield coalition in the global and/or regional hierarchies.[28] Senior political and military visits – meaning, visits by general officers, senior civilian officials (e.g., undersecretary of defence) or high-ranking diplomats (e.g., country ambassador) – confer positional status. Given states have limited resources and cannot send high-level government representatives to all exercises, leadership visits create stratifications among exercises.[29] High-status exercises – with at least one major media report and a senior-level political or military visit – typically indicate a change in strategic relationships, such as forging new partnerships and deepening military cooperation. The participating countries choose to associate themselves very publicly with other participants, effectively interweaving their national identities and interests. Conversely, low-status exercises receive neither major media coverage nor leadership visits, because they are routine and consistent with existing parameters of the security partnership. Status recognition thus matters because it establishes exercise uniqueness – identifying group membership or the standing and responsibility of a security commitment.

Combining these two dimensions produce four ideal-types of MMEs (Figure 3.1). 'Preparing' MMEs involve a high degree of combined military integration and high prestige. 'Partnering' exercises involve a low degree of combined military integration but a high level of prestige. 'Pruning' exercises entail a high degree of combined military integration but low prestige. Finally, 'probing' exercises involve both low levels of combined military integration and prestige.

		Combined Military Integration	
		High	Low
Status Recognition	High	*Preparing*	*Partnering*
	Low	*Pruning*	*Probing*

Figure 3.1 A Typology of Multinational Military Exercises.

3.3 A Theory of Exercise Signals

We theorise that four ideal-type MMEs send distinct signals about collective military capabilities and the extent and depth of political alignments within a battlefield coalition. Importantly, each type of exercise and its associated signals has a different effect on strategic stability. 'Preparing' exercises tend to foster crisis instability, because they create temporary, localised military advantages and raise the dangers of first-strike advantages and surprise attacks. 'Partnering' exercises tend to destabilise peacetime strategic security signalling a potential strategic realignment and an impending power shift. In contrast, 'pruning' and 'probing' exercises generally have positive or neutral effects on strategic stability, because individually, they have marginal effects on the balance of power. Whether multinational military exercises deter military conflict or provoke escalation thus depends critically on the exercise type.

As depicted in Figure 3.2, these effects occur through two key mechanisms. A first mechanism through which exercise type may impact strategic stability is by demonstrating a new or improving capacity to wage coalition warfare. MMEs can be used to showcase new technology or new combined capabilities, like integrated command structures or common doctrine. A more complex exercise with greater operational breadth and/or tactical depth can also demonstrate the capacity of the battlefield coalition to coordinate and combine operational and tactical effects across multinational units for waging high-end conventional warfare.

A second mechanism through which exercise type may impact strategic stability is through a change in perception of alignments. Exercises are visible representations of relationships and are therefore a vehicle for demonstrating changes in relationships between states. Exercises between states without previous military ties can be seen as evidence of the formation of a new battlefield coalition, while exercises that occur more frequently may indicate a deepening commitment to a common mission. Conversely, exercises that occur less often than in the past, or lapse entirely, could indicate a battlefield coalition that is beginning to disintegrate. Taken together, these capability and relationship demonstration mechanisms suggest variation in exercise type matters for strategic stability.

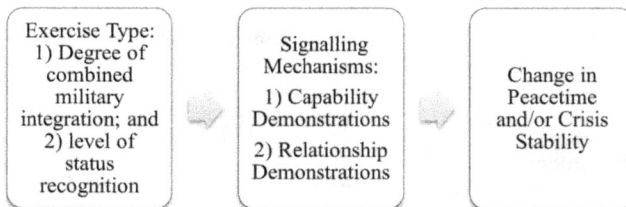

Figure 3.2 Exercise Signalling Mechanisms.

3.3.1 'Preparing' Exercises

Preparing exercises, conducted with high degrees of combined military integration and status, demonstrate most clearly an ability and willingness to wage coalition warfare. High status signals the importance of the exercise both to home and foreign audiences. The high level of combined military integration demonstrates the ability of battlefield coalitions to operate effectively during wartime. In short, preparing exercises come closest to dress rehearsals for combat. For example, NATO's annual *Defender Europe* – a complex exercise with both operational breadth and tactical depth – conducted near Russia, builds interoperability and tests allied plans and procedures. *Defender Europe* also represents a most likely scenario for a conventional military confrontation in Europe, given NATO-Russian tension. Because these exercises are large-scale and costly demonstrations of military power, they cause a temporary shift in the local military balance, provoking fears of surprise attacks and balancing behaviours, including counter exercises or military demonstrations that increase crisis instability. A single preparing exercise negatively affects crisis stability, but the magnitude of the effect is likely cumulative: each additional exercise exacerbates security dilemma dynamics.

3.3.2 'Partnering' Exercises

Partnering exercises, conducted with high status but low levels of combined military integration, signal new partnerships among states, who do not have an existing alliance or partnership. These exercises have a low level of integration because participants do not have prior experience fighting together in major battles and therefore lack common doctrine, command and control structures, and/or equipment interoperability. The exercise receives significant attention from media and government leaders, however, due to the importance the new relationship. In short, by exercising, the states signal increasing military collaboration that often precedes a battlefield coalition. During *Peace Mission 2005*, for example, China and Russia exercised together for the first time in 40 years, conducting parallel manoeuvers in the same locations. Though the level of combined integration was low, this high-profile exercise led pundits to wonder if it was a precursor to a future defence agreement or alliance.[30] Because these exercises generate fears of strategic realignments, they tend to worsen peacetime stability, provoking balancing behaviour from other states, such as efforts to divide the exercising states, boost their own military readiness, and find or bolster new allies and/or partners. The effects of partnering exercises are cumulative, as each additional exercise strengthens the signal of strategic realignment and intensifies security dilemma dynamics.

3.3.3 'Pruning' Exercises

A pruning exercise flips these dimensions: low prestige but high combined military integration. Pruning exercises generally demonstrate close and deep military cooperation required of an effective battlefield coalition, but they generate little concern because the exercise falls within the expected parameters of a long-standing partnership.[31] NATO command post exercises, involving small numbers of headquarters and staff personnel, for example, exhibit significant tactical depth in testing common procedures and interoperability, even if they rarely make international headlines or attract senior leadership visits. Such exercises, even if held with some frequency, typically do little to shift other states' understanding of the status quo, with minimal effect on strategic stability.

3.3.4 'Probing' Exercises

Finally, probing exercises display little combined military integration and low status. Such exercises often indicate an exploration of prospective battlefield coalitions without any security commitments. Because the exercising states have little, if any, prior shared military experience, they involve low combined military integration. Simultaneously, these exercises are low status, as neither the exercising states nor other powers understand them as portending a change in strategic relations. *Open Spirit* 2004, involving NATO navies and the Russian navy, for example, cleared world war era mines and other explosives in the Baltic Sea. While the exercise was large and involved Russia, it was not a high-status exercise, garnering little media coverage and no senior-level visits.[32] Because these exercises signal neither a buildup of combined military capability nor a potential strategic realignment, they tend, even in aggregate, to have minimal effects on strategic stability.

3.4 Case Study: US and NATO Exercises with Ukraine, 1994–2021

This section tests our theory's predictions against the case of US and NATO exercises conducted with Ukraine between 1994 and 2021. This historically important case features significant variation in the theory's independent variables and in the outcomes we seek to explain – whether exercises bolster or harm strategic stability. Before 2014, the United States and NATO conducted a series of high-status, low-integration trainings, or partnering exercises. After 2014, however, our first independent variable – combined military integration – changes in value, while our second independent variable – the status of the exercise – changes in value for a subset of exercises.[33] The empirical evidence in each case permits us to process trace how shifts in exercise status and capability integration affected Russia's behaviour and, in turn, strategic stability.[34]

We find that partnering exercises conducted before 2014 negatively affected peacetime stability by increasing Moscow's fears of strategic realignment and provoking balancing behaviour. In contrast, after 2014, preparing exercises with their high-status and high-capability integration negatively affected crisis stability, typically triggering an immediate and aggressive Russian response. During that same period, however, low-profile but high-integration pruning exercises generated little if any observable response from Moscow. These findings suggest the alignment and capability concerns posited by our theory drove Russian behaviour.

3.4.1 Partnering Exercises, 1994–2014

The United States and NATO conducted partnering exercises with Ukraine in the period from 1994 until the Russian invasion of eastern Ukraine in 2014. Our theory predicts that these high-status but low-integration exercises would have made Russia wary of a strategic realignment – a potential future battlefield coalition – generating more intense Russian balancing behaviour over the long term. Consistent with our theory, these exercises led Moscow to try to drive a 'wedge' between Ukraine and NATO and build up other security partnerships, intensifying security cooperation and worsening peacetime stability.

After joining the NATO Partnership for Peace program in February 1994, Ukraine participated in a series of partnering exercises with the United States and NATO.[35] These exercises were high-status events, garnering a great deal attention from the media and government officials. The first such exercise, codenamed *Cooperative Bridge* 94, saw the NATO Supreme Allied Commander General George Joulwan, the Polish prime minister, several defence ministers, and some 300 journalists in attendance.[36] Ukraine continued to participate in similar high-status US and NATO exercises over the next 20 years, including the *Peace Shield* exercise series, conducted annually between 1995 and 2005, which became the annual *Rapid Trident* exercise series after 2006. Acknowledging the exercise series conveyed high status, the Russian News Agency (TASS), commented, in 1996, 'The political significance of the exercise was underscored by the fact that the opening ceremony was attended by the defence ministers of 12 countries.'[37]

While high status, the exercises involved low levels of combined military integration. They were about building trust and relationships, not meaningful interoperability, and thus lacked operational breadth and tactical depth. In *Cooperative Bridge* 94, US soldiers compared weapons and equipment with Ukrainian and former-Warsaw Pact soldiers; evening social gatherings were a main focal point.[38] Exercise scenarios focused on less operationally demanding search and rescue, humanitarian assistance, and peacekeeping missions – or what the Russian media termed a 'Balkan scenario.'[39] 'Militarily, these were very low-level exercises, comprised mainly of platoon and company-level forces,' General Helge Hansen, commander of NATO's Allied Forces Center Europe, explained, 'But the political importance of these exercises was high.'[40] In other words, the high status of the exercises imbued them with a political and strategic significance well beyond their immediate military value.

Both Ukraine and NATO often used these exercises to signal the potential for the strategic partnership to develop into a formal invitation for Ukraine to join the alliance.[41] The *Sea Breeze* exercise series, an annual multinational naval exercise in and around the Black Sea, co-hosted by the United States and Ukrainian navies since 1997, emerged as a high-status training event, in part, because protests from Crimea's ethnic Russian majority attracted major media coverage.[42] They were also a costly signal from the United States, as Russia made clear its objections to these exercises, including a State Duma statement issued in 1997 and signed by 302 Russian deputies.[43] In persisting with these exercises, both Washington and Kyiv sent a costly signal of their commitment to maintaining a strategic partnership.[44]

Consistent with the theory, these partnering exercises increased Moscow's fears of strategic realignment between Ukraine and NATO, provoking Russian balancing behaviour. The Duma statement characterised Ukraine's participation in NATO exercises as 'a very unfriendly act with regard to Russia.'[45] Russia's foreign ministry consistently voiced its objections over the years, arguing, in 2008, the 'nature of the exercises and attempts them to present them as anti-Russian, as well as the involvement of countries form outside the region cannot but provoke certain question and certain concerns,' and asking, 'why was the Black Sea basin chosen for the drills?'[46] These statements indicate NATO-Ukrainian exercises created a security dilemma for Russia.

In response, Moscow adopted a wedge strategy to try to pull Ukraine out of NATO's orbit and draw it back into the Russian sphere of influence.[47] Russian media propagandised protests by pro-Russian nationalists in Crimea, offering them up as evidence that most Ukrainians opposed joining NATO and calling on Kyiv to hold a referendum on the issue, or risk 'the preservation of political peace in the country.'[48] The Kremlin also made a concerted effort to boost ties with Ukraine. Russian President Boris Yeltsin offered Ukraine security guarantees in the months before the first *Sea Breeze* exercise in 1997 and then arranged a Russian-Ukrainian counter-exercise a few months later.[49] 'It's the business of Black Sea states to hold exercises here,' Russia's Black Fleet commander, Viktor Kravchenko said of the Russian-Ukrainian counter-exercise, 'we will show the whole world and our enemies that there is no need for the presence of third countries in the Black Sea.'[50] As the United States was the only non-Black Sea country to participate in *Sea Breeze*, the message was clear: the United States should pull back its military presence from the Black Sea, and Ukraine should not participate in such exercises.

Russia also strengthened its military position Central Asia and the South Caucasus through bilateral and multilateral military exercises with countries of the Collective Security Treaty Organization (CSTO) – a Russian-led alliance of six former Soviet states formed in 2002. Exercise scenarios ranged from responding to transnational threats to resisting NATO aggression, with Moscow using the organisation to enhance its influence, limit opportunities for Western defence engagement with these countries, and maintain a defensive buffer around its borders.[51] Overall, the evidence suggests that US and NATO partnering exercises with Ukraine shaped Russia's long-term threat perceptions and provoked push back against what it saw as a looming European power realignment.

3.4.2 *Preparing and Pruning Exercises, 2014–2021*

After 2014, when Russia annexed Crimea and pro-Russian separatists seized parts of eastern Ukraine, the United States and NATO significantly stepped up the scale and scope of military exercises with Ukraine. Compared with earlier exercises, post-2014 NATO-Ukraine exercises displayed relatively high levels of combined military integration. During the same period, such exercises varied from low to high status. Our theory predicts that preparing exercises (high-status/high integration) would have provoked an immediate Russian response to the temporary change in the local balance of forces, while pruning exercises (low-status/high integration) would have led to little if any observable response from Moscow. The evidence supports these theoretical propositions: preparing exercises caused greater crisis instability while pruning exercises had a more positive or neutral effect on strategic stability.

NATO's preparing exercises with Ukraine displayed high levels of combined military integration and status. First, the exercises grew in both operational breadth and tactical depth.[52] Whereas *Rapid Trident* 2015 aimed broadly to 'enhance interoperability and military-to-military relationships,' the exercise series grew increasingly ambitious in subsequent years.[53] By 2021, the exercise had grown five times larger, to 6,000 multinational troops, and included combined paratrooper jumps, brigade field exercises, and 'battalion tactical exercises of a multinational battalion with combat shooting in a single combat order.'[54] Similarly, the *Sea Breeze* series more than doubled in size and expanded to include 'amphibious warfare, land manoeuvre warfare, diving operations, maritime interdiction operations, air defence, special operations integration, anti-submarine warfare, and search and rescue operations.'[55] '[T]he number and scope of military exercises grow,' Russian Foreign Minister Sergey Lavrov observed.[56] Exercise scenarios also grew more complex, shifting from a focus primarily on peacekeeping operations to high-end conventional warfare.[57] Russian media paid particular attention to this development, noting in 2016, for example, the addition of an aviation component and expanded amphibious trainings during *Sea Breeze*.[58] In sum, these preparing exercises trained Ukrainian personnel to NATO standard procedures and communications, integrating them into the planning and execution of collective defence operations.

Second, these exercises also displayed high status via media coverage and high-level government representation. Between 2014 and 2021, for example,

Itar-Tass published 53 separate stories on the *Sea Breeze* series, and 17 unique stories on the *Rapid Trident* series.[59] Senior political and military leaders also routinely attended these exercises, including Ukrainian Prime Minister Arseniy Yatsenyuk, President Volodymyr Zelensky and the dual-hatted commander of the US 6th Fleet and NATO's Naval Striking and Support Forces.[60] These preparing exercises not only demonstrated NATO's continuing commitment to Ukraine but also underscored Kyiv's goal of eventually gaining NATO membership.

Consistent with the theory, these preparing exercises negatively affected crisis stability, typically triggering an immediate Russian counterreaction. Because these large-scale exercises shifted the local balance of military power temporarily in NATO and Ukraine's favour, they alarmed Moscow. The fear of surprise attack runs deep in Russian strategic culture, particularly the employment of exercises as a cover for a massive attack on Russian territory from the West.[61] 'The great danger of such training lies in the fact that, under the guise of exercise, they can smoothly turn into a real military offensive,' Yuri Knutov, a military expert, told Russian media, pointing to a surge in NATO exercises with Ukraine.[62] Similarly, in 2017, Capt. Dmitry Litovkin, warned *Sea Breeze* 'drills may have additional goals, including practicing seizures of territories,' adding, 'Amphibious landing indicates that there is an aggressive component, including taking dominance in the region.'[63] Russian President Vladimir Putin even accused NATO of using exercises as a guise for deploying military infrastructure in Ukraine, while the Russian National Security Council warned NATO's exercises in Ukraine posed 'mounting military dangers and military threats to Russia.'[64]

Russia responded to each Ukrainian-NATO exercise, staging coinciding exercises, shadowing or interfering in Ukrainian-NATO trainings, and escalating its rhetoric and threats. First, Russia often responded by staging coinciding exercises, including snap training drills to demonstrate the combat readiness of Russian forces. In 2014, Russia conducted a snap exercise in the Black Sea for the first time during the same period as *Sea Breeze*.[65] Russia conducted its own drills immediately before, after, or during annual *Sea Breeze* and *Rapid Trident* exercises every year between 2014 and 2021, except in 2016.[66] Russia similarly staged exercises immediately before, during, or after annual *Rapid Trident* exercises, except in 2016 and 2019. These exercises highlighted Russia's military modernisation – testing air defence systems and mobilising sizable forces – and were designed to send a none-too-subtle signal of its readiness to confront NATO if need be.

Second, Russia often shadowed, harassed, or otherwise interfered with exercising aircraft and vessels. Russian warplanes and ships followed Ukrainian-NATO vessels during annual *Sea Breeze* exercises between 2014 and 2021, even dangerously sailing into the exercise's line of fire in 2019.[67] Other Russian efforts to disrupt NATO-Ukrainian exercises included conducting cyberattacks and disinformation operations, closing parts of the

Black Sea, jamming communications systems, and aggressively buzzing exercising vessels.[68] An action-reaction spiral set in, whereby NATO-Ukrainian exercise grew larger and Russia adopted ever more dangerous and escalatory countermeasures.[69]

Finally, Russia escalated its rhetoric and threats against Ukraine and NATO countries participating in preparing exercises. In 2015, as *Rapid Trident* kicked off, Moscow warned the exercises risked 'explosive consequences' and threatened to upend the peace process in eastern Ukraine.[70] Even more alarmingly, after calling for *Sea Breeze* 2021 to be cancelled, Russia threatened a 'corresponding reaction' if NATO went ahead with 'combat deployments of such scale near Russian territory.'[71] Upping the ante amid the exercise, Russia threatened to fire on any Ukrainian or NATO vessel that entered what the Kremlin considers its territorial waters.[72] This threat coincided with a particularly harrowing incident at sea, in which the Russians claimed to have fired warning shots and dropped bombs in the path of a British destroyer that crossed briefly into waters off the coast of Crimea.[73] The British denied any such actions took place, but the episode served as stark warning. Overall, NATO's preparing exercises in Ukraine provoked a dangerous game of brinkmanship with the Russians, increasing the dangers of inadvertent or accidental war.

In contrast to preparing exercises, pruning exercises conducted between Ukraine and NATO typically led to no observable response from Moscow. These numerous small-scale and lower-visibility exercises aimed to integrate Ukraine into the alliance's defence operations, down to the technical interoperability of computer systems for exchanging information.[74] They also appeared routine, at least in part, because they took place outside the territory of Ukraine and involved smaller numbers of military personnel.[75] No major media outlets covered these exercises, and no senior political or military leaders attended the opening or closing ceremonies, rendering them low-status events.

Consistent with the theory, these pruning exercises generally had a positive or neutral effect on strategic stability. Whereas preparing exercises frequently triggered a Russian provocation, pruning exercises between NATO and Ukraine resulted in little if any observable response from Moscow – no dueling exercises, public recriminations, or reported exercise interference. Individually, the Russians seem to have viewed pruning exercises as less provocative than other exercises, perhaps reasoning that no single exercise significantly impacted the balance of power, and exercises taking place outside of Ukraine's borders were not meant to signal an upgrade in Ukraine's strategic relationship with the United States or NATO.[76] Whether they shaped Russian threat perceptions in aggregate is harder to determine.

In sum, partnering exercises conducted before 2014 negatively affected peacetime stability by increasing Moscow's fears of strategic realignment and provoking balancing behaviour. After 2014, preparing exercises worsened crisis stability and triggered an immediate and aggressive Russian response, and pruning exercises had a neutral or positive effect on strategic stability. The evidence thus provides support for the causal mechanisms highlighted in our theory.

3.5 Conclusion

MMEs are essential in preparing battlefield coalitions in peacetime, but many pundits fear they are a Catch-22: in exercising, coalitions may raise the dangers of war through accidental or inadvertent escalation; in not exercising, coalitions may lack the capacity to fight together effectively on the battlefield. But not all multinational military exercises are created equal – the degree of combined military integration and the level of status recognition vary across exercises. Through the mechanisms of capability and relationship demonstrations, exercises can send distinct signals about collective military capabilities and the extent and depth of political alignments. Commentators who fear the destabilising effects of MMEs tend to focus mainly on preparing exercises for worsening crisis stability. Partnering exercises, however, may have a subtler but more lasting negative effect on peacetime stability, because they trigger more intense security competition over the long term. In contrast, pruning and probing exercises tend to have positive or neutral effects on strategic stability, suggesting scholars and policymakers alike would benefit from a more nuanced understanding of such exercises.

Building on these findings, future research could explore several theoretical and empirical avenues. First, these arguments require further testing to establish the external validity of our theory and preliminary findings. Second, collecting data on the frequency of different exercises could help to identify whether some coalitions prefer certain types and under what conditions coalitions are likely to employ different kinds of exercises. Finally, a systematic assessment of variation in the target's response to exercises, specifically under what conditions targets are more likely to respond with public statements, capability demonstrations, snap exercises, or some other provocation.

Given the growing importance of MMEs, our findings have several implications for US foreign policy and NATO, especially as Washington and Brussels aim to bolster deterrence and defence on the alliance's eastern flank. First, they may want to place more emphasis on pruning exercises, which enhance allied interoperability without typically provoking an aggressive Russian response. Second, NATO will still need to conduct preparing exercises to both train large-scale complex manoeuvers and signal the alliance has both the capability and commitment to deter and defend against Russian aggression. However, it may want to consider additional measures to mitigate against Russian overreaction, including moving these large-scale exercises farther from Russian territory. Finally, the United States, in particular, should consider carefully the unintended consequences of partnering exercises. Because these exercises involve a low level of combined military integration, it is easy to view them as neither militarily nor strategically significant. But as the NATO-Ukraine-Russia case shows, such exercises can play a significant role in shaping strategic threat perceptions and inadvertently intensify security competition.

Notes

1 'Baltops 22, The Premier Baltic Sea Maritime Exercise, Concludes in Kiel,' Press Release (Naples Italy: Naval Forces Europe and naval Striking and Support Forces NATO Public Affairs, 17 June 2022).
2 'The Russian admiral warned: We will target NATO ships in the Baltic,' Srbin.Info (15 April 2022).
3 Thomas Frear, Ian Kearns, and Lukasz Kulesa, *Preparing for the Worst: Are Russian and NATO Military Exercises Making War in Europe More Likely?* (London: European Leadership Network August 2015); Ralph C. Clem, 'Military Exercises as Geopolitical Messaging in NATO-Russia Dynamic: Reassurance, Deterrence, and (In)Stability,' *Texas National Security Review* 2, no. 1 (November 2018): 130–43, http://dx.doi.org/10.26153/tsw/865.
4 Rosella Cappella Zielinksi and Ryan Grauer, 'A Century of Coalitions in Battle: Incidence, Composition, and Performance, 1900–2003,' Chapter 2, this volume.
5 Vito D'Orazio, 'International Military Cooperation: From Concepts to Constructs,' doctoral dissertation, Pennsylvania State Univ., 2013.
6 Important exceptions are Vito D'Orazio, 'War Games: North Korea's Reaction to US and South Korean Military Exercises,' *Journal of East Asian Studies* 12, no. 2 (2012): 272–94, https://doi.org/10.1017/s1598240800007864; Kyle J. Wolfley, 'Military Statecraft and the Use of Multinational Exercises in World Politics,' *Foreign Policy Analysis* 17, no. 2 (April 2021), https://doi.org/10.1093/fpa/oraa022; Jordan Bernhardt and Lauren Sukin, 'Joint Military Exercises and Crisis Dynamics on the Korean Peninsula,' *Journal of Conflict Resolution* 65, no. 5 (2021): 855–88, https://doi.org/10.1177/0022002720972180; Brian Blankenship and Raymond Kuo, 'Deterrence and Restraint: Do Joint Military Exercises Escalate Conflict?' *Journal of Conflict Resolution* 66, no. 1 (2022): 3–31, https://doi.org/10.1177/00220027211023147.
7 Eric Min, 'Speaking with One Voice: Coalitions and Wartime Diplomacy,' Chapter 10, this volume; Sara Bjerg Moller, 'Learning from Losing: How Defeat Shapes Coalition Dynamics in Wartime,' Chapter 7, this volume; Dan Reiter, 'Command and Military Effectiveness in Rebel and Hybrid Battlefield Coalitions,' Chapter 6, this volume.
8 Beatrice Heuser, 'Reflections on the Purpose, Benefits, and Pitfalls of Military Exercises,' in *Military Exercises: Political Messaging and Strategic Impact,* ed. Beatrice Heuser, Tormod Heier, and Guillaume Lasconjarias (Rome: NATO Defense College, 2018), https://doi.org/10.1515/sirius-2018-3015.
9 Laura Martinho, 'NATO Exercises – Evolution and Lessons Learned,' Report 137 DSCFC 19 E (Brussels: NATO Parliamentary Assembly, 13 October 2019), 3.
10 Glenn Snyder, *Deterrence and Defense: Toward a Theory of National Security* (Princeton, NJ: Princeton University Press, 1961), 15–16; John J. Mearsheimer, *Conventional Deterrence* (Ithaca, NY: Cornell University Press, 1985), 15.
11 Mearsheimer, *Conventional Deterrence*, 23–24.
12 Evan B. Montgomery, 'Signals of Strength: Capability Demonstrations and Perceptions of Military Power,' *Journal of Strategic Studies* 43, no. 2 (2020): 309–30, https://doi.org/10.1080/01402390.2019.1626724.

13 James. D. Fearon, 'Rationalist Explanations for War,' *International Organization* 49, no. 3 (1995): 379–414, https://doi.org/10.1017/s0020818300033324.

14 James D. Fearon, 'Domestic Political Audiences and the Escalation of International Disputes,' *American Political Science Review* 88 (1994): 577–92, https://doi.org/10.2307/2944796; Paul K. Huth, 'Reputations and Deterrence: A Theoretical and Empirical Assessment,' *Security Studies* 7, no. 1 (1997): 72–99, https://doi.org/10.1080/09636419708429334; Vesna Danilovic, 'The Sources of Threat Credibility in Extended Deterrence,' *Journal of Conflict Resolution* 45, no. 3 (2001): 341–69, https://doi.org/10.1177/0022002701045003005.

15 Thomas S. Schelling, *Arms and Influence* (New Haven, CT: Yale University Press, 1966), 35–59.

16 Robert Jervis, 'Cooperation under the Security Dilemma,' *World Politics* 30, no. 2 (1978): 167–214, https://doi.org/10.2307/2009958. MMEs can also present entrapment risks and moral hazard dangers. See Glenn Snyder, *Alliance Politics* (Ithaca, NY: Cornell University Press, 1997); Tongfi Kim, 'Why Alliances Entangle But Seldom Entrap States,' *Security Studies* 20, no. 3 (2011): 350–77, https://doi.org/10.1080/09636412.2011.599201; Roseann W. McManus and Mark David Nieman, 'Identifying the Level of Major Power Support Signaled for Protégés,' *Journal of Peace Research* 56, no. 3 (2019): 364–78, https://doi.org/10.1177/0022343318808842; and Blankenship and Kuo, 'Deterrence and Restraint,' 10–14.

17 Bernhardt and Sukin, 'Joint Military Exercises and Crisis Dynamics on the Korean Peninsula,' 870–81.

18 Barbara Starr, '2 Russian Aircraft Make Unsafe Intercept of US Air Force B-52 Bomber,' CNN, 29 August 2020.

19 Scott D. Sagan, *The Limits of Safety: Organizations, Accidents, and Nuclear Weapons* (Princeton, NJ: Princeton University Press, 1993), 12.

20 Nate Jones, ed., *Able Archer 83: The Secret History of the NATO Exercise That Almost Triggered Nuclear War* (New York: The New Press, 2016).

21 Simon Miles, 'The War Scare That Wasn't: Able Archer 83 and the Myths of the Second Cold War,' *Journal of Cold War Studies* 22, no. 3 (2020): 86–118, https://doi.org/10.1162/jcws_a_00952.

22 There is no single agreed upon definition of strategic stability in the extant literature. This chapter defines the term broadly, as a combination of crisis stability and peacetime stability – two manifestations of the same phenomenon at different time points. Crisis stability refers to the likelihood of immediate crises and conflicts, whereas peacetime stability denotes a long-range build-up in tensions and intensifying competition between states. Our conception of peacetime stability draws from Edward Warner, quoted in Acton. To measure peacetime stability, we examine three key indicators of escalating tensions and intensifying strategic competition: (1) adversary statements (leaders/government documents) about the exercises, assessing whether they express fears of strategic realignment and issue coercive threats; (2) increase capabilities and trainings and/or pursue new security partners or build up existing security partnerships; and (3) attempt to divide the exercising states. To measure crisis stability, we examine two key indicators: (1) adversary statements (leaders/government documents) about the exercises, assessing whether they express fears of surprise attack or issue coercive threats; (2) engage in force demonstrations (e.g., reciprocal exercises, weapons tests); or (3) attempt to disrupt the exercise (e.g., shadow exercising aircraft or vessels). See Todd S. Sechser, Neil Narang, and Caitlin Talmadge, 'Emerging Technologies and Strategic Stability in Peacetime, Crisis, and War,' *Journal of Strategic Studies* 42, no. 6 (2019): 727–35, https://doi.org/10.1080/01402390.2019.1626725; James M. Acton, 'Reclaiming Strategic Stability,' in *Strategic Stability:*

Contending Interpretations, ed. Elbridge A. Colby and Michael S. Gerson (Carlisle, PA: Strategic Studies Institute and U.S. Army War College Press, 2013), 117–46, https://doi.org/10.21236/ada572928.

23 Grieco, Kelly A., 'War by Coalition: The Effects of Coalition Military Institutionalization on Coalition Battlefield Effectiveness,' doctoral dissertation, Massachusetts Institute of Technology, 2016, 37–39.

24 Grieco, 'War by Coalition,' 33–5.

25 Pål Røren, 'On the Social Status of the European Union,' *Journal of Common Market Studies* 58, no. 3 (2020): 706–22, https://doi.org/10.1111/jcms.12962.

26 Jonathan Renshon, *Fighting for Status: Hierarchy and Conflict in World Politics* (Princeton: Princeton University Press, 2017), 43–5.

27 Major media outlets refers to the leading news services (e.g., Al Jazeera, Associated Press, Reuters, BBC) and the top daily national newspapers of the countries making up the reference group.

28 Tristen Naylor, *Social Closure and International Society: Status Groups from the Family of Civilized Nations to the G20* (New York: Routledge, 2018); Deborah Welch Larson, 'New Perspectives on Rising Powers and Global Governance: Status and Clubs,' *International Studies Review* 20, no. 2 (2018): 247–54, https://doi.org/10.1093/isr/viy039.

29 Marina G. Duque, 'Recognizing International Status: A Relational Approach,' *International Studies Quarterly* 62, no. 3 (2018): 577–92, https://doi.org/10.1093/isq/sqy001.

30 'First China-Russia War Games Begin,' *The Guardian*, 18 August 2005; Richard Weitz, 'Assessing Russian-Chinese Military Exercises,' *Small Wars Journal* (30 September 2009).

31 Pruning is a key part of tending a garden, encouraging healthy growth and flowering. Similarly, pruning exercises are an important piece of regular maintenance to keep a battlefield coalition effective; these exercises keep cooperative capabilities maintained.

32 'Intl Mine-Clearing Training at Sea Open Spirit 2004 Starting in Lithuania,' *Baltic News Service*, 3 September 2004.

33 The reference group is the major regional powers – France, Germany, Russia, the United Kingdom, and the United States. As the most important countries in the European system, this group decides the European status hierarchies. Major media outlets are the leading news services and daily national newspapers of the reference group: Associated Press, Agence-France Presse, BBC, Deutsche Presse-Agentur, Le Figaro, *Financial Times*, Itar-Tass, Izvestia, Le Monde, Frankfurter Rundschau, *The Moscow Times*, *The New York Times*, *The Washington Post*, and Reuters, Süddeutsche Zeitung, and Ria Novosti.

34 Alexander L. George and Andrew Bennett, *Case Studies and Theory Development in the Social Sciences* (Cambridge, MA: MIT Press, 2004).

35 Jeffrey Simon, 'Partnership for Peace: Stabilizing the East,' *Joint Forces Quarterly* 5 (1994): 36–45.

36 Sergei Anisimov, 'Cooperative Venture '94,' *Current Digest of the Post-Soviet Press* 46, no. 16 (16 November 1994): 22; Rick Atkinson, 'Poland Hosts Mission Improbable – NATO Games with Ex-Warsaw Pact,' *The Washington Post*, 13 September 1994; Jane Perlez, 'The Cold War Armies Meet, Just to Link Arms,' *The New York Times*, 15 September 1994.

37 'Multinational Military Exercise Ends in Ukraine,' TASS, 7 June 1996.

38 Jane Perlez, 'The Cold War Armies Meet, Just to Link Arms.'

39 Kori N. Schake, 'NATO Chronicle: New World Disorder,' *Joint Forces Quarterly* (1999): 21; Galina Nekrasova 'Ukraine-Army-Drill. Ukraine Invites 50 States to Take Part in Peacemakers' Drill,' Itar-Tass, 23 February 2001.

40 Gen Helge Hansen, 'Training and Exercises for Partnership for Peace' (1995).
41 NATO membership itself was never more than a vague promise. Daniel Dombey, 'US Gives Way on NATO for Georgia and Ukraine,' *Financial Times*, 26 November 2008.
42 Lev Ryabchinkov, 'Ukraine-Protest-Exercise,' Itar-Tass, 7 November 1997; Valery Rzhevsky, 'Ukraine Seaside City Aroused by US Warplanes,' Itar-Tass, 30 May 2006; 'Deputies of Odessa Council against Sea Breeze drill,' TASS, 9 June 2008; 'Left-Wing Parties Try to Upset Sea Breeze-2011 Exercise Opening Ceremony,' Itar-Tass (6 June 2011).
43 'Duma Says Sea Breeze Exercise Violates Cooperation Rules,' Itar-Tass, 3 September 1997.
44 Natasha Lisova, 'Anti-NATO Protests Hurt Ukraine's Image but Not Kiev's NATO Aspirations, Official Says,' Associated Press, 13 June 2006.
45 'Duma Says Sea Breeze Exercise Violates Cooperation Rules,' Itar-Tass, 3 September 1997.
46 'Russia Concerned over US-Ukraine Black Sea Military Exercises,' Ria Novosti, 18 July 2008.
47 Timothy W. Crawford, *Power to Divide: Wedge Strategies in Great Power Competition* (Ithaca, NY: Cornell University Press, 2021).
48 'Russian Parliament Blasts Ukraine's Plans to Joint NATO,' Ria Novosti, 7 June 2006; 'Most Residents of Ukraine Oppose Integration in NATO,' Itar-Tass, 5 June 2006.
49 'Ukrainian Official Unhappy with Russian Visit,' Associated Press, 17 August 1997.
50 Steve Gutterman, 'Russian Fleet Leader Denounces Foreign Exercises in Black Sea,' Associated Press, 26 September 1997.
51 Richard Weitz, *Assessing the Collective Security Treaty Organization: Capabilities and Vulnerabilities* (Carlisle, PA: United States Army Strategic Studies Institute, 2018).
52 Joint Statement of the NATO-Ukraine Commission, Press Release, 124, 4 September 2014.
53 'Exercise Rapid Trident 2015 Begins in Ukraine,' Press Release (Wiesbaden, Germany: US Army Europe Public Affairs 21 July 2015).
54 Menegay and Valles, 'US, NATO, Ukraine Enhance Operability with Rapid Trident Exercise.'
55 'US Sixth Fleet announces Sea Breeze 2021 participation,' Press Release (Naples, Italy: US Sixth Fleet, 21 June 2021).
56 'Moscow Concerned about NATO's Policy of Containing Russia – Lavrov,' Itar-Tass, 2 November 2016.
57 Ministry of Defense and the General Staff of the Armed Forces of Ukraine, *The White Book 2016: The Armed Forces of Ukraine* (Kyiv: Ministry of Defense of Ukraine, 2017), 50.
58 'Ukrainian-US Military Exercises Sea Breeze Begin in Black Sea,' Itar-Tass, 18 July 2016.
59 These numbers generated from a search of Itar-Tass reports available from East View Information Services.
60 'NATO Membership among Ukraine's Priorities – Yatsenyuk,' Itar-Tass, 1 September 2015; 'President at the Opening Ceremony of Rapid Trident 2020,' Press Release (Kyiv, Ukraine: President of Ukraine, 17 September 2020); 'Partner Nations Fly Together during Exercise Sea Breeze 2021,' Press Release (Naples, Italy: US Naval Forces Europe-Africa/US Sixth Fleet, 5 July 2021).
61 Jean-Christophe Romer, *La pensée stratégique russe au XXᵉ siècle* (Paris: Economica 1997), 8–9.

62 Taran and Komarova, 'Provocative Approach.'
63 'US, Ukraine "May Practice Seizing Territories" during Joint Drills in Black Sea,' Sputnik News, 10 July 2017.
64 Robyn Dixon, 'The US-Ukraine Sea Breeze Exercises, Explained,' *The Washington Post*, 2 July 2021; 'NATO's Military Activity Leads to Increased Military Threats for Russia – Security Council,' Itar-Tass, 14 July 2021.
65 Joshua Kucera, 'Dueling NATO, Russia Naval Exercises on Black Sea,' Eurasianet, 11 July 2014.
66 Russia also did not hold dueling drills amid *Rapid Trident* 2019.
67 Ryan Pickrell, 'A Russian Destroyer Allegedly Sailed into the Line of Fire during Multinational Shooting Drills in the Black Sea,' Business Insider, 12 July 2019.
68 Sam LaGrone, 'NATO Ship in Black Sea Byzzed by Russian Planes, Russia Disputes Account,' USNI News, 9 September 2014; Damien Sharkov, 'Russia Reportedly Closes Quarter of Black Sea till 19 August,' BBC, 30 July 2019; Yuri Lapaiev, 'Russian Disinformation Shadows Ukrainian-British-US Joint Endeavor 2020 Exercise,' Eurasia Daily Monitor 17/140 (2020); Mark Episkopos, 'The Seabreeze Military Exercise Are Stoking Tension in the Black Sea,' *The National Interest*, 8 July 2021.
69 Clem, 'Military Exercise as Geopolitical Messaging in the NATO-Russia Dynamic,' 139.
70 Julian Robinson, 'Russia Warns US-Led Military Exercises May Have "Explosive Consequences" as 1,800 Troops from 18 Countries Being "Morale Boosting" Trailing Drills in Western Ukraine,' MailOnline, 20 July 2015.
71 'Sea Breeze Exercise Has Outspokenly Anti-Russian Implications – Zakharova,' TASS, 1 July 2021.
72 'Sea Breeze 21 Begins in the Black Sea after Russia Threatens to Fire on "Intruding" Warships,' *Navy Times*, 28 June 2021.
73 Episkopos, 'The Seabreeze Military Exercise Are Stoking Tension in the Black Sea.'
74 Interfax-Ukraine, 'Ukrainian Military Personnel Take Part in NATO Multinational Exercises CWIS 2019,' *Kyiv Post*, 25 June 2019, available at https://www.kyivpost.com/ukraine-politics/ukrainian-military-personnel-take-part-in-nato-multinational-exercises-cwix-2019.html.
75 '300 Romanian and Foreign Servicemen Participate in "Platinum Eagle 19.1,"' Agerpres, 13 March 2019; 'Exercise Platinum Eagle 19.1 at the End,' *Tactica Magazine*, 27 March 2019, available at https://www.tacticamagazine.com/2019/03/27/exercise-platinum-eagle-19-1-at-the-end/; Eduard Pasco, 'Multinational Exercise in Tulcea with Javelin and Anti-Tank Missiles,' *Defense Romania*, 19 September 2019.
76 Stephen Watts et al., *Deterrence and Escalation in Competition with Russia* (Washington, DC: Rand Corporation, 2022).

Bibliography

Acton, James M. 'Reclaiming Strategic Stability.' In *Strategic Stability: Contending Interpretations*, edited by Elbridge A. Colby and Michael S. Gerson, 117–46. Carlisle, PA: Strategic Studies Institute and U.S. Army War College Press, 2013. 10.21236/ada572928.

Anisimov, Sergei. 'Cooperative Venture '94.' *Current Digest of the Post-Soviet Press* 46, no. 16 (16 November 1994): 22.

Atkinson, Rick. 'Poland Hosts Mission Improbable – NATO Games with Ex-Warsaw Pact.' *The Washington Post*, 13 September 1994.

'Baltops 22, The Premier Baltic Sea Maritime Exercise, Concludes in Kiel.' Press Release. Naples Italy: Naval Forces Europe and naval Striking and Support Forces NATO Public Affairs, 17 June 2022.

Bernhardt, Jordan, and Lauren Sukin. 'Joint Military Exercises and Crisis Dynamics on the Korean Peninsula.' *Journal of Conflict Resolution* 65, no. 5 (2021): 855–88. 10.1177/0022002720972180

Blankenship, Brian, and Raymond Kuo. 'Deterrence and Restraint: Do Joint Military Exercises Escalate Conflict?' *Journal of Conflict Resolution* 66, no. 1 (2022): 3–31. 10.1177/00220027211023147

Cappella Zielinski, Rosella, and Ryan Grauer. 'Organizing for Performance: Coalition Effectiveness on the Battlefield.' *European Journal of International Relations* 26, no. 4 (2020): 953–78.

Clem, Ralph C. 'Military Exercises as Geopolitical Messaging in NATO-Russia Dynamic: Reassurance, Deterrence, and (In)Stability.' *Texas National Security Review* 2, no. 1 (November 2018): 130–43. 10.26153/tsw/865.

Crawford, Timothy W. *Power to Divide: Wedge Strategies in Great Power Competition*. Ithaca, NY: Cornell University Press, 2021.

Danilovic, Vesna. 'The Sources of Threat Credibility in Extended Deterrence.' *Journal of Conflict Resolution* 45, no. 3 (2001): 341–69. 10.1177/0022002701045003005

'Deputies of Odessa Council against Sea Breeze drill.' TASS, 9 June 2008.

Dixon, Robyn. 'The US-Ukraine Sea Breeze Exercises, Explained.' *The Washington Post*, 2 July 2021.

Dombey, Daniel. 'US Gives Way on NATO for Georgia and Ukraine.' *Financial Times*, 26 November 2008.

D'Orazio, Vito. 'International Military Cooperation: From Concepts to Constructs.' Doctoral dissertation, Pennsylvania State Univ., 2013.

D'Orazio, Vito. 'War Games: North Korea's Reaction to US and South Korean Military Exercises.' *Journal of East Asian Studies* 12, no. 2 (2012): 272–94. 10.1017/s1598240800007864

'Duma Says Sea Breeze Exercise Violates Cooperation Rules,' Itar-Tass, 3 September 1997.

Duque, Marina G. 'Recognizing International Status: A Relational Approach.' *International Studies Quarterly* 62, no. 3 (2018): 577–92. 10.1093/isq/sqy001.

Episkopos, Mark. 'The Seabreeze Military Exercise Are Stoking Tension in the Black Sea.' *The National Interest*, 8 July 2021.

'Exercise Rapid Trident 2015 Begins in Ukraine.' Press Release. Wiesbaden, Germany: US Army Europe Public Affairs, 21 July 2015. https://www.army.mil/article/152586/exercise_rapid_trident_2015_begins_in_ukraine

Fearon, James D. 'Domestic Political Audiences and the Escalation of International Disputes.' *American Political Science Review* 88 (1994): 577–92. 10.2307/2944796

Fearon, James D. 'Rationalist Explanations for War.' *International Organization* 49, no. 3 (1995): 379–414. 10.1017/s0020818300033324

Fearon, James D. 'Signaling Foreign Policy Interests.' *Journal of Conflict Resolution* 41, no. 1 (1997): 68–90.

'First China-Russia War Games Begin.' *The Guardian*, 18 August 2005.

Foggo, Adm James G. 'On the Horizon: Navigating the European and African Theatres.' https://www.c6f.navy.mil/Media/transcripts/Article/2043822/on-the-horizon-navigating-the-european-and-african-theatres-episode-11/

Frear, Thomas, Ian Kearns, and Lukasz Kulesa. *Preparing for the Worst: Are Russian and NATO Military Exercises Making War in Europe More Likely?*. London: European Leadership Network, August 2015. https://www. europeanleadershipnetwork.org/wp-content/uploads/2017/10/Preparing-for-the-Worst.pdf

Friberg, John. 'Sea Breeze 2021 – An Exercise in the Black Sea.' SOF News, 8 July 2021. https://sof.news/exercises/sea-breeze-2021/

George, Alexander L., and Andrew Bennett. *Case Studies and Theory Development in the Social Sciences*. Cambridge, MA: MIT Press, 2004.

Grieco, Kelly A. 'War by Coalition: The Effects of Coalition Military Institutionalization on Coalition Battlefield Effectiveness.' Doctoral dissertation, Massachusetts Institute of Technology, 2016.

Gutterman, Steve. 'Russian Fleet Leader Denounces Foreign Exercises in Black Sea.' Associated Press, 26 September 1997.

Hansen, Gen Helge. 'Training and Exercises for Partnership for Peace' (1995). https://www.csdr.org/95Book/Hansen.htm

Heuser, Beatrice, 'Reflections on the Purpose, Benefits, and Pitfalls of Military Exercises.' In *Military Exercises: Political Messaging and Strategic Impact*, edited by Beatrice Heuser, Tormod Heier, and Guillaume Lasconjarias, 9–26. Rome: NATO Defense College, 2018. 10.1515/sirius-2018-3015

Hockstader, Lee. 'Suspicions Lurks at Summit – Military Exercise Alarms Russians Already Wary of NATO Plan.' *The Washington Post*, 20 March 1997.

Huth, Paul K. 'Reputations and Deterrence: A Theoretical and Empirical Assessment.' *Security Studies* 7, no. 1 (1997): 72–99. 10.1080/09636419708429334

Interfax-Ukraine. 'Ukrainian Military Personnel Take Part in NATO Multinational Exercises CWIS 2019.' *Kyiv Post*, 25 June 2019. Available at https://www. kyivpost.com/ukraine-politics/ukrainian-military-personnel-take-part-in-nato-multinational-exercises-cwix-2019.html

'Intl Mine-Clearing Training at Sea Open Spirit 2004 Starting in Lithuania.' Baltic News Service, 3 September 2004.

Jervis, Robert. 'Cooperation under the Security Dilemma.' *World Politics* 30, no. 2 (1978): 167–214. 10.2307/2009958

Joint Statement of the NATO-Ukraine Commission, Press Release (2014) 124 (4 September 2014). https://www.nato.int/cps/en/natohq/news_112695.htm

Jones, Nate (ed.). *Able Archer 83: The Secret History of the NATO Exercise That Almost Triggered Nuclear War*. New York: The New Press, 2016.

Kim, Tongfi. 'Why Alliances Entangle but Seldom Entrap States.' *Security Studies* 20, no. 3 (2011): 350–77. 10.1080/09636412.2011.599201

Kucera, Joshua. 'Dueling NATO, Russia Naval Exercises on Black Sea.' Eurasianet, 11 July 2014.

LaGrone, Sam. 'NATO Ship in Black Sea Buzzed by Russian Planes, Russia Disputes Account.' USNI News, 9 September 2014.

Lapaiev, Yuri. 'Russian Disinformation Shadows Ukrainian-British-US Joint Endeavor 2020 Exercise.' *Eurasia Daily Monitor* 17, no. 140 (2020). https://jamestown.org/ program/russian-disinformation-shadows-ukrainian-british-us-joint-endeavor-2020-exercise/

Larson, Deborah Welch. 'New Perspectives on Rising Powers and Global Governance: Status and Clubs.' *International Studies Review* 20, no. 2 (2018): 247–54. 10.1093/isr/viy039.

'Left-Wing Parties Try to Upset Sea Breeze-2011 Exercise Opening Ceremony.' Itar-Tass, 6 June 2011.

Lisova, Natasha. 'Anti-NATO Protests Hurt Ukraine's Image but Not Kiev's NATO Aspirations, Official Says.' Associated Press, 13 June 2006.

Martinho, Laura. 'NATO Exercises – Evolution and Lessons Learned.' Report 137 DSCFC 19 E. Brussels: NATO Parliamentary Assembly, 13 October 2019. https://docslib.org/doc/8987068/report-137-dscfc-19-e-nato-exercises-evolution-and-lessons-learned-pdf

Mearsheimer, John J. *Conventional Deterrence*. Ithaca, NY: Cornell University Press, 1985.

Menegay, Sgt. 1st Class Chad and Capt. Aimee Valles. 'US, NATO, Ukraine Enhance Interoperability with Rapid Trident Exercise,' Press Release. Fort Bragg, NC: 22nd Mobile Public Affairs Detachment, 20 September 2021. https://www.army.mil/article/250444/us_nato_ukraine_enhance_interoperability_with_rapid_trident_exercise

McManus, Roseann W., and Mark David Nieman. 'Identifying the Level of Major Power Support Signaled for Protégés.' *Journal of Peace Research* 56, no. 3 (2019): 364–78. 10.1177/0022343318808842

Miles, Simon. 'The War Scare That Wasn't: Able Archer 83 and the Myths of the Second Cold War.' *Journal of Cold War Studies* 22, no. 3 (2020): 86–118. 10.1162/jcws_a_00952

Ministry of Defense and the General Staff of the Armed Forces of Ukraine. *The White Book 2016: The Armed Forces of Ukraine*. Kyiv: Ministry of Defense of Ukraine, 2017.

Ministry of Defense and the General Staff of the Armed Forces of Ukraine. *White Book 2014: The Armed Forces of Ukraine*. Kyiv: Ministry of Defense of Ukraine, 2015.

Montgomery, Evan B. 'Signals of Strength: Capability Demonstrations and Perceptions of Military Power.' *Journal of Strategic Studies* 43, no. 2 (2020): 309–30. 10.1080/01402390.2019.1626724

'Moscow Concerned about NATO's Policy of Containing Russia – Lavrov.' Itar-Tass, 2 November 2016.

'Most Residents of Ukraine Oppose Integration in NATO.' Itar-Tass, 5 June 2006.

'Multinational Military Exercise Ends in Ukraine.' TASS, 7 June 1996.

'NATO Allies and Partners Ready for Exercise SEA BREEZE 21.' Press release. Mons, Belgium: SHAPE Public Affairs Office, 25 June 2021. https://shape.nato.int/news-archive/2021/nato-allies-and-partners-ready-for-exercise-sea-breeze-21.aspx

'NATO Membership among Ukraine's Priorities – Yatsenyuk.' Itar-Tass, 1 September 2015.

'NATO's Military Activity Leads to Increased Military Threats for Russia – Security Council.' Itar-Tass, 14 July 2021.

Navy Times staff. 'Sea Breeze 21 Begins in the Black Sea after Russia Threatens to Fire on "Intruding" Warships.' *Navy Times*, 28 June 2021. https://www.navytimes.com/news/your-navy/2021/06/28/sea-breeze-21-begins-in-the-black-sea-after-russia-threatens-to-fire-on-intruding-warships/

Naylor, Tristen. *Social Closure and International Society: Status Groups from the Family of Civilized Nations to the G20*. New York: Routledge, 2018.

Nekrasova, Galina. 'Ukraine-Army-Drill. Ukraine Invites 50 States to Take Part in Peacemakers' Drill.' Itar-Tass, 23 February 2001.

'Partner Nations Fly Together during Exercise Sea Breeze 2021.' Press Release. Naples, Italy: US Naval Forces Europe-Africa/US Sixth Fleet, 5 July 2021. https://www.c6f.navy.mil/Press-Room/News/News-Display/Article/2681955/partner-nations-fly-together-during-exercise-sea-breeze-2021/

Pedi, Revecca. 'Small States in Europe as a Buffer between East and West.' In *Handbook on the Politics of Small States*, edited by Geoffrey Baldacchino and Anders Wivel, 168–88. Northampton, MA: Edwards Elgan Publishing Ltd, 2020.

Perlez, Jane. 'The Cold War Armies Meet, Just to Link Arms.' *The New York Times*, 15 September 1994. https://www.nytimes.com/1994/09/15/world/biedrusko-journal-the-cold-war-armies-meet-just-to-link-arms.html

Pickrell, Ryan. 'A Russian Destroyer Allegedly Sailed into the Line of Fire during Multinational Shooting Drills in the Black Sea.' Business Insider, 12 July 2019.

'Poland, Britain, Ukraine hold Joint Military Maneuver.' Xinhua News Agency, 26 August 2005.

'President at the Opening Ceremony of Rapid Trident 2020.' Press Release. Kyiv, Ukraine: President of Ukraine, 17 September 2020.

'Rapid Trident 19 Demonstrates Multinational Proficiency in Ukraine.' US Fed News, 30 September 2019.

Renshon, Jonathan. *Fighting for Status: Hierarchy and Conflict in World Politics*. Princeton: Princeton University Press, 2017.

'Representatives of the ICRC Participate in Rapid Trident 2019 for the First Time.' Ukrainian Government News, 21 September 2019.

Robinson, Julian. 'Russia Warns US-Led Military Exercises May Have "Explosive Consequences" as 1,800 Troops from 18 Countries Bebeing "Morale Boosting" Trailing Drills in Western Ukraine.' MailOnline, 20 July 2015.

Romer, Jean-Christophe. *La pensée stratégique russe au XXe siècle*. Paris: Economica, 1997.

Røren, Pål. 'On the Social Status of the European Union.' *Journal of Common Market Studies* 58, no. 3 (2020): 706–22. 10.1111/jcms.12962

'The Russian Admiral Warned: We Will Target NATO Ships in the Baltic.' Srbin. Info, 15 April 2022. https://srbin.info/en/svet/ruski-admiral-upozorio-drzacemo-na-nisanu-brodove-nato-a-u-baltickom-moru/?lang=lat

'Russian Parliament Blasts Ukraine's Plans to Joint NATO.' Ria Novosti, 7 June 2006.

Ryabchinkov, Lev. 'Ukraine-Protest-Exercise.' *Itar-Tass*, 7 November 1997.

Rzhevsky, Valery. 'Ukraine Seaside City Aroused by US Warplanes.' Itar-Tass, 30 May 2006.

'Saber Junction 19 Media Day.' Grafenwoehr, Germany: US 7th Army Training Command, 13 September 2019. https://www.europeafrica.army.mil/ArticleView PressRelease/Article/1959454/saber-junction-19-media-day/

Sagan, Scott D. *The Limits of Safety: Organizations, Accidents, and Nuclear Weapons*. Princeton, NJ: Princeton University Press, 1993.

Schafer, Susanne M. 'US Defense Chief Observes First US-Ukrainian Military Exercise.' Associated Press, 25 May 1995.

Schake, Kori N. 'NATO Chronicle: New World Disorder.' *Joint Forces Quarterly* 21 (1999): 18–24.

Schelling, Thomas S. *Arms and Influence*. New Haven, CT: Yale University Press, 1966.

'Sea Breeze 2021 Increases Risk of Military Incidents in Black Sea – Russian Embassy to U.S.' *Ukraine General Newswire*, 23 June 2021.

'Sea Breeze Exercise Has Outspokenly Anti-Russian Implications – Zakharova.' TASS, 1 July 2021.

Sechser, Todd S., Neil Narang, and Caitlin Talmadge. 'Emerging Technologies and Strategic Stability in Peacetime, Crisis, and War.' *Journal of Strategic Studies* 42, no. 6 (2019): 727–35. 10.1080/01402390.2019.1626725

Sharkov, Damien. 'Russia Reportedly Closes Quarter of Black Sea till 19 August.' BBC, 30 July 2019.

Simon, Jeffrey 'Partnership for Peace: Stabilizing the East.' *Joint Forces Quarterly* 5 (1994): 36–45.

Snyder, Glenn. *Deterrence and Defense: Toward a Theory of National Security.* Princeton, NJ: Princeton University Press, 1961.

Snyder, Glenn. *Alliance Politics*. Ithaca, NY: Cornell University Press, 1997.

Starr, Barbara. '2 Russian Aircraft Make Unsafe Intercept of US Air Force B-52 Bomber.' CNN, 29 August 2020. https://www.cnn.com/2020/08/29/politics/russian-aircraft-us-bomber-black-sea/index.html#:~:text=(CNN)%20Two%20Russian%20aircraft%20made,Air%20Forces%20Africa%20Public%20Affairs

'Statement on Exercise Rapid Trident 2014,' *Targeted News Service*, 3 September 2014; 'Press Release – Rapid Trident Exercise in Ukraine.' Press Release. Wiesbaden, Germany: U.S. Army Europe and Africa, 16 September 2021. https://www.europeafrica.army.mil/ArticleViewPressRelease/Article/2776365/press-release-rapid-trident-exercise-in-ukraine/

Taran, Irina, and Elizaveta Komarova. 'Provocative Approach: How Kiev and NATO Countries Are Increasing the Number of Military Exercises in Ukraine.' *Russia Today*, 4 January 2022. https://russian.rt.com/ussr/article/945278-ukraina-nato-ucheniya-2022-god-rossiya

'The Rapid Trident 2020 Exercise Officially Kicks Off.' *DVIDS*, 17 September 2020. https://www.dvidshub.net/news/379818/rapid-trident-2020-exercise-officially-kicks-off

'Ukraine, NATO launch Sea Breeze 2011 Military Exercise.' *Xinhua News Agency*, 6 June 2011.

'Ukraine's Def Minister to Discuss Military Cooperation in London.' TASS, 10 September 2001.

'Ukrainian, British Military Officials Pleased with Joint Exercises.' BBC, 10 September 2003.

'Ukrainian Defense Minister Attends Military Exercise in UK.' BBC, 6 September 2004.

'Ukrainian Official Unhappy with Russian Visit.' Associated Press, 17 August 1997.

'Ukrainian-US Military Exercises Sea Breeze Begin in Black Sea.' Itar-Tass, 18 July 2016.

'US Sixth Fleet Announces Sea Breeze 2021 Participation.' Press Release. Naples, Italy: US Sixth Fleet, 21 June 2021. https://www.navy.mil/DesktopModules/ArticleCS/Print.aspx?PortalId=1&ModuleId=523&Article=2664699

'US, Ukraine "May Practice Seizing Territories" during Joint Drills in Black Sea.' Sputnik News, 10 July 2017.

Watts, Stephen, Bryan Rooney, Gene Germanovich, Bruce McClintock, Stephanie Pecard, Clint Reach, and Melissa Shostak. *Deterrence and Escalation in Competition with Russia*. Washington, DC: Rand Corporation, 2022.

Weitz, Richard. 'Assessing Russian-Chinese Military Exercises.' *Small Wars Journal*, 30 September 2009. https://smallwarsjournal.com/jrnl/art/assessing-russian-chinese-military-exercises

Weitz, Richard. *Assessing the Collective Security Treaty Organization: Capabilities and Vulnerabilities*. Carlisle, PA: United States Army Strategic Studies Institute, 2018. https://publications.armywarcollege.edu/pubs/3661.pdf

Wolfley, Kyle J. 'Military Statecraft and the Use of Multinational Exercises in World Politics.' *Foreign Policy Analysis* 17, no. 2 (April 2021). 10.1093/fpa/oraa022

Zaks, Dmitry. 'Russians Bristle at NATO Sea Breeze.' *The Moscow Times*, 26 August 1997.

'Zelensky: Combat-Ready Navy amid Russian Aggression One of Main Components of Ukraine's Security in Black Sea.' *Ukraine General Newswire*, 5 July 2021.

'Zelensky Visits Naval Forces in Odesa.' UATV, 8 July 2019. https://uatv.ua/en/zelensky-visits-naval-forces-odesa/

4 When the Coalition Determines the Mission

NATO's Detour in Libya

Stéfanie von Hlatky and Thomas Juneau

War by committee has become the dominant form of warfare since the 1990s. As noted in the introduction to this volume, the majority of major battles fought since the end of the Cold War have been carried out as a collective effort.[1] How do these battlefield coalitions plan and execute their combined operations?

In this chapter, we engage in a theory-building exercise and, drawing on insights from various existing literatures, we argue that the composition of the battlefield coalition and the respective capabilities and interests of its members significantly shape the evolution of a mission's design and conduct – meaning how mandates are interpreted by the coalition and what military actions ensue. We develop this claim through a close examination of the war in Libya in 2011, and specifically Operation Unified Protector (OUP), carried out by the North Atlantic Treaty Organisation (NATO). This time-bound mission is particularly useful for theory development because it involves a diverse coalition of democratic and non-democratic states, as well as non-state actors on the ground and powerful external stakeholders like Russia.

Given such diversity, there is reason to believe that the mission dynamics uncovered by this case study may be illustrative of collective operations writ large, whether other NATO-led operations or non-NATO battlefield coalitions: many actors vie for influence, but each is faced with specific opportunities and constraints that limit their ability to drive the direction and thrust of collective action. Our argument and findings thus point to a critical limitation in existing literature on battlefield coalitions: it is as important to examine how the coalition shapes the mission as it is to study how the mission determines the coalition.

To develop and begin assessing our argument, this chapter collapsed into civil war. The scope of this article, however proceeds as follows. The first section introduces the relevant literature on alliance and coalition operations, and leads to a second section, where we provide an overview of our key hypotheses. The third section presents our case study, which relies on interviews with policymakers and military officials who were involved in OUP as well as the mining of policy documents and secondary sources. In doing so, we shift the focus from coalition formation and battlefield performance, which has received much scholarly attention, to the

DOI: 10.4324/9781003399896-6

process of mission design and execution in battlefield coalition operations. In the final section, we discuss our findings and conclude with policy recommendations and avenues for future research.

4.1 Understanding Mission Design and Execution

The reason coalition formation and performance have received significant scholarly attention is rooted in the fact that, despite the challenges of coordination under fire, actors in the international system have recognised that fighting with partners is worth the hassle.[2] That is, whether they form for normative reasons like political legitimacy or more tangible benefits like resource pooling, basing rights, and airspace access, collectives created for the purpose of fighting together seem to give belligerents a better chance at success on the battlefield.[3] What happens between the formation of a collective and victory, however, is just as important as the book-end phenomena, as there are many ways in which the path connecting the two may fail. Warfighting collectives, whether loosely aligned ad hoc coalitions or deeply institutionalised alliances, rarely have a clear mandate and mission template to guide their actions when they enter combat. When mandates do exist, they are noteworthy for being couched in vague language and the result of a lowest-common-denominator approach. To make such vague instructions concrete, and maximise the collective's likelihood of battlefield success, members must thus often negotiate over the guidance to be given to fighting forces. The fact of negotiation means that belligerents' previous experiences, available capabilities, and, to put it bluntly, power and interests are all likely to shape the final form of group's mission and its execution. Critically, therefore, studying a collective's composition is essential for understanding how it conceives of and implements its operations.

To study this relationship, two clearer conceptualisations are needed. First, while the collectives we discuss are drawn solely from formal alliances, they need not be. Our goal is to develop a framework to uncover the dynamics and operations of battlefield coalitions writ large, and preexisting treaty obligations are not a necessary condition for all such groups. Second, we define the term 'mission' here to include any mandate given to a battlefield coalition and the range of military tasks that are conducted in support of that guidance. In other words, a mission can be understood as having two distinct phases: the design, wherein military planners and others draw up plans interpreting the mandate, and the execution, which entails carrying out the plan through specific military tasks and can range from air strikes to special forces raids to full ground operations. And again, because mandates are often unclear or vague, identifying the process by which a mission goes from design to execution is critical for our understanding of how battlefield coalitions operate.

Existing literature provides some guidance in thinking about coalitions' mission design and execution, but is limited in important ways. For example, the literature on alliance management and politics offers some important insights, including that institutionalisation is likely to amplify the voices of all members of the group.[4] However, the dynamics of mission design and execution in a group that has been bound together by a treaty is likely to look quite different from those in a group that has not. Consider, for example, Saudi Arabia's military intervention in Yemen in March 2015, under a vague mandate based on United Nations Security Council Resolution 2216 and at the head of an ad hoc coalition of ten members. Not all of the ten mostly Arab and Sunni states in the coalition contributed equally; some have dropped out and only the United Arab Emirates (UAE) made more than a limited contribution to the fighting. This imbalance in contribution is reflected in the mission design and execution: the coalition's approach and effort has been shaped primarily by the constant negotiation between the coalition's two most powerful members, Saudi Arabia and the UAE.[5]

Developing a theory of collective mission design and execution that is relevant for battlefield coalition operations writ large thus requires tying together different strands of existing literature on both alliances and coalitions to understand the internal dynamics of the groups. Accordingly, in addition to the aforementioned work on alliances, the literature on how coalitions come together is helpful for our purposes. It has identified domestic and international factors, such as ideology or alliance commitments, that explain which states participate and which states opt out of certain collective efforts, with important empirical findings about how potential members perceive coalition objectives before making any formal commitments. These perceptions are often indicative of broad areas of agreement and disagreement in the national preferences of states comprising a coalition, and are likely to endure beyond the initial decision to opt in or out.[6] Another segment of the literature has highlighted the impact of domestic constraints, such as regime type and the corresponding autonomy of the executive, when comparing coalition members' varied contributions to military operations.[7] The particular domestic constraints at play in any given coalition, whether material or political, determine what is feasible or not in a very practical sense, independent of the group's overarching objectives. Finally, other scholars have examined how coalition members interpret strategic directives in distinct ways, which can cause variation in how different members' forces operate at the tactical level.[8] As this literature collectively makes clear, a coalition's composition has operational consequences. While tying together the coalition-specific factors that influence operations is difficult, it can improve our understanding of how intra-coalition dynamics explain the design and execution of collective missions, delivering important theoretical insights in the process.[9]

4.2 Missions Designed in a Multinational Setting

To better probe the links between coalitions' composition and their missions, we introduce five hypotheses rooted in the insights articulated in existing scholarship. Our hypotheses emphasise the way in which a diverse array of coalition attributes may impact mission design and execution.

Hypothesis 1: The coalition's most powerful members set the key parameters of the mission.

This first hypothesis aligns with traditional realist arguments about how power yields influence in a multinational context: powerful states set the parameters of the mission and ultimately have a veto over how the mission is designed and carried out. In NATO, for example, the powerful states would include the United States, and, secondarily, the United Kingdom, France, and, in some circumstances, Germany. Similar power imbalances may also be present in coalitions operating outside an alliance framework. In the case of the Global Coalition against the Islamic State, for example, the United States, alongside a second tier of powers including France, the United Kingdom, and Turkey are significantly more powerful than other members.[10] In both of these types of cases, we might expect these big players to be disproportionately influential in mission design and execution by virtue of their preponderance. Crucially, members' relative power may be variable depending on the context: a smaller state providing a large number of troops may gain influence, while a powerful state choosing to forego a large contribution may lose some burden-sharing matters. The observable implication of this hypothesis is that the powerful members, however strength is measured, will meet informally in small groups, excluding other coalition members, and make decisions on mission design and execution, despite objections that might be voiced by other coalition partners.

While power shapes individual coalition members' influence on mission design and conduct, it is a crude variable. As Schmitt notes, the classical power narrative 'only tells a limited aspect of the story.'[11] Thus, it is likely insufficient on its own to account for mission specificities. Our next three hypotheses therefore offer more granularity by exploring how other aspects of internal coalition dynamics shape mission design and conduct.

Hypothesis 2: Intra-coalition processes, such as common operational planning doctrines and standard operating procedures, drive mission design and execution.

Militaries follow fixed operational planning processes and rigid chains of command. In a multinational setting, armed forces typically follow their national chain of command, but may also be subject to the lead state's chain of command. If an institution is lending its leadership to a mission, NATO being the most obvious example, then there may be common doctrine and alliance procedures shaping the design of the mission. In a non-institutionalised coalition, the lead state provides this military guidance, which may also impact mission design and execution in significant ways.

The observable implication of this hypothesis is that power matters, of course, but it is constrained by the procedures of the alliance or the coalition; mission design and execution are driven by power, but also by pre-established beliefs and processes.

Hypothesis 3: Coalition partners with the least restrictive caveats are able to decisively shape the mission.

Another way in which the composition of the coalition may impact the mission has to do with military caveats–restrictions on what forces can and cannot do. Caveats are determined by members' domestic political and military considerations. The presence of caveats influences the mission's pecking order: states with the fewest caveats will want more say when it comes to interpreting the mandate, since they will ultimately be responsible for carrying out a broader range of tasks, including the riskier, and often more scrutinised, elements of a mission. This dynamic was on full display in the context of the fluctuating NATO-led coalition in Afghanistan and of the US-led coalition against the Islamic State.

The observable implication of this hypothesis is that countries with heavy caveats will be partially or completely sidelined from important military discussions at key decision points.

Hypothesis 4: The coalition member tasked with commanding operations will shape mission design.

Another factor that might influence mission design is leadership. The state tasked with commanding the mission, whether it is under an institutional umbrella or not, gains access to the decision-making table, which translates into more opportunities to influence mission design and execution. This is likely to be the case whether or not the state commanding the mission is the top troop contributor, as was the case with the ISAF mission in Afghanistan, which saw a host of non-US commanders, namely from the United Kingdom, Turkey, Germany, Canada, France, and Italy.

The observable implication of this hypothesis is that command roles translate into access that may not otherwise be possible, with the country in command exercising more discretion when it comes to making certain decisions regarding the content of the mission and the coalition's core tasks.

These four hypotheses emphasise dynamics internal to a coalition and how they might affect mission design and execution. It is conceivable, however, that forces external to the coalition shape mission design and conduct.

Hypothesis 5: External actors, like local partners or rival powers, shape mission design and conduct.

Many actors external to a coalition may want to shape its evolution, with local partners and coalition rivals as the most likely influencers. Consistent with moral hazard logic, local partners like rebel groups may act more decisively on the ground if they expect the coalition to support them.[12] This kind of behaviour can pressure the coalition to expand its mission and operations beyond what it had initially planned. At the very least, such possibilities are likely to factor into coalition decision-making leading up to operations. Rival great or regional powers, by contrast, can be spoilers, imposing new or unanticipated constraints on the execution of the mission that stall, sabotage, or simply narrow the options available to the coalition. Russia, for example, tried to impose constraints on NATO-led operations in the former Yugoslavia in the 1990s and on the operations of the US-led coalition against the Islamic State, especially in Syria.

The observable implication is that these external actors, at any stage of the mission, can deploy political, military, or financial resources or articulate objections that force the hand of coalition members with respect to the design and execution of the collective mission.

4.3 Case Study Analysis: NATO's Operation Unified Protector in Libya

To assess the relative merit of our five hypotheses generated from existing scholarship, we explore the design and execution of NATO's Operation Unified Protector in 2011. In doing so, we rely on primary documents generated by governments, the United Nations (UN), and NATO, as well as ten virtual interviews with civilian and military officials from contributing coalition members and from NATO.[13] To analyse the collected data, we performed a qualitative content analysis on the official documents, testimonies, and statements to identify the distinct preferences of coalition partners. The interviews were used to fill in data gaps left by the documents and to validate information from primary and secondary sources.

4.3.1 Background

In January 2011, to Libya's west, Tunisia's long-serving President, Zine El-Abedine Ben Ali, fled the country after a few weeks of protests. Shortly after, to Libya's east, Egypt's president of 30 years, Hosni Mubarak resigned after weeks of demonstrations. Protests then rapidly spread throughout the Middle East, including to Yemen, Bahrain, and Syria.[14] Libya proved no exception: decades of accumulated anger and frustration in the face of repression, corruption, and poverty exploded into large-scale demonstrations throughout the country.[15] The opposition scored some early gains in February and March, seizing key towns. But pro-Qaddafi forces regrouped and started moving east along the coast. Loyalist forces approached Benghazi as the rebellion was accumulating defeats. This sparked mounting fears, especially in Europe, of Srebrenica- or Sarajevo-scale massacres.

Responding to the violence, on 26 February the UN Security Council (UNSC) adopted Resolution 1970, which called for 'steps to fulfill the legitimate demands of the Libyan population' and imposed sanctions on the Libyan government, including an asset freeze, a travel ban on senior officials, and an arms embargo. As the violence further spread, in early March the US, UK, France, and other NATO allies began, or intensified, work on contingency plans.[16] On 10 March, US Secretary of Defence Robert Gates met NATO's secretary-general and told him that any American role would be contingent on a new UNSC resolution and on regional endorsement. This last condition was met on 12 March, when the Arab League said it would support a no-fly zone (with only Syria and Algeria opposing). The final piece in the puzzle came on 17 March, when the Security Council adopted Resolution 1973, authorising member states 'to take all necessary measures, to protect civilians' short of a foreign occupation. Two days later, on 19 March, a coalition of the willing that included the United States, France, and the United Kingdom began striking Libya. Operation Odyssey Dawn, initially run under separate commands but coordinated under the auspices of US Africa Command, succeeded in halting Qaddafi's advance on Benghazi and destroying his regime's air defences as well as much of its airpower and heavy weapons. From the start, the mandate's vague terms fostered intense discussions at the United Nations, but also at NATO.

Despite significant internal disagreement, on 23 March, NATO agreed to enforce the arms embargo on Libya and, on 24 March, also the no-fly zone. Many allies did not intend to participate, notably Germany, but none openly opposed these steps. On the following day, Lt. Gen. Charles Bouchard, a Canadian who until then was deputy commander of Allied Joint Force Command in Naples, was chosen to take over command of the NATO mission.[17] Finally, on 27 March, NATO announced that it would implement all aspects of UNSCR 1973. As a result, on 31 March, NATO formally assumed command of the campaign and renamed it Operation Unified Protector.[18]

The situation rapidly evolved into a stalemate: after holding off pro-Qaddafi units in the early days of the intervention, NATO-backed rebels made only limited progress. As frustration in allied capitals started to mount, NATO extended the mission for an additional three months on 1 June.[19] In August, the situation started improving rapidly once Tripoli fell to the rebels. Two months later, on 20 October, Qaddafi was killed by rebel forces. The UNSC then terminated its authorisation for the use of force with Resolution 2016, effective 31 October.[20]

There are two key features of this sequence related to the definition and execution of the OUP mandate that require further investigation. First, OUP's civilian protection mandate was not clearly defined in UNSCR 1973.

There was no doubt that it involved stopping Qaddafi forces' assault on Benghazi, but there was an absence of strategic guidance beyond this point. As a result, the early days and weeks of OUP were marked by debate, among military planners and at the political level, over OUP's desired end-state. Through NATO's institutional channels, the coalition members eventually settled on a more precise meaning at the 14 April foreign ministerial meeting, where it was agreed that operations would continue until:

- Attacks against civilians have ended
- Pro-regime forces have withdrawn to bases
- The regime permits full humanitarian access to people in Libya

Second, and related to the previous point, the mission definition did not explicitly call for regime change. As many interviewees emphasised, however, there are inevitable and necessary consequences to the decision to implement a no-fly zone to protect civilians in the pursuit of these three objectives ('it is not as simple as flying planes around to prevent the Libyan air force from flying'). Some interviewees specified that it is not clear how these objectives could have been achieved without the Qaddafi regime falling. Others went further and clearly stated that regime change was an implicit but necessary consequence of these goals. Below, we show that factors internal to the coalition largely explain how this mission was designed and carried out, from the interpretation of the mandate through to the execution of specific military tasks, while external factors had a marginal impact.

4.3.2 The Coalition and the Mission

Considering the various drivers of mission design and execution highlighted in our five hypotheses, we find that the role of coalition leaders and the absence of caveats explain the bulk of the variation in terms of influence over the coalition's mandate and core tasks. That the most powerful members shape the design and conduct of a mission is not surprising in and of itself, but accounting for other intra- and extra-coalitional dynamics is necessary to explain the more granular aspects of the combined operations, as summarised in Table 4.1.

The first point of nuance our analysis brings is that it is not power alone that matters, but the nature and scope of the contributions made, combined with the US imposing caveats on its contribution (H3), which allowed for France and the United Kingdom to drive the mission in ways that would not have been possible otherwise. Moreover, Germany, a powerful state, had political influence through the North Atlantic Council, but its status did not translate into influence on mission design and execution given its very limited military contribution. As Schmitt and Adler-Nissen and Pouliot have pointed out, opportunities for influence must not only exist, but they must also be competently exploited by states through their diplomatic and military

Table 4.1 Evidence across the Five Hypotheses

	Supporting Evidence	Disconfirming Evidence
H1 Powerful members	Usual suspects: US, France, UK	Germany is a marginal player
H2 Intra-coalition processes	NATO-specific interpretation of mandate: 0 casualty guideline	Many of the processes stood up were ad hoc
H3 Caveats	UK and France have joint lead role, given few caveats	US retains high influence through prominent positions within NATO structure
H4 Command authority	Lead country benefits from increased access to key meetings	Chain of command still flows through more powerful states
H5 External actors	Evidence of coordination (rebels) and public criticism (rivals)	No unity of effort/voice (rebels), no concrete actions taken to back up criticism (rivals)

channels.[21] With OUP, the most powerful and influential members of the NATO-led coalition were the United States, France, and the United Kingdom, which were instrumental in both interpreting the mandate and identifying the key military tasks that would be deployed in support of that mandate. As one interviewee put it: 'the more a nation brings to the table, the more influence it is going to have.' We should add that those same countries were part of Operation Odyssey Dawn, the initial mission in Libya, and also had a first mover advantage when it came to standing up OUP.

Despite the famous claim that it was 'leading from behind,' first, the United States was a top contributor (a table will be provided in an online annex which details the assets of top NATO contributors).[22] In total, American forces flew more than 7,100 sorties, or about 27% of total sorties flown during OUP.[23] Yet American combat assets were mostly involved in the early phases of the mission, in March under Operation Odyssey Dawn and then in the first days of OUP, under NATO command. By early April, the United States withdrew many of its combat assets and focused more on provisioning key enablers to allow its allies to sustain their operations, notably refueling and ISTAR (intelligence, surveillance, target acquisition, and reconnaissance). The operational planning doctrine was also heavily influenced by the United States, especially when it came to targeting, as American officers held key positions in the hierarchy, notably the Supreme Allied Commander Europe (SACEUR). This poses a challenge for Hypothesis 2, which posits that multinational command structures can impose constraints on great power influence.[24]

The United States, in contrast to past coalition operations, had important caveats. President Obama himself was initially reluctant to commit American assets to Libya, having campaigned in 2008 on disengaging the United States from Iraq and Afghanistan. He insisted that the intervention should be as multilateral as possible to avoid the perception of American unilateralism.[25] He also pushed hard for NATO to be actively involved.[26] Similarly, the Obama Administration sought buy-in from regional organisation, especially the Arab League and pushed its regional partners to contribute.[27]

Unlike the United States, France was looking for a leadership role and initially preferred the mission not to be led by NATO, not because of constraints imposed by its command structure, as hinted at by Hypothesis 2, but because it preferred the EU route. Despite France's initial reluctance to accept transferring the mission under NATO command, it gave in when it became clear that there was no other viable option as far as the United Kingdom was concerned, but made sure that some political oversight would be delegated to the Libya Contact Group, outside NATO.[28] Ultimately, France's contribution to the intervention in Libya, Opération Harmattan, represented its largest air and naval engagement since the war in Kosovo in 1999. French aircraft were responsible for about 5,600 sorties, or a quarter of coalition sorties, 35% of offensive missions, and 20% of strikes.[29] France was also one of the only contributors (with the United States and the United Kingdom) able to provide essential enablers such as tankers and ISTAR assets (e.g., unmanned aerial vehicles [UAVs], reconnaissance planes, space imagery, and human intelligence).[30] By the end of the mission in October, France was responsible for the largest share of operations, especially in terms of sorties and strike missions, and at the top of the mission's pecking order along with the United Kingdom.

When international debate on the possibility of establishing a no-fly zone gathered steam in late February 2011, France quickly emerged as one of the most vocal proponents of the idea. In speeches, senior French leaders repeatedly sought to portray France's role as one of pro-active leadership, emphasising that France 'fought without respite' to ensure the protection of Libyan civilians under the responsibility to protect.[31] President Sarkozy, in particular, was very active in trying to convince counterparts on the UN Security Council to support, or at least not oppose, what eventually became UNSCR 1973, authorising air operations beyond a no-fly zone.[32] With respect to its position on the viability of the Libyan regime, France called for Qaddafi's departure as soon as early March, as this was the only path to ensure the protection of Libyan civilians.[33]

The United Kingdom conducted one-fifth of all airstrikes.[34] It was not as forceful as France in wanting a strong leadership role from the start, given the politically sensitive nature of the mission back home, but did insist that the mission be under NATO command.[35] The interplay between the two countries was thus key in setting the earliest parameters of the mission. The justification of the United Kingdom's role was carefully crafted, with Prime Minister David Cameron, articulating the conditions under which military intervention is necessary to defend humanitarian principles, echoing UN language but cuing other players towards a particular interpretation of the mission's mandate.[36] First, he emphasised the need for such an intervention both in terms of the potential for a humanitarian catastrophe and the United Kingdom's national interests, chiefly instability in Southern Europe and Libya being a country that harboured terrorists that perpetrated attacks on British soil in 1988. Second, the United Kingdom (like the United States and France) stressed the importance of regional support, primarily through seeking Arab League and African Union endorsement and contributions from regional partners. The third condition was having a clear legal basis from the UN. So, whereas France was pushing for intervention no matter what (like in Kosovo), the United Kingdom wanted to clarify the UN mandate publicly and on its terms. Early on, Prime Minister Cameron also emphasised that the intervention was about protecting people, not regime change, pushing back on accusations of mission creep stemming from parliamentary debates, questions in defence committee meetings, and in the media.[37] Over time, however, the United Kingdom's position veered closer to France's, and by April, the British Defence Secretary, Liam Fox, appeared comfortable with the idea of taking sides.[38] How the mandate was interpreted was ultimately the result of a UK-France compromise.

The United States, France, and the United Kingdom were dominant because of their power status and material contributions, but caveats on the US side made France and the United Kingdom's leadership more visible. Caveats, following the logic from Hypothesis 3, also determined who the other significant players were. In general, caveats were handled like any other NATO mission: tasks were assigned based on what individual countries agreed to do. Within OUP, the least caveated allies formed a 'striker group,' a core club of allies who participated in airstrikes and made key operational decisions. The striker group included the eight alliance members (Britain, France, Canada, Belgium, Denmark, Norway, the United States, and Italy; Norway withdrew its planes in June) who employed their aircraft for ground attack missions. There were, in total, only 130 combat aircraft, of which only 55 could carry out air-to-ground operations.[39]

Membership in the striker group undeniably provided its members with more influence on the mission, like the translation of strategic goals into operational ones. The striker group's role was to discuss 'most decisions informally prior to deliberation with the full complement of allies and in the process decided many of the important issues.'[40] In practice, the North Atlantic Council was a 'secondary framework' for decision-making; what mattered more was small group of especially committed nations.[41]

More than half of the bombing and ground-attack sorties were flown by British and French pilots. These two countries, most active in airstrikes, both in terms of number of sorties and strikes and as the two members of the striker group with the fewest restrictions, were the ones pushing most aggressively for an ambitious/expansive interpretation of the mandate. Although one interviewee noted that 'the risk threshold was different, and this played out in targeting,' suggesting that the British were more risk-averse than the French. Additionally, British and French special forces on the ground – acting outside the NATO chain of command – were able to pass on information to help guide targeting discussions.

By contrast, there was no consensus among interviewees, or in the literature, on the precise extent of American influence in OUP. Some argued that the United States had slightly less influence because it 'led from behind' while others argued that its influence was just as dominant because it brought niche capabilities that were vital for the mission, such as ISTAR, medical evacuation, and intelligence. According to this second view, the United Kingdom and France were content with the narrative of the United States leading from behind, as this gave them more space politically. One interviewee, for example, emphasised that

> even though (the Americans) were less involved than they normally would have been, (they still) had huge influence over targeting ... Clearly, (it is) just the way NATO works, from NAC, to (the Military Committee), to SHAPE (the Supreme Headquarters Allied Powers Europe), to op and tactical commanders, the whole system is full of US officers and civilians.

This stands in sharp contrast to Germany, which is a powerful player in NATO but had relatively less influence in the OUP context because it was not a military contributor. Still, Germany and Turkey were able to gain influence at early critical junctures by raising or threatening to raise objections. Arab states were also involved, even if they did not bring much to the table in terms of air capabilities. They were important politically for the success of the mission, but not its design and conduct.

When it came to the most controversial aspects of the mission, notably political and military actions that were seen to stretch the UN mandate, 'the US, UK and France tried to massage the language to recognise the legal constraints they were under, saying the mission was legally to protect civilians on the ground, however, in the course of doing that, you may have to take out Qaddafi and his forces to achieve that.' That was a constant challenge for the military legal advisers because politically all leaders wanted to keep talking about 'Qaddafi must go.'

The NATO context was also a permissive environment for the protection of civilians' mandate to be interpreted in this way, but not precisely as predicted by Hypothesis 2. Instead, our findings suggest that NATO, as an institutional actor, mattered for how the mandate of the mission was interpreted, which is more consistent with an argument rooted in a strategic culture logic. Indeed, NATO's own interpretation of what the protection of civilians means, in general terms, shaped the application of this concept in the context of OUP operations. To begin with, the mission's objectives were not clearly defined by UNSCR 1973, which gave 'quite a bit of latitude' to NATO, as one interview put it, and this happened through the commander, who drew up the initial campaign plan. Indeed, the civilian protection part of the mandate was the least clearly defined in the Resolution in terms of how it should be achieved, putting the burden of its interpretation onto NATO, so that it could be translated into precise military guidance and tasks. NATO's own take on protecting civilians is different, organisationally speaking, from the UN or the EU, and is best exemplified in the Libya case by its zero collateral damage directive. Rather than actively protecting civilians, operations were designed to avoid killing them.[42] Moreover, given the compressed time horizon that NATO was operating under, the normal operational planning process gave way to more ad hoc actions, a point also raised by interviewees. This ultimately serves to bolster the influence of the most influential players. As one interviewee put it when asked about the planning process: 'it was utter chaos, it was horrible.'

Among those influential players was Canada, given that a Canadian officer commanded OUP. LGen Bouchard had experience with air campaigns and considerable autonomy when designing the campaign plan, given the loose guidance from the UNSC mandate or from the NAC's interpretation of it. Moreover, because headquarters were initially understaffed, LGen Bouchard was able to rapidly bring Canadian officers and civilian analysts as augmentees. Thus interviewees agreed that LGen Bouchard, along with this Canadian staff, was able to leave his imprint on the mission, with one going as far as to say Naples 'became a mini-Canada.' Ultimately, however, he was still part of the NATO chain of command and had to answer to the United States, France, and the United Kingdom primarily and 'Sam Locklear, the 4-star US commander, was playing a pivotal role in guiding, almost commanding Bouchard.' While the interviews we conducted support the view that commanding the mission provided Canada with some influence, it can mostly be measured in terms of prestige and visibility. Canada could not have prioritised its national preferences through this command role, nor could LGen Bouchard have made the choices he did without the acquiescence of the United States, the United Kingdom, and France, which undermines support for Hypothesis 4.

Finally, we did not find evidence to support Hypothesis 5, which offers a set of competing explanatory factors focused on the role of external actors. The United States, the United Kingdom, France, as well as non-NATO partners (the UAE and Qatar in particular) had relations with Libyan rebels, largely on a bilateral basis. There was no direct political channel between NATO and anti-Qaddafi forces and no NATO footprint on the ground in Libya. The information they collected was channelled back to headquarters through the NATO Ground Effects Cell.[43] However, OUP's alignment with the Libyan rebels did not appear to have an impact in terms of steering the alliance towards a more expansive interpretation of its mandate. If anything, there was much frustration on the side of the rebels, who early on had unrealistically high expectations regarding OUP's role.[44] Finally, fractionalisation of actors on the ground also worked to lessen their influence/significance beyond the provision of intelligence to the alliance.[45]

As for rival actors, Russia and (to a lesser extent) China frequently criticised the NATO mission in Libya, arguing that it had mutated from being one mandated with protecting civilians into a regime change operation. Interviewees involved in discussions with diplomats from these two countries reported being told by Russian and Chinese counterparts that 'they had not signed up for this,' referring to their vote on UNSCR 1973. Beyond somewhat chilly discussions, however, OUP was not a major point of discussion in bilateral relations. Russian and Chinese frustration or public criticism of OUP for allegedly overreaching its mandate did not have an impact on the evolution of the mission, even if OUP is and will continue to be an important reference point for future objections to UN-sponsored, NATO-led missions.

To summarise our key findings, we should first emphasise that France and the United Kingdom are, after the United States, typically the most influential actors in defining any NATO mission (H1). It is therefore to be expected that they would have had significant influence in shaping the design and implementation of OUP. In the event, their influence over the direction of the mission was greater than might have been expected: when the United States limited its role once NATO took over, France and the United Kingdom conducted the most airstrikes and were hampered by few caveats (H3), they provided most enablers not provided by the United States, and they doubled down with special operations forces on the ground, outside the NATO framework, that could relay intelligence and provide training to the Libyan opposition forces. As such, France and the United Kingdom played an outsized role in steering the mission from its initially vague mandate – even more than a reading solely based on the distribution of power within the alliance would suggest. At the same time, other coalition members, notably Germany, who had more reservations with respect to the mission took more of a backseat; as such, their influence on mission design was more limited and almost absent on execution. Canada gained more visibility by assuming the command role and had a direct role when it came to mission design and conduct, but only within the bounds of what France, the United Kingdom and the United States would permit.

Another feature of coalition operations was important: NATO was a consequential institutional actor, especially for the interpretation of the mandate (H2). The organisation's take on protecting civilians, narrowly defined as a zero civilian casualty directive, was aligned with what France and the United Kingdom wanted to achieve and, thus, this directive was the lens through which the mandate was interpreted. NATO's rapid set-up and execution of the mission further circumvented the normal course of the operational planning process, however, which signals that the logic of ad hoc coalition processes might not be all that different from this more institutionalised setting. This increased the influence of France and the United Kingdom, with the full support of the United States.[46] Interviewees noted that allies were united in wanting to see this mission over quickly and that NATO had a rather full plate with concurrent missions in Afghanistan, Kosovo, and Iraq, translating into a rather high operational tempo.

4.4 Conclusion

Our investigation of the importance of a coalition's configuration suggests that, in the case of the NATO-led OUP, the most powerful members certainly had significant influence over mission design and conduct, but that other factors internal to the coalition also shaped the interactions that led to the interpretation of the mandate and the identification of the core military tasks that would be executed. This demonstrates that a reading solely based on power, while not inaccurate, is incomplete. Our argument that the coalition shapes the mission is further supported by the fact that external actors did not matter much for mission design and conduct: neither rival great powers (in this case, Russia and China) nor local actors (rebel groups backed by the members of the coalition) were able to generate incentives or constraints strong enough to shape the mission in a significant way.

A strong framework for analysis of coalition mission design is one that is generalisable and replicable to other cases. The next step in formulating such a framework would be to further test our hypotheses to validate whether they are specific to the particular case of OUP or whether they can also explain the evolution of mission design and conduct in other cases of coalition warfare, whether NATO-led or in other settings. We suspect that our hypotheses are useful for explaining the evolution of mission design and implementation in other cases of NATO-led operations, for example in the Balkans or in Afghanistan. In those instances, the United States was similarly dominant, but other factors, notably the distribution of caveats and the role of intra-alliance planning and operational processes, also played an important role in shaping outcomes.

The case of the Global Coalition against the Islamic State, led by the United States since 2014 and operating in Iraq and Syria, would offer an appropriate testing ground for assessing mission design in coalition operations not operating under NATO or another formalised institutional setting. Clearly, the United States is by far the dominant member in the coalition, followed by a second tier made of the usual suspects, notably France and the United Kingdom. Our findings also hold promise in the case of ad hoc multinational settings: it is likely in such instances that a reading based solely on the intra-coalition distribution of power will provide a useful but incomplete initial analysis of the evolution of mission design and implementation. To gain a finer understanding, other aspects of intra-coalition politics will need to be brought in. In such cases of non-NATO, ad hoc battlefield coalitions, some of our hypotheses will likely play out differently. The absence of a formal institutional umbrella, in particular, might remove an important constraint on the ability of the lead state to shape the mission.

That said, other intra-coalition dynamics, such as the relative distribution of caveats and internal planning and operational processes, are likely to still have an important impact, albeit a modified one, even absent a formal institutional setting.

To conclude, our findings should encourage more theory development about and testing of the ways in which the composition of military coalitions shapes their missions. Our framework holds promise in terms of explaining coalition operations under NATO and non-NATO settings by introducing variables tied to intra-collective politics. Only when considering all coalition actors, those with deeply institutionalised ties but also partners of convenience, can one truly understand battlefield dynamics and processes.

Notes

1 See also Rosella Cappella Zielinski and Ryan Grauer, 'A Century of Coalitions in Battle: Incidence, Composition, and Performance, 1900–2003,' Chapter 2, this volume.

2 See, for example, Michael Barnett and Jack Levy, 'Domestic Sources of Alliances and Alignments: The Case of Egypt, 1962–73,' *International Organization* 45, no. 3 (Summer 1991): 369–95, https://doi.org/10.1017/S0020818300033142; Stephen Walt, *The Origins of Alliances* (Ithaca: Cornell University Press, 1987); Alex Weisiger, 'Exiting the Coalition: When Do States Abandon Coalition Partners During War?,' *International Studies Quarterly* 60, no. 4 (December 2016): 753–65, https://doi.org/10.1093/isq/sqw029. For more background, see also the introduction to this volume.

3 See Cappella Zielinski and Grauer, 'A Century of Coalitions in Battle,' Chapter 2, this volume. Martha Finnemore and Kathryn Sikkink, 'International Norm Dynamics and Political Change,' *International Organization* 52, no. 4 (Autumn 1998): 887–917, https://doi.org/10.1162/002081898550789; Stéfanie von Hlatky, *American Allies in Times of War: The Great Asymmetry* (Oxford: Oxford University Press, 2013).

4 See, for example, Brett Ashley Leeds, 'Alliance Reliability in Times of War: Explaining State Decisions to Violate Treaties,' *International Organization* 57, no. 4 (Autumn 2003): 801–27, https://doi.org/10.1017/S0020818303574057; von Hlatky, *American Allies in Times of War*; Mark Crescenzi, Jacob Kathman, Katja Kleiberg, and Reed Wood, 'Reliability, Reputation, and Alliance Formation,' *International Studies Quarterly* 56, no. 2 (June 2012): 259–74, https://doi.org/10.1111/j.1468-2478.2011.00711.x.

5 May Darwich, 'Escalation in Failed Military Interventions: Saudi and Emirati Quagmires in Yemen,' *Global Policy* 11, no. 1 (February 2020): 103–12, https://doi.org/10.1111/1758-5899.12781; Thomas Juneau, 'The UAE and the War in Yemen: From Surge to Recalibration,' *Survival* 62, no. 4 (August-September 2020): 183–208, https://doi.org/10.1080/00396338.2020.1792135.

6 J. Davidson, *America's Allies and War: Kosovo, Afghanistan, and Iraq* (New York: Palgrave Macmillian, 2011); von Hlatky, *American Allies in Times of War*; Justin Massie and Stéfanie von Hlatky, 'Ideology, Ballots, and Alliances: Canadian Participation in Multinational Military Operations,' *Contemporary Security Policy* 40, no. 1 (August 2018): 101–15, https://doi.org/10.1080/135232 60.2018.1508265; Stéfanie von Hlatky and Jessica N. Trisko, 'Cash or Combat? America's Asian Alliances during the War in Afghanistan,' *Asian Security* 11, no. 1 (March 2015): 31–51, https://doi.org/10.1080/14799855.2015.1006360.

7 Benjamin Zyla, *Sharing the Burden? NATO and Its Second-Tier Powers* (Toronto: University of Toronto Press, 2015); James Sperling and Mark Webber, 'NATO: From Kosovo to Kabul,' *International Affairs* 85, no. 3 (May 2009): 491–511, https://doi.org/10.1111/j.1468-2346.2009.00810.x.

8 Bastian Giegerich and Stéfanie von Hlatky, 'Experiences May Vary: NATO and Cultural Interoperability in Afghanistan,' *Armed Forces & Society* 46, no. 3 (July 2020), 495–516, https://doi.org/10.1177/0095327X19875490; H. Christian Breede, 'Defining Success: Canada in Afghanistan 2006–2011,' *American Review of Canadian Studies* 44, no. 4 (December 2014): 483–501, https://doi.org/10.1080/02722011.2014.973425; Aaron Ettinger and Jeffrey Rice, 'Hell Is Other People's Schedules: Canada's Limited-Term Military Commitments, 2001–2005,' *International Journal* 71, no. 3 (September 2016): 371–92, https://doi.org/10.1177/0020702016662797.

9 Martha Finnemore and Kathryn Sikkink, 'International Norm Dynamics and Political Change'; Heidi Hardt, *NATO's Lesson in Crisis: Institutional Memory in International Organizations* (Oxford: Oxford University Press, 2018); Marina Henke, *Constructing Allied Cooperation: Diplomacy, Payments, and Power in Multilateral Military Coalitions* (Ithaca: Cornell University Press, 2019).

10 Andrew Mumford, *The West's War against Islamic State: Operation Inherent Resolve in Syria and Iraq* (London: I.B. Tauris, 2021).

11 Olivier Schmitt, 'International Organization at War: NATO Practices in the Afghan Campaign,' *Cooperation and Conflict* 52, no. 4 (2017): 502, https://doi.org/10.1177/0010836717701969.

12 Alan J. Kuperman, 'The Moral Hazard of Humanitarian Intervention: Lessons from the Balkans,' *International Studies Quarterly* 52, no. 1 (March 2008): 49–80, https://doi.org/10.1111/j.1468-2478.2007.00491.x.

13 We promised strict anonymity to each. Both authors obtained the required ethics clearances from their universities. The full list of interview questions is available upon request.

14 Stéphane Lacroix and Jean-Pierre Filiu, eds., *Revisiting the Arab Uprisings: The Politics of a Revolutionary Moment* (Oxford: Oxford University Press, 2018).

15 Ethan Chorin, *Exit the Colonel: The Hidden History of the Libyan Revolution* (PublicAffairs, 2012).

16 Jeffrey Michaels, 'Able but Not Willing: A Critical Assessment of NATO's Libya Intervention,' in *The NATO Intervention in Libya: Lessons Learned from the Campaign*, ed.Kjell Engelbrekt, Marcus Mohlin, and Charlotte Wagnsson (Milton Park, Abingdon, Oxon and New York: Routledge, 2014), 19.

17 'Canadian to Lead NATO's Libya Mission,' CBCNews, 25 March 2011, https://www.cbc.ca/news/world/canadian-to-lead-nato-s-libya-mission-1.1046678.

18 Press Briefing by NATO Spokesperson Oana Lungescu, 31 March 2011, https://www.nato.int/cps/en/natohq/opinions_71897.htm?selectedLocale=en.

19 The extension was renewed a second time in September.

20 Libya has, since then, collapsed into civil war. The scope of this chapter, however, ends with the termination of the NATO mission. For more, see Frederic Wehrey, *The Burning Shores: Inside the Battle for the New Libya* (New York: Farrar, Straus and Giroux, 2018).

21 Schmitt, 'International Organization at War: NATO Practices in the Afghan Campaign'; Rebecca Adler-Niessen and Vincent Pouliot, 'Power in Practice: Negotiating the International Intervention in Libya,' *European Journal of International Relations* 20, no. 4 (): 889–911, https://doi.org/10.1177/1354066113512702.

22 Ryan Lizza, 'The Consequentialist: How the Arab Spring Remade Obama's Foreign Policy,' *The New Yorker*, 2 May 2011.

23 Deborah Kidwell, 'The U.S. Experience: Operational,' in *Precision and Purpose: Airpower in the Libyan Civil War*, ed. Karl P. Mueller (Santa Monica: RAND Corporation, 2015), 146.

24 The importance of holding key positions was also mentioned in Schmitt's study of NATO in Afghanistan, where he identified several mechanisms of US domination in an alliance setting. Beyond the material contributions that the US makes, procedural practice can also deliver some benefits. See Schmitt, 'International Organization at War.'

25 In an analysis, Jonathan Paquin found that the coded theme 'multilateralism' was found in 20 of 21 official statements dealing with the crisis in Libya. Jonathan Paquin, 'Is Ottawa Following Washington's Lead in Foreign Policy? Evidence from the Arab Spring,' *International Journal* 67, no. 4 (Fall 2012): 1023, https://doi.org/10.1177/002070201206700409.

26 Matt Negrin, 'Obama: Allies to Take Libya Lead Soon,' *Politico*, 21 March 2011. https://www.politico.com/story/2011/03/obama-allies-to-take-libya-lead-soon-051680.

27 'We had … an international mandate for action, a broad coalition prepared to join us, the support of Arab countries, and a plea for help from the Libyan people themselves.'
   Office of the Press Secretary, The White House. 'President Obama's Speech on Libya,' 28 March 2011.

28 'Libye: Que signifie la prise de commandement de l'OTAN?' *Le Monde*, 28 March 2011, https://www.lemonde.fr/afrique/article/2011/03/28/libye-que-signifie-la-prise-de-commandement-de-l-otan_1499512_3212.html.

29 Camille Grand, 'The French Experience: Sarkozy's War,' in *Precision and Purpose: Airpower in the Libyan Civil War*, ed. Karl Mueller (Santa Monica, CA: RAND Corp., 2015), 188.

30 Grand, 'The French Experience: Sarkozy's War,' 199.

31 In French, 'la France s'est battue sans relâche.' See, for example, Prime Minister François Fillon in a debate in the National Assembly on 22 March 2011, http://www.assemblee-nationale.fr/13/cri/2010-2011/20110144.asp.

32 Philippe Gros. *De Odyssey Dawn à Unified Protector: Bilan transitoire, perspectives et premiers enseignements de l'engagement en Libye*. Fondation pour la recherche stratégique 2011. Avril; see, for example, Prime Minister François Fillon the National Assembly on 22 March 2011, http://www.assemblee-nationale.fr/13/cri/2010-2011/20110144.asp.

33 Senior officials repeatedly emphasised that Qaddafi 'has to go.' See, for example, Foreign Minister Alain Juppé in a debate in the National Assembly on 9 March 2011, http://questions.assemblee-nationale.fr/q13/13-3048QG.htm.

34 UK Parliament (Defence Committee), 'Minutes of Evidence,' 12 October 2011, https://publications.parliament.uk/pa/cm201012/cmselect/cmdfence/950/1110120 1.htm.

35 Domestically, questions were constantly raised about the nature of British involvement in light of the country's controversial participation in the 2003 Iraq War.

36 UK Parliament Debate (Commons Chamber), 'UN Security Council Resolution (Libya),' Volume 525 (18 March 2011), https://hansard.parliament.uk/Commons/2011-03-18/debates/11031850000007/UNSecurityCouncilResolution (Libya).

37 UK Parliament Debate (Commons Chamber), 'UN Security Council Resolution (Libya),' Volume 525 (18 March 2011), https://hansard.parliament.uk/

Commons/2011-03-18/debates/11031850000007/UNSecurityCouncilResolution (Libya).
38 UK Parliament (Defence Committee), 'Minutes of Evidence – HC950,' 26 April 2011, https://publications.parliament.uk/pa/cm201012/cmselect/cmdfence/950/11042701.htm.
39 Frederic Wehrey, 'NATO's Intervention,' in *The Libyan Revolution and its Aftermath*, ed. Peter Cole and Brian McQuinn (New York: Oxford University Press, 2015), 112.
40 Chivvis (2014), 100.
41 Alastair Cameron, 'The Channel Axis: France, the UK and NATO,' in *Short War, Long Shadow: The Political and Military Legacies of the 2011 Libya Campaign*, ed. Adrian Johnson and Saqeb Mueen (London: RUSI, 2012), 18.
42 Stian Kjeksrud, Jacob Aasland Ravndal, Andreas Øien Stensland Cedric de Coning, Walter Lotze, and Erin A. Weir, *Protection of Civilians in Armed Conflict – Comparing Organisational Approaches* (Norwegian Defence Research Establishment [FFI] – FFI-rapport 2011/01888), 2011, https://publications.ffi.no/nb/item/asset/dspace:2236/11-01888.pdf.
43 Wehrey (2015), 114–5.
44 Wehrey (2015).
45 This was highlighted in our interviewees and is also discussed in the academic literature; see, for example, Chivvis 'Libya and the Future of Liberal Intervention,' 74.
46 See the mid-April, Obama, Sarkozy, and Cameron joint statement: the mission's humanitarian objectives could only be reached Qadhafi was no longer in power; https://www.bbc.com/news/world-africa-13090646.

Bibliography

Adler-Niessen, Rebecca, and Vincent Pouliot. 'Power in Practice: Negotiating the International Intervention in Libya.' *European Journal of International Relations* 20, no. 4 (2014): 889–911. 10.1177/1354066113512702

Barnett, Michael, and Jack Levy. 'Domestic Sources of Alliances and Alignments: The Case of Egypt, 1962–73.' *International Organization* 45, no. 3 (Summer 1991): 369–95. 10.1017/S0020818300033142.

Baron, Kevin. 'For the U.S., War against Qaddafi Cost Relatively Little: $1.1 Billion.' *The Atlantic*, 21 October 2011. https://www.theatlantic.com/international/archive/2011/10/for-the-us-war-against-qaddafi-cost-relatively-little-11-billion/247133/

Breede, H. Christian. 'Defining Success: Canada in Afghanistan 2006–2011.' *American Review of Canadian Studies* 44, no. 4 (2014): 483–501. 10.1080/02722011.2014.973425

Cameron, Alastair. 'The Channel Axis: France, the UK and NATO.' In *Short War, Long Shadow: The Political and Military Legacies of the 2011 Libya Campaign*, edited by Adrian Johnson and Saqeb Mueen. London: RUSI 2012.

'Canadian to Lead NATO's Libya Mission.' CBCNews, 25 March 2011. https://www.cbc.ca/news/world/canadian-to-lead-nato-s-libya-mission-1.1046678

CBC News. 'Canada's military contribution in Libya.' 20 October 2011. https://www.cbc.ca/news/world/canada-s-military-contribution-in-libya-1.996755

Chorin, Ethan. *Exit the Colonel: The Hidden History of the Libyan Revolution*. New York: PublicAffairs, 2012.

Crescenzi, Mark, Jacob Kathman, Katja Kleiberg, and Reed Wood. 'Reliability, Reputation, and Alliance Formation.' *International Studies Quarterly* 56, no. 2 (June 2012): 259–74. 10.1111/j.1468-2478.2011.00711.x.

Darwich, May. 'Escalation in Failed Military Interventions: Saudi and Emirati Quagmires in Yemen.' *Global Policy* 11, no. 1 (February 2020): 103–12. 10.1111/1758-5899.12781

Davidson, Jason. *America's Allies and War: Kosovo, Afghanistan, and Iraq.* New York: Palgrave Macmillan, 2011.

Ettinger, Aaron, and Jeffrey Rice. 'Hell Is Other People's Schedules: Canada's Limited-Term Military Commitments, 2001–2005.' *International Journal* 71 (September 2016): 371–92. 10.1177/0020702016662797.

Finnemore, Martha, and Kathryn Sikkink. 'International Norm Dynamics and Political Change.' *International Organization* 52, no. 4 (Autumn 1998): 887–917. 10.1162/002081898550789.

Giegerich, Bastian, and Stéfanie von Hlatky. 'Experiences May Vary: NATO and Cultural Interoperability in Afghanistan.' *Armed Forces & Society* 46, no. 3 (July 2020): 495–516. 10.1177/0095327X19875490.

Grand, Camille. 'The French Experience: Sarkozy's War.' In *Precision and Purpose: Airpower in the Libyan Civil War*, edited by Karl Mueller. Santa Monica, CA: RAND Corporation 2015.

Gros, Philippe. *De Odyssey Dawn à Unified Protector: Bilan transitoire, perspectives et premiers enseignements de l'engagement en Libye.* Paris: Fondation pour la recherche stratégique, 2011.

Hardt, Heidi. *NATO's Lesson in Crisis: Institutional Memory in International Organizations.* Oxford: Oxford University Press, 2018.

Henke, Marina. *Constructing Allied Cooperation: Diplomacy, Payments, and Power in Multilateral Military Coalitions.* Ithaca, NY: Cornell University Press, 2019.

Juneau, Thomas. 'The UAE and the War in Yemen: From Surge to Recalibration.' *Survival* 62, no. 4 (August-September 2020): 183–208. 10.1080/00396338.2020. 1792135.

Kidwell, Deborah. 'The U.S. Experience: Operational.' In *Precision and Purpose: Airpower in the Libyan Civil War*, edited by Karl P. Mueller. Santa Monica, CA: RAND Corporation, 2015.

Kjeksrud, Stian, Jacob Aasland Ravndal, Andreas Øien Stensland Cedric de Coning, Walter Lotze, and Erin A. Weir. *Protection of Civilians in Armed Conflict – Comparing Organisational Approaches.* Norwegian Defence Research Establishment (FFI) – FFI-rapport 2011/01888, 2011. https://publications.ffi.no/nb/item/asset/dspace:2236/11-01888.pdf

Kuperman, Alan J. 'The Moral Hazard of Humanitarian Intervention: Lessons from the Balkans.' *International Studies Quarterly* 52, no. 1 (March 2008): 49–80. 10.1111/j.1468-2478.2007.00491.x

Lacroix, Stéphane, and Jean-Pierre Filiu, eds. *Revisiting the Arab Uprisings: The Politics of a Revolutionary Moment.* Oxford: Oxford University Press, 2018.

Leeds, Brett Ashley. 'Alliance Reliability in Times of War: Explaining State Decisions to Violate Treaties.' *International Organization* 57, no. 4 (Autumn 2003): 801–27. 10.1017/S0020818303574057.

'Libye: Que signifie la prise de commandement de l'OTAN?' *Le Monde*, 28 March 2011. https://www.lemonde.fr/afrique/article/2011/03/28/libye-que-signifie-la-prise-de-commandement-de-l-otan_1499512_3212.html.

Lizza, Ryan. 'The Consequentialist: How the Arab Spring Remade Obama's Foreign Policy.' *The New Yorker*, 2 May 2011.

Massie, Justin, and Stéfanie von Hlatky. 'Ideology, Ballots, and Alliances: Canadian Participation in Multinational Military Operations.' *Contemporary Security Policy* 40, no. 1 (August 2018): 101–15. 10.1080/13523260.2018.1508265

Michaels, Jeffrey. 'Able but Not Willing: A Critical Assessment of NATO's Libya Intervention.' In *The NATO Intervention in Libya: Lessons Learned from the Campaign*, edited by Kjell Engelbrekt, Marcus Mohlin, and Charlotte Wagnsson. Milton Park, Abingdon, Oxon and New York: Routledge, 2014.

Mueller, Karl P., ed. *Precision and Purpose: Airpower in the Libyan Civil War*. Santa Monica, CA: RAND Corporation, 2015.

Mumford, Andrew. *The West's War against Islamic State: Operation Inherent Resolve in Syria and Iraq*. London: I.B. Tauris, 2021.

Negrin, Matt. 'Obama: Allies to take Libya lead soon,' *Politico*, 21 March 2011. https://www.politico.com/story/2011/03/obama-allies-to-take-libya-lead-soon-051680

Paquin, Jonathan. 'Is Ottawa Following Washington's Lead in Foreign Policy? Evidence from the Arab Spring.' *International Journal* 67, no. 4 (Fall 2012): 1001–28. 10.1177/002070201206700409.

Perry, David. *Leading from Behind Is Still Leading: Canada and the International Intervention in Libya*. Ottawa: CDA Institute, 2011. https://cdainstitute.ca/wp-content/uploads/2012/06/vimypaper2012-libya.pdf 11.

Schmitt, Olivier. 'International Organization at War: NATO Practices in the Afghan Campaign.' *Cooperation and Conflict* 52, no. 4 (2017): 502–518. 10.1177/001083 6717701969.

Sperling, James, and Mark Webber. 'NATO: From Kosovo to Kabul.' *International Affairs* 85, no. 3 (May 2009): 491–511. 10.1111/j.1468-2346.2009.00810.x.

Vampouille, Thomas. 'Guerre en Libye: la France a dépensé 300 millions d'euros.' *Le Figaro*, 21 October 2011. https://www.lefigaro.fr/international/2011/10/21/01003-20111021ARTFIG00508-la-guerre-en-libye-a-coute-300-millions-d-euros-a-la-france.php

von Hlatky, Stéfanie. *American Allies in Times of War: The Great Asymmetry*. Oxford: Oxford University Press, 2013.

von Hlatky, Stéfanie, and Jessica N. Trisko. 'Cash or Combat? America's Asian Alliances during the War in Afghanistan.' *Asian Security* 11, no. 1 (March 2015): 31–51. 10.1080/14799855.2015.1006360.

Walt, Stephen. *The Origins of Alliances*. Ithaca, NY: Cornell University Press, 1987.

Wehrey, Frederic. 'NATO's Intervention.' In *The Libyan Revolution and Its Aftermath*, edited by Peter Cole and Brian McQuinn. Oxford: Oxford University Press, 2015.

Wehrey, Frederic. *The Burning Shores: Inside the Battle for the New Libya*. New York: Farrar, Straus and Giroux, 2018.

Weisiger, Alex. 'Exiting the Coalition: When Do States Abandon Coalition Partners During War?.' *International Studies Quarterly* 60, no. 4 (December 2016): 753–765. 10.1093/isq/sqw029.

Zyla, Benjamin. *Sharing the Burden? NATO and Its Second-Tier Powers*. Toronto: University of Toronto Press, 2015.

Part III
Organisation

5 Battlefield Coalitions as International Institutions

A Conceptual Framework

Casey Mahoney

What explains when battlefield coalitions win or lose?[1] Why do they come together in the first place, and what circumstances lead their members to part ways? Foundational questions to the study of battlefield coalitions like these aim to discover the conditions under which states and other groups cooperate in armed conflict. Scholarship included in this volume and elsewhere explains these outcomes with reference to the basic interests, capabilities, identities, and security and technological environments of coalition members. In turn, these factors lead coalitions to undertake a variety of additional activities – combining command and control structures, developing shared operational plans, conducting exercises, or sharing intelligence – variation in which explains outcomes of coalition cooperation.

This chapter contends that many of these activities can be conceptualised and analysed as international institutions in ways useful to explaining when cooperation in battlefield coalitions succeeds or fails. Accounts of alliance politics and coalition outcomes more generally draw important explanatory power from descriptions of the institutional contexts in which interactions among coalition members occur.[2] Left unstated in many research works on battlefield coalitions specifically, however, is a recognition that many activities, like prewar operational planning or integrating command and control systems, involve establishing behavioural standards that in fact constitute international institutions: sets of rules that are 'persistent and connected,' 'formal [or] informal,' and 'prescribe behavioural roles, constrain activity, and shape expectations.'[3]

At the core of all battlefield coalitions is a shared understanding among coalition members of one such rule: that the enemy of my enemy will act as if my friend. But which behaviours,behaviours, precisely, are expected of battlefield partners or 'friends'? Though rarely conceptualised as such, these behavioural expectations can emerge from the more complex sets of sub-rule coalitions institutionalised to clarify standards for how co-belligerents ought to enact a battlefield partnership. How will coalition military units respond if attacked from the west? If from the east? Who will decide if and how an operational plan will be updated when circumstances change? How will information about those circumstances be gathered?

DOI: 10.4324/9781003399896-8

Under what conditions will coalition partners begin – or stop – contributing to an operation?

When formalised in writing (or agreed upon in other, more tacit ways), the answers coalitions provide to these questions constitute international institutions. Such institutions, in turn, shape whether battlefield coalition cooperation succeeds or fails. In this chapter, I argue that explicitly articulating the implicit rule-setting functions of institutions in battlefield coalitions directs attention to the political bargains and normative standards of these collectives, factors that scholars of international institutions have long known to be important variables in cooperation processes.[4] As the conceptual framework I develop here shows, thinking about certain battlefield coalition activities as institutions allows important features of their design, like their formalisation, scope, or centralisation, to be measured and understood as variables that serve as causes or consequences in models of battlefield coalition behaviour.

Identifying and measuring key sources of variation in battlefield coalition institutions promises to yield a number of analytic payoffs by linking the study of battlefield coalitions to literatures on institutions in international politics more generally. By articulating these payoffs, this chapter makes three key contributions. First, I translate concepts developed in other empirical contexts to the battlefield coalition context, thus providing a framework researchers can use to describe how coalitions vary with greater consistency. To date, researchers have tended to adopt conceptual frameworks well-suited to their specific analytic purposes to examine phenomena of interest. As the study of battlefield coalitions progresses, however, greater convergence around conceptual conventions can facilitate the accumulation of new insights in this field.

Second, the chapter demonstrates how this conceptual framework allows for more systematic comparisons within and across coalitions in ways that can generate new lines of inquiry about relationships among these variables. By providing a new lens through which to observe differences in the rules that coalitions do or do not adopt, the framework directs attention to sources of novel, potentially unexplained variance in coalition cooperation.

Third, conceptualising battlefield coalitions as institutions provides a language in which to make research into the causes and consequences of their variation accessible to scholars outside security studies. Related literatures on alliance treaties and other international security institutions provide models of how institutionalist concepts can be usefully applied to questions related to conflict outcomes, security cooperation, and change processes related to coalitions.[5] Scholarship on coalitions at the battlefield level, however, has yet to fully exploit the potential that these approaches offer. Given that the conditions under which institutions constrain powerful actors remains a persistent question in international relations literatures,[6] utilising the framework proposed here provides one opportunity to bring insights from the study of battlefield coalitions into dialogue with central debates in the field.

The chapter proceeds as follows. It begins with a discussion of how I identify institutions in battlefield coalitions. It then turns to introduce five dimensions of international institutions – formalisation, membership, scope, control, and centralisation – by defining them in the context of battlefield coalitions and introducing the range of empirical indicators that researchers can use to measure them. I make the case that using these concepts can help open new lines of inquiry that existing work has yet to fully explore by illustrating the ways in which these concepts capture substantive variation in a range of empirical cases. The discussion draws from examples from the early twentieth century from the Boxer War through the world wars, from Cold War conflicts like the Yom Kippur War, and from more recent US- and NATO-led coalition operations. The chapter concludes by suggesting ways this framework can guide future research.

5.1 Identifying Institutions in Battlefield Coalitions

Scholars recognise that the domestic institutional contexts in which individual militaries act and how these contexts shape military institutions themselves serve as important explanatory variables for their strategic choices and operational effectiveness.[7] Military organisations tend to adhere to clear hierarchies, maintain strict orders of discipline, inculcate a strong sense of shared identity, and enforce norms that support these characteristics. It follows, then, that the extensive rule-setting that makes militaries and other armed actors rather institutionalised social groups themselves might also extend to battlefields where collectives of these groups operate side by side.

The coalition battlefield is a context in which powerful incentives exist for co-belligerents to establish ways of reducing transaction costs of cooperation.[8] In the face of the complexity of modern warfare, the uncertainty the fog of war creates, and the diverse motivations that drive battlefield coalition members to fight, institutions that establish standards for intra-coalition relations provide one means of making combined operations more efficient and effective. Before investigating causal questions, like whether these incentives are what drive battlefield coalitions to establish institutions or whether these institutions have the effects their creators intend, it is necessary to establish a basis for identifying institutions in battlefield coalitions where they exist, as I do in this section, and for measuring and comparing them, as I do in the next.

To define battlefield coalition institutions, I follow Robert Keohane's canonical definition as sets of rules that are 'persistent and connected,' 'formal [or] informal,' and 'prescribe behavioural roles, constrain activity, and shape expectations' of the members of battlefield coalitions.[9] In his typology, institutions can be classified as one of three different types: formal international organisations, regimes comprised of explicitly negotiated rules, and conventions comprised of informal or implicit rules that emerge through practice. The following examples illustrate how various examples of battlefield coalition institutions span the range of each of these categorisations.

Some battlefield coalitions create intra-coalition institutions by establishing *organisations*. Arrangements that grant personnel access to coalition organisations, like liaison officers assigned to a foreign military unit, or like non-state militia members embedded in the governance bodies of rebel groups, meet a minimal definition of an international organisation. More robust forms of institutionalised organisations can be found in the integrated, multinational command-and-control (C2) structures that direct coalition operations. NATO members operating in Kosovo and Afghanistan established such organisations, as did the United States in other ad hoc coalitions it assembled to fight in conflicts from the Persian Gulf War to the campaign against the Islamic State of Iraq and the Levant (ISIL) in Syria and Iraq. In each of these cases, explicit rules written in agreements assigned roles and duties to commanders, subordinate officers, and other actors who convened in standing, multinational staff organisations, thus forming international institutions.

Other institutions in battlefield coalitions lack standing organisations but nonetheless have explicit rules that constitute international *regimes* among coalition members. Logistics, technology-sharing, or intelligence-sharing agreements, like the British-US Communication Agreement that grew into the Five Eyes Alliance among the United States, United Kingdom, Canada, Australia, and New Zealand, establish regimes that may facilitate cooperation in coalition battlefield operations. Schedules of multinational military exercises (MMEs) involving the forces of battlefield coalitions can also constitute international regimes. As Boehlefeld and Grieco show in the context of the NATO-Ukraine *Peace Shield* and *Rapid Trident* exercises through the mid-2000s, MMEs can occur under the auspices of agreements that institutionalise the frequency, locations, invited actors, and other parameters of these activities. As their analysis suggests, the status that institutionalising an MME can confer on the relationship among MME participants impacts the probability conflict involving a coalition or its members occurs and, thus, on the effectiveness of coalition cooperation.[10]

In other cases, the rules that govern practices of battlefield coalitions related to personnel exchanges, command and control, technology- and intelligence-sharing, or multinational exercises are less explicit. Without organisational structures or formally negotiated agreements, they might instead resemble *conventions* that have emerged through practice among international partners. For example, a practice reported by veterans of NATO coalition forces operating in Afghanistan was to share important, tactically relevant intelligence across national units.[11] Though unit commanders did so without formal authorisation or explicit political agreement among the states involved, practices like this can become norms in battlefield coalitions. Behavioural norms and practices like this that emerge at the operational and tactical levels of warfare among militaries of battlefield coalitions approximate other types of conventions that define international practices among states.

The examples of battlefield coalition institutions discussed here do not comprise an exhaustive list, nor does the organisation-regime-convention typology necessarily provide a means for describing all varieties of such institutions. For example, written operational plans for a military campaign of a battlefield coalition, which sets rules for the roles the units of a multi-national force are to play in operations, meet the definitional standard of an international regime. Yet, the fact that scholars tend to associate international regimes with complexes of layered international agreements that emerge and endure over longer periods of time – like the Bretton Woods regime – suggests that conceptual tools with greater precision than broad typologies offer would be better suited to comparing the types of institutions that tend to emerge among battlefield coalition members. How can scholars interested in investigating the causes and consequences of variation in institutions in battlefield coalition contexts draw these comparisons?

The remainder of this chapter proposes one approach to comparing such institutions across time and space by considering several dimensions along which battlefield coalition institutions of any type may vary. It illustrates the range of variation among these institutions by drawing on empirical examples of one broad category of such institutions: C2 systems within battlefield coalitions during wartime. Moller defines C2 as 'the exercise of authority and direction of forces in battle.'[12] In a single military, C2 systems establish rules that define which officers and units must follow orders of which superior commanders, and they delegate authority to make certain types of decisions to officers of different ranks.[13] In coalition contexts, C2 systems perform the same function but must also identify circumstances in which officers and units are to comply with orders issued by those of *other* nationalities.[14] In doing so, coalition C2 systems establish explicit and implicit rules and norms – institutions – that assign roles across the international membership of a battlefield coalition.

Such rules, or C2 systems, thus constitute international institutions by the definition adopted above and bear similar traits to institutions in other contexts. Like any other international legal or political agreement, coalitions can establish C2 systems through explicit negotiation and written agreement among state officials. Sometimes, such arrangements are inscribed in peacetime alliance treaties, such as the North Atlantic Treaty or the Arab League treaty, which identify a supreme commander role and establish the authorities of this individual. Other times, C2 arrangements are established in political agreements negotiated before or during wartime, like those the American and British political and military leadership concluded during the Arcadia and subsequent bilateral war-planning conferences of the Second World War.[15]

Like other institutions, coalition C2 institutions can establish standing organisations that operate on the battlefield. Just as any government might appoint officials to serve in an international organisation secretariat, coalition members can appoint military officers to lead a multinational force from a combined headquarters and send personnel to serve on C2 staffs. Alternatively, just as with other international regimes and conventions, other instances of coalition C2 may lack such standing organisations and are established as less-centralised self-enforcing arrangements instead. Still others may lack many traits of institutions at all: on some battlefields, rules for intra-coalition decision-making may be very sparse or absent altogether, meaning no or few coalition-specific institutions govern command relations in such cases of anarchy.

As these examples suggest, battlefield coalitions have established institutions concerning command and control that have taken a variety of forms. What unites these phenomena is the possibility of describing them as international institutions. Next, I turn to presenting a method by which to systematically describe their differences.

5.2 Dimensions of Battlefield Coalition Institutions

Coalition institutions serve a diverse array of functions on the battlefield, which raises the question of how to compare institutions with differing purposes, forms, or other features. Characterising these sources of variation requires careful thinking about the dimensions along which institutions in battlefield coalitions may vary. This section identifies five distinct conceptual dimensions that can apply to the range of phenomena that can be identified as institutions: *formalisation, membership, scope, control,* and *centralisation.*

I highlight these five dimensions as a starting point for further research, given their importance in the study of institutions in other contexts. Formalisation, which relates to degree to which institutions are explicitly encoded rather than left implicit in habitual practice, is a foundational characteristic of institutions that captures one of their important constitutive elements: given that 'rules' that comprise institutions tend to have some degree of non-trivial formality or rigidity to them, formalisation describes the 'institutionness' of an institution. The remaining four dimensions are those to which Barbara Koremenos, Charles Lipson and Duncan Snidal pointed in their 2001 agenda-setting study of institutionalist literature as key variables central to the field.[16]

It is important to note that this list is far from exhaustive. The analysis I conduct might instead have focused on other concepts institutionalist scholars use, like Keohane's institutional *commonality* (i.e. the coherence of members' beliefs), *specificity* (i.e. the precision or detailedness of rules), and *autonomy* (i.e. an organisation's ability to act without explicit direction of institutional members).[17] The five dimensions I do examine here, however, accomplish the goal of exploring variation that scholars of battlefield coalitions have found important in their models of coalition outcomes and thus provide a basis for additional conceptual work in the future.

Each of the following subsections defines the dimension under consideration; identifies the range of values each dimension might take in the context of formal battlefield coalition institutions; and uses empirical examples to illustrate the relevance of each concept to questions of substantive interest. As with the selection of the five specific dimensions themselves, the specific parameters by which I define each of these dimensions are not definitive either. Instead, the discussion aims to show that greater clarity about the bases of comparison across instances of institutions in battlefield coalitions can provide a more solid foundation for interrogating relations between these features and their causes and consequences.

5.2.1 Formalisation: Which Rules?

In the coalition context, *formalisation*, or *institutionalisation*, describes how coalition members articulate the rules by which they agree to abide.[18] One question we can ask about how rules are articulated focuses on measuring formalisation as a static feature of an institution at a point in time: Are the rules to which battlefield coalitions agree explicitly written down, left as explicit but unwritten verbal agreements, or otherwise left implicit as emergent habit or practice?[19] Highly institutionalised coalitions, like the NATO forces that fought in Afghanistan or Libya, operate on the basis of extensive written documentation of procedures for assigning operational and decision-making roles to commanders across different nationalities. Other coalitions, including some that are comprised of non-state actors or of hybrid state and non-state actors, may engage primarily in tacit, informalised coordination where lines of authority across state groups do not exist, are not explicitly negotiated, or are never written down.[20]

A second way of assessing institutional formalisation focuses on the processes by which the status of formal or informal rules as socially valuable institutions emerges: How have battlefield institutions come to be respected as legitimate by coalition members? In the context of establishing battlefield coalition C2, such processes include ratification, political agreement (e.g. executive orders, diplomatic notes), and military orders. The most formal battlefield coalition institutions are those that form as a result of a legal ratification procedure. Egyptian command of the Arab coalition that fought the 1973 Yom Kippur War, for instance, was formalised in the Arab League treaty ratified by Egypt, Syria, and other participants.[21]

Short of legal ratification instruments, other processes with legitimacy of their own can confer formal authority on coalition institutions. In the Second World War, it was only by executive agreement that British Prime Minister Winston Churchill and US President Franklin D. Roosevelt subordinated their own forces to the authority of one another's commanders in the Mediterranean, Southeast Asian, and Western European theatres of war. Though the US Congress had yet to ratify any treaty of alliance with Britain or act otherwise to formally legitimate Roosevelt's agreement with Churchill, combined C2 in the war was formalised in the sense that the US Constitution provides the president authority to conduct foreign policy and to act as commander-in-chief of US military forces.

What are the implications of these distinctions among coalition C2 systems, and battlefield institutions more generally? A substantial body of evidence establishes that the formalisation of alliance agreements at the inter-state level impacts outcomes, from the onset of war with third-party states to levels of conflict within alliances, even if the direction, magnitude, or causal significance of those impacts remains under debate.[22] Scholars have done far less to examine if variation in how (or whether) battlefield-level institutions are formalised also plays a role in determining outcomes. Measuring formalisation as I suggest here can advance research on and understanding of the effects of formalisation in at least two ways.

First, the static and procedural indicators of formalisation discussed here do not necessarily correlate. In the Boxer War, a major-power coalition coalesced to counter an uprising in Peking by the initiative of Vice Admiral Edward Seymour of the British Royal Navy, who convened 27 'Councils of Senior Naval Commanders' of the eight nations' navies with ships ashore at Peking to discuss the response options to the impending conflict. These officers established written rules to govern relations among national C2 chains of command and set rules of engagement, thus establishing a formal battlefield coalition institution in the static sense.[23] Yet, the process by which they came to that agreement lacked a greater degree of formality that, for instance, a ratified treaty negotiated among their states' diplomats would have conferred. But did that lack of formality make for a less effective coalition? Understanding more generally the circumstances under which different types of formalisation arise or affect outcomes of concern requires first identifying the modes of formalisation across and within cases as I suggest.

Second, measurements of *de jure* formalisation of institutions offers an important baseline against which to compare *de facto* practice in battlefield coalitions as a means of understanding the causes and consequences of institutionalisation. The Alliance Treaty Obligations and Provisions data set, for instance, tracks whether peacetime and wartime alliance treaties include (formal) provisions for the circumstances under which their members are to subordinate their forces to some form of combined coalition C2 in the case of war.[24] But, to fully understand the battlefield impacts of formalising C2 arrangements, it is necessary to know whether wartime coalition partners concluded other types of agreements – like Anglo-American arrangements for C2 during the Second World War discussed above – that formalised C2. It is further necessary to know the extent to which coalitions implemented or complied with such arrangements.

Investigating how formal or informal coalition C2 and other functional coalition institutions within formal or informal alliances can optimise coalition coherence, longevity, or performance requires measuring these features of institutions. As policymakers grapple with questions of whether and how to institutionalise international security partnerships to deter future conflict and ensure that collective defences are effective, examining formalisation across a wider range of battlefield coalition institutions than scholars have to date is an important task for future research.

5.2.2 Membership: Who Belongs?

An institution's *membership* is comprised of the actors who respect the rules the institution sets as legitimate. Though a seemingly obvious point, descriptions of coalitions' institutional context benefit greatly from measurements of how the constituent *identities* and *size* (i.e. counts of nations, counts of forces) of coalition actors subject to an institutional agreement changes, or stays static, over time and space.[25] Naming or counting the types of constituent actors in an institution enables analysts to also measure second-order indicators, like the degree of *overlap* (i.e. set intersection as a subset, identity, or superset) between the membership of one institution with others. Measuring when the identities or number of nations, military branches, or specific units in an institution change over time, or where there is a relatively high or low degree of overlap in which coalition members participate across multiple battlefield institutions is an important baseline step in analysing the causes and consequences of battlefield coalition institutions.

Coalition C2 systems establish rules for how commanders and military units ought to behave in relation one another, making these commanders and the forces they lead the main indicator of the membership of C2 institutions. As states or other actors add new forces to battlefield coalitions, as existing forces attrite in battle, and commanders reorganise units under their command, the membership of any given C2 system can ultimately be a highly dynamic variable over time. Moreover, within a single wartime coalition, the membership of C2 organisations across theatres and across coalition functions varies.

The example of membership in allied institutions during the Second World War illustrates this. Though some 26 states made up the United Nations coalition that fought against the Axis powers, not all of them counted among the members of the various institutions that structured intra-coalition interactions on the battlefield. For instance, the Combined Chiefs of Staff, which coordinated the strategy and logistics systems of the Western Allies, formally included only the United States and United Kingdom as members. Yet, this organisation oversaw operational C2 systems that included those of several other member states and forces. Commanders from Free France and Nationalist China served in Allied C2 structures, and division-sized units from Brazil and Poland, among many others, executed strategy the CCS determined and British and American high-level commanders implemented. A key step to understanding how the interests of all the members of a coalition are represented, or not, in the institutional politics of coalition operations requires careful attention to which subsets of actors do and do not have seats at the negotiating table.

Characterising variation in the identities, numbers, or overlap of institutional membership is also important for assessing any number theories about the factors and mechanisms that account for ways C2 structures impact effectiveness. Some theories argue particular leader-, unit-, or state-level features explain coalition success and failure. If so, to continue with the Second World War examples, it is important to know whether it was Eisenhower, Montgomery, or Patton who commanded an operation, or to know whether it was the US 1st or 20th Armoured Division that was included in a coalition C2 chain of command. Measuring membership is also important to test theories of collective action that posit more systemic, structural explanations for outcomes. If C2 organisations overseeing coalitions drawn from *many* national militaries tend to perform well because they have access to more warfighting or financial resources, it is important to measure membership in order to distinguish relatively diverse coalitions, like the Allied force that fought in Italy from 1943, from less diverse collectives, like the predominantly British force assembled against Japan in Burma in 1942 and 1943. Making these distinctions is also necessary to test whether, conversely, it is institutions with fewer members that have advantages that lead them to cooperate more effectively, make decisions more nimbly, or reduce coordination problems arising from less diverse forces with greater interoperability.

Understanding how logics of collective action operate at the level of battlefield coalitions – and whether institutions ameliorate those that do exist – is a foundational issue for their study. In a statistical analysis of coalition wars since 1816, Weisiger does not find evidence of a collective action logic at work in battlefront-level coalition member decisions to defect from fighting.[26] Indeed, battlefield coalitions derived from many-member alliances and coalitions, from the Allies of the Second World War to the 85-member Global Coalition against Daesh, do succeed. But, research that pays careful attention to how the membership of warfighting collectives vary from battlefield to battlefield can make progress in squaring the supposed benefits of large coalitions with the potential challenges for military effectiveness that large groups are theorised to create.[27]

5.2.3 Scope: Which Issues Are Decided?

Scope characterises the issue set that falls within an institution's authorities. Whereas considering the membership of an institution corresponds to what is often referred to as the 'breadth' of an institution, *scope* refers to the 'depth' of cooperation an institution facilitates by prescribing that its members make joint decisions and, more precisely, around which they would not otherwise have cooperated in the absence of the institution.[28] In the context of battlefield coalitions, a description of an institution's scope refers to the military functions over which it is authorised to make decisions.

The scope of coalition C2 institutions can be defined by domain or function. If defined by domain, the scope of a C2 institution includes the types of forces it has authority to command in ground, sea, air, space, or cyber domains. The institutional scope of NATO C2 in Libya, for example, grew over time from including primarily air forces to implement a no-fly zone to later including a larger range of forces to support a 'maritime embargo, a no-fly zone, [and] a civilian-protection [ground] mission.'[29] Alternatively, scope can be measured in terms of the number or type of functions in an institution's competency to control. The scope of coalition C2 systems often includes decision-making authority over some combination of planning, operations, and logistics functions. The scope of the Anglo-American Combined Chiefs of Staff system, for instance, was only over long-term strategic plans and inter-theatre logistics, whereas the scope of operational C2 institutions was to devise operational plans and intra-theatre logistics. In another example, during the NATO mission to Afghanistan, some coalition members placed caveats on the participation of their forces in coalition operations, effectively shrinking the scope of the authority commanders in the multinational chain of command had over time.[30]

Each of these approaches to comparing institutional scope can be used to assess the relative *breadth* or *narrowness* of similar institutions. As with the other dimensions discussed thus far, doing so can support the assessment of important cross-variable relationships. One example of this is how attention to these institutional dimensions sheds light on an unresolved debate over whether institutions tend to face trade-offs between breadth (membership or size) and depth (scope).[31] As institutional scope increases, the argument goes, the more difficult it is to achieve as large or diverse a membership.

Assessing this argument in the context of battlefield coalitions requires a systematic approach to defining and measuring scope (and membership) such as those presented here. In at least three of the empirical cases of battlefield coalitions examined during research for this chapter, evidence of this trade-off was identified: C2 institutions with more limited scope over theatre-specific operations had larger memberships, whereas C2 institutions with broader scope over strategic planning for multiple theatres or functions had smaller memberships. In the Second World War, Allied theatre-level C2 had larger memberships than the more broadly-scoped, but purely Anglo-American CCS. In the Yom Kippur War of 1973, Iraq, Jordan, and several other Arab League members participated in battlefield fighting and placed forces under direct command of or in coordination with the Egyptian or Syrian armed forces, but it was an exclusively Egyptian and Syrian military committee that decided the overall strategy for the Arab attack. Last, in devising campaign strategy in Iraq and Syria for the counter-Daesh coalition, from the mid-2010s onward,

Table 5.1 Conceptual Dimensions of Institutions and Potential Measures, Part I

Dimension	Indicators	Potential Indicator Values
Formalisation	Are institutional rules written down or not?	Yes, no/verbal, *or* no/implicit
	By which processes has the legitimacy of an institution been established?	Political ratification, executive agreement, diplomatic note, *or* commander's orders
Membership	Which states, non-state entities, or military units are members of an institution?	Identities of actors at various levels of abstraction; count variable
	How does an institution's membership compare to (overlap with) that of other institutions in same coalition?	Identities of actors; indicator of identity, subset, *or* superset
Scope	About which military functions or objectives does an institution have authority to make decisions (e.g. tactics, logistics and resupply, strategic-level movement)?	Relatively broad (i.e. 'deep cooperation') *or* relatively narrow ('shallow cooperation'); by military function, domain, operational level

US leaders convened meetings to review strategic progress within a coalition 'Small Group,' which comprised of the ministers of only a subset of the 85 total members of the coalition that otherwise participated in high-level institutionalised political consultations and contributed forces under coalition C2.

Whether this pattern generalises to broader populations of battlefield coalitions is an open question. Yet, this pattern shows it is plausible that considering battlefield coalitions as institutions of varied scope might bring work on operational warfighting collectives into dialogue with scholarship on other types of institutions in which breadth-depth trade-offs or other institutional dynamics may play out.

Table 5.1 summarises the indicators and potential values they can take on for the dimensions of formalisation, membership, and scope.

5.2.4 Control: Which Rules Structure Decision-Making?

The dimension of *control* captures the processes by which institutional rules provide decisions will be made. Indicators of control type describe the procedural rules by which the members of an institution make and act on decisions.[32]

Control in battlefield coalition institutions can be of two types. What I refer to as *unstructured* control exists when ad hoc coordination or consensus-based decision-making mean that anarchy is the ordering rule for operational decision-making. Battlefield coalitions with C2 structures that operate in 'parallel' (e.g. M-shaped), which lack any subordination of a nation's forces to the commander of another, have unstructured institutional control.[33] For instance, British and Chinese Nationalist ground forces all operated in the Burma theatre during the Second World War from mid-1942 to mid-1943 in an institutional framework in which national commanders laboured under few rules governing their battlefield interactions. Each operational commander had autonomy to control the forces under his authority and no institutional requirement to accept the orders of others, thus comprising an unstructured control environment.

What I refer to as *structured* control exists in institutions where their rules define roles for members in decision-making. Two sub-types of structured control, which vary according to the equality of coalition members' formal power, are *hierarchic* and *pluralistic* control. In a coalition C2 context, hierarchic institutional control occurs when rules differentiate coalition actors' roles in decision-making on the basis of vertical lines of authority. These institutions correspond to U-shaped command structures.[34] In the summer of 1943 after initial failed attempts at winning territory back from Japanese control in Burma, Churchill, Roosevelt, and Chiang Kai-Shek agreed to establish a C2 system with structured, hierarchic control known as the South East Asia Command. Organising British, American, and Chinese Nationalist officers, staffs, and units under a British supreme commander mirrored how the Allied Forces Headquarters structure put Allied units under American supreme command in the Mediterranean. Doing so improved the Allies' performance in Burma in 1944 compared with the previous fighting season.[35] In other cases of structured control, in principle, battlefield coalition C2 institutions could define more pluralistic decision-making processes based on an equitable power distribution among members, as a committee-style majoritarian voting rule might, in order to avoid granting sole decision-making power to a single commander.[36]

C2 institutions can also blend these features of control. As one example, in the secretive Egyptian-Syrian strategic planning council that developed the dual-front attack on Israel that began the Yom Kippur War in 1973, military leaders operated under a rule of consensus (i.e. unstructured control) that nevertheless granted the Egyptian army chief of staff a final say on key issues (i.e. structured, hierarchic control).[37] Or, in the anti-Iraq coalition in the Persian Gulf War, C2 was blended between structured and unstructured types. French forces in Iraq operated under a separate, national chain of command that operated *without* structured means of operational decision-making vis-à-vis the multinational C2 organisation that operated, internally, with hierarchic control among the United States, Saudi, and other commanders.[38]

Analysing the institutional arrangements by which collectives of states or other armed actors agree to subordinate their forces to the control of battlefield coalition C2 systems cuts to the heart of key puzzles of cooperation under anarchy. Understanding the circumstances under which coalition members agree to sacrifice autonomy or sovereignty in this way requires using concepts like those proposed here. These concepts draw distinctions among various forms of institutional control and, as discussed in the next section on centralisation, avoid conflating control with other features of C2 systems that in fact signal the presence of distinct institutional dimensions.

5.2.5 Centralisation: Where Does Communication Occur?

The dimension of *centralisation* answers the question: does a single organisational node (or do many dispersed nodes) perform key institutional functions to facilitate communication and information-exchange? As a dimension of an international institution, centralisation refers to the degree to which an institution's rules grant it corporate competencies to organise and process information.[39] When enacted, centralised institutions engage in internally coordinated activities, like gathering reports from members (coordination and monitoring) or issuing reports or decisions (dispute resolution), which would otherwise be functions that members themselves would have to execute in a dispersed, uncoordinated fashion, if at all.[40] In battlefield coalitions, centralised institutions set rules for how transmission of information related to shared threats, capabilities, and coalition strategic and operational plans occur. These rules prescribe how coalition members organise as standing staffs or through regularised meetings or communications, thus determining how members can access the information these processes produce.

The frequency and regularity with which decision-makers, commanders, and functional experts formally convene serve as leading indicators of centralisation. Institutions with high degrees of centralisation establish standing combined military staffs and permanent multinational liaisons to coordinate activities to execute their given operational, planning, logistics, or intelligence-gathering functions. In the context of multinational C2, archetypal 'integrated' coalition command systems with structured, hierarchical control will establish standing multinational staffs to coordinate and implement operations. Many examples from the Anglo-American Allied Forces Headquarters in the Mediterranean from the Second World War through the multinational commands established in Iraq and Syria during operations against Daesh reflect a high degree of centralisation in this way. Yet, high degrees of centralisation do not imply high degrees of control. In the Persian Gulf War, though the French ground force, division *Daguet*, was not subject to coalition control, it nevertheless participated in the Coalition Coordination, Communication, and Integration Center that served as a highly centralised information hub.[41]

C2 institutions with modest degrees of centralisation may instead institutionalise regularised meetings of key individuals or implement reporting schedules. The Allied Supreme War Council and the subordinate Board of Military Representatives during the First World War are examples of such institutions that met monthly and weekly, respectively, under formalised processes to exchange information.[42] Or, a battlefield coalition can institutionalise such communications and meetings on an ad hoc basis. The Councils of Senior Naval Commanders of the Eight Nation Alliance in the Boxer War provided one such form. C2 systems might lack any substantive institutionalised centralisation, making no or little provisions for convening personnel or exchanging information. Syrian, Jordanian, and Iraqi commanders on the northern front of the Yom Kippur war lacked agreed-upon means of exchanging information or coordinating operations for most of the brief conflict with Israel and thus exemplify a battlefield coalition with low centralisation.

As these examples suggest, highly centralised C2 structures sometimes have hierarchical control, but not always. Some typologies of coalition C2 categorise systems along a unidimensional spectrum from 'integrated' or 'combined,' to 'mixed,' to 'parallel,' or 'independent.'[43] Such schematics offer a useful heuristic for gauging the depth of cooperation coalition partners are able to achieve when they agree to rules to subordinate military forces to coalition partners, to share information, to exchange liaison officers, and so forth. However, they fail to draw distinctions among what are ultimately independent dimensions of C2 systems. Even if institutions with hierarchical control and those with high degrees of centralisation do tend to correlate with one another,[44] each of these features represent discrete sets of decisions reflecting specific bargains among coalition members. One coalition in one context may agree to centralise information sharing, but not agree to a structured form of control; another may agree to structured control in C2, but fail to formalise any form of centralisation.

Each of these bargains, in turn, has the potential to shape outcomes through how they institutionalise a redistribution of power in a given domain of coalition activity. Distinguishing variation in a C2 system's distribution of authority in decision-making procedures – control – from other processes, like those that determine how information is gathered and shared among coalition members – centralisation – is therefore an important task for research on battlefield coalitions. Isolating how a battlefield coalition institution mediates control over decision-making from how it does or does not facilitate the centralisation of the information that may go into decision-making processes is key to analysing the relationship between these features, the reasons why institutions with these features arise in the first place, and which of these features promote coalition performance or other outcomes of interest.

Table 5.2 summarises the indicators and potential values they can take on for the dimensions of control and centralisation.

Table 5.2 Conceptual Dimensions of Institutions and Potential Measures, Part II

Dimension	Indicators	Potential Indicator Values
Control	Is institutional decision-making over a given military function unstructured (i.e. anarchic/ 'parallel'/consensus-driven) or structured?	Unstructured *or* Structured
	If structured, is decision-making hierarchic (i.e. authoritative) or pluralistic (i.e. majoritarian or other voting rule)?	Hierarchic *or* pluralistic
Centralisation	Which procedures that govern how information-sharing occurs in an institution?	Standing staff or bureaucracy, regular meetings, *or* irregular or rare meetings

5.3 Conclusion

Understanding the ways battlefield coalitions overcome cooperation problems is a central task in studying these collectives. Assessing the extent to which operational level institutions facilitate or complicate cooperation processes and outcomes requires both identifying these institutions where they exist and applying rigorous methods of conceptualising and observing variation among them. The analysis presented here finds that a range of rule-making phenomena in coalition contexts meet definitional standards as international institutions and bear many features that resemble other types of institutions recognisable as international organisations. As such, I have argued that explicit attention to features of battlefield coalitions – like formalisation, membership, scope, control, and centralisation – holds promise for making progress in understanding them. Doing so can bring findings of research on battlefield coalitions into dialogue with the study of institutions found in other international contexts.

Basic concepts and measurement approaches used in institutionalist scholarship therefore provide a foundation on which researchers of battlefield coalitions can continue to build. This chapter suggests at least a few ways institutionalist analysis can point the way to new discoveries. First, rigorous efforts to draw descriptive inferences can sometimes lead to new puzzles and research questions on their own. The suggestive evidence of a breadth-depth trade-off in the membership and scope of coalition C2 institutions presented above, for instance, raises the question of why states would accept delegating higher-order, strategy-making functions to a small group of coalition members to which their military forces will be expected to submit. Using the conceptual dimensions presented above as a basis for observations like this is an important step towards uncovering new associations or trade-offs that may exist among these variables across levels of analysis, time, and context. Does this breadth-depth trade-off exist across all cases of coalitions? Does increased formalisation come at the cost of control? Do control and centralisation tend to be independent, sufficient factors in promoting coalition performance, or are they both necessary components of strong, effective C2? The starting point for both asking and answering questions like these is attending to the types of measurement strategies recommended here.

Second, problematising dominant conceptual frameworks can generate useful insights into important features of battlefield coalitions like their formation, performance, and longevity by helping to reveal the role of omitted variables and look at problems in the new light the refraction of alternative analytic lenses shed. In their study of how unitary and multi-divisional C2 structures interact with belligerents' resource endowments and affect coalition performance, for instance, Cappella Zielinski and Grauer largely focus on different patterns of institutional control, as defined in this chapter. While their theory notes information transmission tends to be less efficient in unitary structures than in others, it does not consider whether the quality of institutional measures that enhance or degrade centralisation might independently affect battlefield performance.[45] Doing so raises additional questions for research about how battlefield coalitions that effectively use unitary C2 (i.e. hierarchical control) overcome the information centralisation problems such structures bring. Efforts to assess centralisation and other dimensions discussed here as explanatory variables for C2 effectiveness or any number of other important outcomes can strengthen the credibility of inferences drawn about the relative importance of these factors.

Last, the framework proposed here provides a launching-off point from which to bridge the security studies literatures at the basis of most scholarship on battlefield coalitions with other political science subfields. Approaching 'hard security' empirical settings like military coalitions from the neoliberal institutionalist or political-economy perspectives on which this framework draws leads to questions examined in other contexts. These include questions about the political origins of institutions, compliance, institutional stability and path dependencies, and the unintended consequences to which institutional politics give rise. Whether the answers to these 'big questions' in the study of coalitions mirror or diverge from the answers in other contexts can be of general interest that extends beyond international relations scholars steeped in military affairs.

Some researchers may worry that introducing additional concepts to the study of battlefield coalitions comes at the price of parsimony that other typologies or concepts provide, or that new concepts risk cluttering the language used to communicate about coalitions.[46] The validity of measurement strategies always depends in part on the purposes for which they are put to use.[47] So, the contention of this chapter has been not that all analyses of battlefield coalitions will benefit from the approaches presented here. Rather, these approaches underscore that conceptualising institutions as such is one tool of many that future research can usefully apply to advance knowledge about battlefield coalitions beyond its current state.

Notes

1 The author is grateful for comments from the editors, Alex Weisiger and Sara Bjerg Moller, which improved the quality of this chapter and the research on which it draws. Any errors remain his own.

2 Mancur Olson and Richard Zeckhauser, 'An Economic Theory of Alliances,' *The Review of Economics and Statistics* 48, no. 3 (1966): 266–79; Ronald R. Krebs, 'Perverse Institutionalism: NATO and the Greco-Turkish Conflict,' *International Organization* 53, no. 2 (1999): 343–77, https://doi.org/10.1162/00208189955 0904; Helga Haftendorn, Robert O. Keohane, and Celeste A. Wallander, eds., *Imperfect Unions: Security Institutions over Time and Space* (Oxford: Oxford University Press, 1999), https://doi.org/10.1111/1468-2346.00198; Patricia A. Weitsman, *Dangerous Allies: Proponents of Peace, Weapons of War* (Stanford CA: Stanford University Press, 2004); Brett Ashley Leeds and Sezi Anac, 'Alliance Institutionalization and Alliance Performance,' *International Interactions* 31, no. 3 (2005): 183–202, https://doi.org/10.1080/03050620500294135; Andrew G. Long, Timothy Nordstrom, and Kyeonghi Baek, 'Allying for Peace: Treaty Obligations and Conflict between Allies,' *Journal of Politics* 69, no. 4 (2007): 1103–17, https://doi.org/10.1111/j.1468-2508.2007.00611.x; Brett Ashley Leeds, Michaela Mattes, and Jeremy S. Vogel, 'Interests, Institutions, and the Reliability of International Commitments,' *American Journal of Political Science* 53, no. 2 (2009): 461–76, https://doi.org/10.1111/j.1540-5907.2009.00381.x; Marina Henke, *Constructing Allied Cooperation* (Ithaca NY: Cornell University Press, 2020), https://doi.org/1 0.1515/9781501739705.

3 Robert O. Keohane, 'Neoliberal Institutionalism: A Perspective on World Politics,' in *International Institutions and State Power: Essays in International Relations Theory*, ed. Robert O. Keohane (Boulder, CO: Westview Press, 1989), 3.

4 Erik Voeten, 'Making Sense of the Design of International Institutions,' *Annual Review of Political Science* 22 (2019): 147–63, https://doi.org/10.1146/annurev-polisci-041916-021108.

5 Brett Ashley Leeds, 'Alliance Treaty Obligations and Provisions, 1815–1944,' *International Interactions* 28, no. 3 (2002): 237–60, https://doi.org/10.1080/03 050620213653; Celeste A. Wallander and Robert O. Keohane, 'Risk, Threat, and Security Institutions,' in *Imperfect Unions*, ed. Haftendorn et al. (Oxford; Oxford University Press, 1999) 21–47, https://doi.org/10.1111/1468-2346.00198; Ana Arjona, 'Wartime Institutions: A Research Agenda,' *Journal of Conflict Resolution* 58, no. 8 (2014): 1360–89, https://doi.org/10.1177/002200271454 7904.

6 Robert Jervis, 'Realism, Neoliberalism, and Cooperation: Understanding the Debate,' *International Security* 24, no. 1 (1999): 42–63, https://doi.org/10.1162/ 016228899560040.

7 Peter D. Feaver, *Armed Servants: Agency, Oversight, and Civil-Military Relations* (Cambridge MA: Harvard University Press, 2005), https://doi.org/1 0.4159/9780674036772; Stephen Biddle, *Military Power: Explaining Victory and Defeat in Modern Battle* (Princeton, NJ: Princeton University Press, 2006), https://doi.org/10.2307/j.ctt7s19h; Caitlin Talmadge, *The Dictator's Army: Battlefield Effectiveness in Authoritarian Regimes* (Ithaca, NY: Cornell University Press, 2015), https://doi.org/10.7591/9781501701764; Elizabeth Kier, *Imagining War: French and British Military Doctrine between the Wars*

(Princeton, NJ: Princeton University Press, 2017), https://doi.org/10.1515/9781400887477.

8 Nora Bensahel, 'International Alliances and Military Effectiveness: Fighting Alongside Allies and Partners,' in *Creating Military Power: The Sources of Military Effectiveness*, ed. Risa A. Brooks and Elizabeth A. Stanley (Stanford, CA: Stanford University Press, 2007), https://doi.org/10.11126/stanford/9780804753999.003.0008.

9 Keohane, 'Neoliberal Institutionalism,' 3–4.

10 Kathryn M.G. Boehlefeld and Kelly Grieco, 'Exercising Escalation: Do Multinational Military Exercises Provoke Interstate Security Crises?,' Chapter 3, this volume.

11 Bensahel, 'International Alliances,' 198.

12 Sara Bjerg Moller, 'Learning from Losing: How Defeat Shapes Coalition Command Arrangements in Wartime,' Chapter 7, this volume; Martin Van Creveld, *Command in War* (Cambridge: Harvard University Press, 1985), 9.

13 Ryan Grauer, *Commanding Military Power: Organizing for Victory and Defeat on the Battlefield* (Cambridge: Cambridge University Press, 2016), https://doi.org/10.1017/CBO9781316670170.

14 Rosella Cappella Zielinski and Ryan Grauer, 'Organizing for Performance: Coalition Effectiveness on the Battlefield,' *European Journal of International Relations* 26, no. 4 (2020): 953–78, https://doi.org/10.1177/1354066120903369.

15 Combined Chiefs of Staff, 'Washington War Conference: Post-Arcadia Collaboration,' Memorandum ABC-4/CS4 (14 January 1942), https://www.jcs.mil/Portals/36/Documents/History/WWII/Arcadia3.pdf.

16 This analysis excludes the fifth, flexibility, included in the study by Barbara Koremenos, Charles Lipson, and Duncan Snidal, 'The Rational Design of International Institutions,' *International Organization* 55, no. 4 (2001): 763, https://doi.org/10.1162/002081801317193592. As the following analysis of other institutional dimensions shows, coalition institutions vary in their propensity to change or remain static over time, that is, their flexibility. The baseline framework presented here can serve as a template for future work to identify general indicators and empirical examples of institutional flexibility, as a second-order institutional feature with respect to the others I discuss here.

17 Keohane, 'Neoliberal Institutionalism,' 4–5.

18 James D. Morrow, 'Alliances: Why Write Them Down?,' *American Review of Political Science* 3 (2000): 63–83, https://doi.org/10.1146/annurev.polisci.3.1.63.

19 Keohane's definition of institutions, which I adopt here, includes those that are unformalised. Compare this with Koremenos et al., 'Rational Design,' where institutions are defined as only those that are formalised as 'explicit arrangements, negotiated among international actors, that prescribe, proscribe, and/or authorize behaviour.'

20 Martin C. Steinwand and Nils W. Metternich, 'Who Joins and Who Fights? Explaining Tacit Coalition Behavior among Civil War Actors,' *International Studies Quarterly* 66, no. 4 (2022): 1–14, https://doi.org/10.1093/isq/sqac077; Dan Reiter, 'Command and Military Effectiveness in Rebel and Hybrid Battlefield Coalitions,' Chapter 6, this volume.

21 Notably, institutions with high procedural formality need not imply actors comply with their provisions. Despite bearing the formal title of supreme commander, Egyptian Field Marshal Ahmad Ismail did not exert command in a meaningful operational sense during the conflict. Saad El Shazly, *The Crossing of the Suez* (San Francisco, CA: American Mideast Research, 1980); Abraham Sela, 'The 1973 Arab War coalition: Aims, Coherence, and Gain-Distribution,'

Israel Affairs 6, no. 1 (1999): 36–69, https://doi.org/10.1080/1353712990871 9546.

22 Leeds and Anac, 'Alliance Institutionalization'; Long, Nordstrom, and Baek, 'Allying for Peace'; Michael R. Kenwick, John A. Vasquez, and Matthew A. Powers, 'Do Alliances *Really* Deter?,' *Journal of Politics* 77, no. 4 (2015): 943–54, http://dx.doi.org/10.1086/681958; Brett Ashley Leeds and Jesse C. Johnson, 'Theory, Data, and Deterrence: A Response to Kenwick, Vasquez, and Powers,' *Journal of Politics* 79, no. 1 (2016): 335–40, http://dx.doi.org/10.1086/687285; James D. Morrow, 'When Do Defensive Alliances Provoke Rather than Deter?,' *Journal of Politics* 79, no. 1 (2016): 341–45, http://dx.doi.org/10.1086/686973.

23 Umio Otsuka, 'Coalition Coordination during the Boxer Rebellion: How Twenty-Seven "Councils of Senior Naval Commanders" Contributed to the Conduct of Operations,' *Naval War College Review* 71, no. 4 (2018): 111–30.

24 Leeds, 'Alliance Treaty Obligations and Provisions, 1815–1944.'

25 Cappella Zielinski and Grauer make a significant contribution to this end in their Belligerents in Battle data set by identifying not only which states and non-state actors are present on different coalition battlefields, but also the numbers and types of forces they contribute. Future scholarship can focus on characterising the institutional links that bind these actors together. Rosella Cappella Zielinski and Ryan Grauer, 'A Century of Coalitions in Battle: Incidence, Composition, and Performance, 1900–2003,' *Journal of Strategic Studies* 45, no. 2 (2022): 186–210, https://doi.org/10.1080/01402390.2021.2011233.

26 Alex Weisiger, 'Exiting the Coalition: When Do States Abandon Coalition Partners during War?,' *International Studies Quarterly* 60, no. 4 (2016): 753–65, https://doi.org/10.1093/isq/sqw029.

27 Mancur Olson, *The Logic of Collective Action: Public Goods and the Theory of Groups* (Cambridge, MA: Harvard University Press, 1965); Sarah E. Kreps, *Coalitions of Convenience: United States Military Interventions after the Cold War* (New York: Oxford University Press, 2011).

28 George W. Downs, David M. Rocke, and Peter N. Barsoom, 'Is the Good News about Compliance Good News about Cooperation?,' *International Organization* 50, no. 3 (1996): 379–406, https://doi.org/10.1017/S0020818300033427.

29 Christopher S. Chivvis, 'Strategic and Political Overview of the Intervention,' in *Precision and Purpose: Airpower in the Libyan Civil War*, ed. Karl P. Mueller (Santa Barbara, CA: RAND Corporation, 2015), 26. See also Stéfanie Von Hlatky and Thomas Juneau, 'When the Coalition Determines the Mission: NATO's Detour in Libya,' Chapter 4, this volume.

30 Stephen M. Saideman and David P. Auerswald, 'Comparing Caveats: Understanding the Sources of National Restrictions upon NATO's Mission in Afghanistan,' *International Studies Quarterly* 56, no. 1 (2012): 67–84, https://doi.org/10.1111/j.1468-2478.2011.00700.x.

31 Michael J. Gilligan, 'Is There a Broader-Deeper Trade-off in International Multilateral Agreements?,' *International Organization* 58, no. 3 (2004): 459–84, https://doi.org/10.1017/S0020818304583029.

32 Koremenos et al., 'Rational Design,' 772–3. It is important to distinguish the use of the term *control* as a conceptual dimension of an institution here from its use in military jargon to connote 'operational control' or 'tactical control' as functions of operational or tactical command. The United States and allied forces distinguish between the 'command' (e.g. authority to assign missions or tasks) and the 'control' (e.g. authority to execute such missions or tasks) functions of a C2 system. See Sara Bjerg Moller, 'Fighting Friends: Institutional Cooperation and

Military Effectiveness in Multinational War' (PhD dissertation, Columbia University 2016), 490–5, for a useful summary. Here, I conceptualise a C2 system as an institution with a type of *institutional control* determined by how the procedural rules say decisions pertaining to both 'command' and 'control' are to be made with respect to a given set of coalition military forces.

33 Cappella Zielinski and Grauer, 'Organizing for Performance.'
34 Ibid.
35 Charles F. Romanus and Riley Sunderland, *Stilwell's Mission to China* (Washington, DC: Department of the Army, 1953); Allied Forces Headquarters, *History of Allied Forces Headquarters: Part 1, August – December 1942* (1945).
36 Whether C2 systems with control rules institutionalised as such arise frequently or behave differently than those with unstructured control is an empirical question for future work.
37 El Shazly, Crossing of the Suez.
38 US Department of Defense, *Conduct of the Persian Gulf War: Final Report* (April 1992), Annex K.
39 The use of the term centralisation here differs from others where it references how high in a C2 hierarchy certain types of operational or tactical decisions are made. In those contexts, the 'centralisation' or 'decentralisation' of a command structure refers to the scope of authorities delegated to lower, or 'less central,' levels in a decision-making structure.
40 Koremenos et al. discuss centralisation in terms of functions 'to disseminate information, to reduce bargaining and transaction costs, and to enhance enforcement,' 'Rational Design,' 771. Aaron Schneider, 'Decentralization: Conceptualization and Measurement,' *Studies in Comparative International Development* 38, no. 3 (2003): 32–56, https://doi.org/10.1007/BF02686198, describes centralisation as how much institutions prescribe the transfer of power and resources to actors higher in a hierarchy or more centrally located in a dependency network.
41 US Department of Defense, *Conduct of the Persian Gulf War*, K-25.
42 Meighen McCrae, *Coalition Strategy and the End of the First World War: The Supreme War Council and War Planning, 1917–1918* (Cambridge: Cambridge University Press, 2019), 12–26, https://doi.org/10.1017/9781108566711.
43 Moller, *Fighting Friends*; Daniel S. Morey, 'Centralized Command and Coalition Victory,' *Conflict Management and Peace Science* 37, no. 6 (2020): 716–34; Kelly Grieco, 'War by Coalition: The Effects of Coalition Military Institutionalization on Coalition Battlefield Effectiveness' (PhD dissertation, Massachusetts Institute of Technology, 2016).
44 Koremenos et al. claim, 'While Centralization May Reduce Control in Some Cases, the Two Dependent Variables Generally Vary Independently,' 'Rational Design,' 772. Only by distinguishing them conceptually is it possible to evaluate how frequently they do vary independently empirically.
45 'Organizing for Performance,' 958–60.
46 Giovanni Sartori, 'Concept Misformation in Comparative Politics,' *American Political Science Review* 64, no. 4 (1970): 1033–53, https://doi.org/10.2307/1958356; John Gerring, 'What Makes a Concept Good? A Criterial Framework for Understanding Concept Formation in the Social Sciences,' *Polity* 31, no. 3 (1999): 357–93, https://doi.org/10.2307/3235246.
47 Robert Adcock and David Collier, 'Measurement Validity: A Shared Standard for Qualitative and Quantitative Research,' *American Political Science Review* 95, no. 3 (2001): 529–46, https://doi.org/10.1017/S0003055401003100.

Bibliography

Adcock, Robert, and David Collier. 'Measurement Validity: A Shared Standard for Qualitative and Quantitative Research.' *American Political Science Review* 95, no. 3 (2001): 529–46. 10.1017/S0003055401003100

Arjona, Ana. 'Wartime Institutions: A Research Agenda.' *Journal of Conflict Resolution* 58, no. 8 (2014): 1360–89. 10.1177/0022002714547904

Bensahel, Nora. 'International Alliances and Military Effectiveness: Fighting Alongside Allies and Partners.' In *Creating Military Power: The Sources of Military Effectiveness*, edited by Risa A. Brooks and Elizabeth A. Stanley. Stanford, CA: Stanford University Press, 2007. 10.11126/stanford/9780804753999.003.0008

Biddle, Stephen. *Military Power: Explaining Victory and Defeat in Modern Battle*. Princeton, NJ: Princeton University Press, 2006. 10.2307/j.ctt7s19h

Cappella Zielinski, Rosella, and Ryan Grauer. 'Organizing for Performance: Coalition Effectiveness on the Battlefield.' *European Journal of International Relations* 26, no. 4 (2020): 953–78. 10.1177/1354066120903369

Cappella Zielinski, Rosella, and Ryan Grauer. 'A Century of Coalitions in Battle: Incidence, Composition, and Performance, 1900–2003.' *Journal of Strategic Studies* 45, no. 2 (2022): 186–210. 10.1080/01402390.2021.2011233

Chivvis, Christopher S. 'Strategic and Political Overview of the Intervention.' In *Precision and Purpose: Airpower in the Libyan Civil War*, edited by Karl P. Mueller. Santa Barbara, CA: RAND Corporation, 2015.

Downs, George W., David M. Rocke, and Peter N. Barsoom. 'Is the Good News about Compliance Good News about Cooperation?' *International Organization* 50, no. 3 (1996): 379–406. 10.1017/S0020818300033427

El Shazly, Saad. *The Crossing of the Suez*. San Francisco, CA: American Mideast Research, 1980.

Feaver, Peter D. *Armed Servants: Agency, Oversight, and Civil-Military Relations*. Cambridge, MA: Harvard University Press, 2005. 10.4159/9780674036772

Gerring, John. 'What Makes a Concept Good? A Criterial Framework for Understanding Concept Formation in the Social Sciences.' *Polity* 31, no. 3 (1999): 357–93. 10.2307/3235246

Gilligan, Michael J. 'Is There a Broader-Deeper Trade-off in International Multilateral Agreements?' *International Organization* 58, no. 3 (2004): 459–84. 10.1017/S0020818304583029

Grauer, Ryan. *Commanding Military Power: Organizing for Victory and Defeat on the Battlefield*. Cambridge: Cambridge University Press, 2016. 10.1017/CBO9781316670170

Grieco, Kelly. 'War by Coalition: The Effects of Coalition Military Institutionalization on Coalition Battlefield Effectiveness.' PhD dissertation, Massachusetts Institute of Technology, 2016.

Haftendorn, Helga, Robert O. Keohane, and Celeste A. Wallander, eds. *Imperfect Unions: Security Institutions over Time and Space*. Oxford: Oxford University Press, 1999. 10.1111/1468-2346.00198

Henke, Marina. *Constructing Allied Cooperation*. Ithaca, NY: Cornell University Press, 2020. 10.1515/9781501739705

Jervis, Robert. 'Realism, Neoliberalism, and Cooperation: Understanding the Debate.' *International Security* 24, no. 1 (1999): 42–63. 10.1162/016228899560040

Kenwick, Michael R., John A. Vasquez, and Matthew A. Powers. 'Do Alliances *Really* Deter?' *Journal of Politics* 77, no. 4 (2015): 943–54. 10.1086/681958

Keohane, Robert O. 'Neoliberal Institutionalism: A Perspective on World Politics.' In *International Institutions and State Power: Essays in International Relations Theory*, edited by Robert O. Keohane, 1–20. Boulder, CO: Westview Press, 1989.

Kier, Elizabeth. *Imagining War: French and British Military Doctrine between the Wars*. Princeton, NJ: Princeton University Press, 2017. 10.1515/9781400887477

Koremenos, Barbara, Charles Lipson, and Duncan Snidal. 'The Rational Design of International Institutions.' *International Organization* 55, no. 4 (2001): 761–99. 10.1162/002081801317193592

Krebs, Ronald R. 'Perverse Institutionalism: NATO and the Greco-Turkish Conflict.' *International Organization* 53, no. 2 (1999): 343–77. 10.1162/002081899550904

Kreps, Sarah E. *Coalitions of Convenience: United States Military Interventions after the Cold War*. New York: Oxford University Press, 2011. 10.1093/acprof:oso/9780199753796.001.0001

Leeds, Brett Ashley 'Alliance Treaty Obligations and Provisions, 1815–1944.' *International Interactions* 28, no. 3 (2002): 237–60. 10.1080/03050620213653

Leeds, Brett Ashley, and Jesse C. Johnson. 'Theory, Data, and Deterrence: A Response to Kenwick, Vasquez, and Powers.' *Journal of Politics* 79, no. 1 (2016): 335–40. 10.1086/687285

Leeds, Brett Ashley, Michaela Mattes, and Jeremy S. Vogel. 'Interests, Institutions, and the Reliability of International Commitments.' *American Journal of Political Science* 53, no. 2 (2009): 461–76. 10.1111/j.1540-5907.2009.00381.x.

Leeds, Brett Ashley, and Sezi Anac. 'Alliance Institutionalization and Alliance Performance.' *International Interactions* 31, no. 3 (2005): 183–202. 10.1080/0305 0620500294135

Long, Andrew G., Timothy Nordstrom, and Kyeonghi Baek 'Allying for Peace: Treaty Obligations and Conflict between Allies.' *Journal of Politics* 69, no. 4 (2007): 1103–17. 10.1111/j.1468-2508.2007.00611.x

McCrae, Meighen. *Coalition Strategy and the End of the First World War: The Supreme War Council and War Planning, 1917–1918*. Cambridge: Cambridge University Press, 2019. 10.1017/9781108566711

Morey, Daniel S. 'Centralized Command and Coalition Victory.' *Conflict Management and Peace Science* 37, no. 6 (2020): 716–34. 10.1177/073889422 0934884

Morrow, James D. 'Alliances: Why Write Them Down?' *American Review of Political Science* 3 (2000): 63–83. 10.1146/annurev.polisci.3.1.63

Morrow, James D. 'When Do Defensive Alliances Provoke Rather than Deter?' *Journal of Politics* 79, no. 1 (2016): 341–5. 10.1086/686973

Olson, Mancur. *The Logic of Collective Action: Public Goods and the Theory of Groups*. Cambridge, MA: Harvard University Press, 1965.

Olson, Mancur, and Richard Zeckhauser. 'An Economic Theory of Alliances.' *The Review of Economics and Statistics* 48, no. 3 (1966): 266–79.

Otsuka, Umio. 'Coalition Coordination during the Boxer Rebellion: How Twenty-Seven "Councils of Senior Naval Commanders" Contributed to the Conduct of Operations.' *Naval War College Review* 71, no. 4 (2018): 111–30.

Romanus, Charles F., and Riley Sunderland. *Stilwell's Mission to China*. Washington, DC: Department of the Army, 1953.

Saideman, Stephen M., and David P. Auerswald. 'Comparing Caveats: Understanding the Sources of National Restrictions upon NATO's Mission in Afghanistan.' *International Studies Quarterly* 56, no. 1 (2012): 67–84. 10.1111/j.1468-2478.2011.00700.x.

Sartori, Giovanni. 'Concept Misformation in Comparative Politics.' *American Political Science Review* 64, no. 4 (1970): 1033–53. 10.2307/1958356

Schneider, Aaron. 'Decentralization: Conceptualization and Measurement.' *Studies in Comparative International Development* 38, no. 3 (2003): 32–56. 10.1007/BF02686198

Sela, Abraham. 'The 1973 Arab War Coalition: Aims, Coherence, and Gain-Distribution.' *Israel Affairs* 6, no. 1 (1999): 36–69. 10.1080/13537129908719546

Steinwand, Martin C., and Nils W. Metternich. 'Who Joins and Who Fights? Explaining Tacit Coalition Behavior among Civil War Actors.' *International Studies Quarterly* 66, no. 4 (2022): 1–14. 10.1093/isq/sqac077

Talmadge, Caitlin. *The Dictator's Army: Battlefield Effectiveness in Authoritarian Regimes*. Ithaca, NY: Cornell University Press, 2015. 10.7591/9781501701764

Van Creveld, Martin. *Command in War*. Cambridge: Harvard University Press, 1985.

Voeten, Erik. 'Making Sense of the Design of International Institutions.' *Annual Review of Political Science* 22 (2019): 147–63. 10.1146/annurev-polisci-041916-021108

Weisiger, Alex. 'Exiting the Coalition: When Do States Abandon Coalition Partners during War?' *International Studies Quarterly* 60, no. 4 (2016): 753–65. 10.1093/isq/sqw029

Weitsman, Patricia A. *Dangerous Allies: Proponents of Peace, Weapons of War*. Stanford, CA: Stanford University Press, 2004.

6 Command and Military Effectiveness in Rebel and Hybrid Battlefield Coalitions

Dan Reiter

Military historians and political scientists have long studied coalition war-fare, typically focusing on collections of states fighting each other, such as the several coalitions fighting Napoleonic France, the Allies fighting the Central Powers in World War I, and the Axis Powers in World War II, and United Nations coalitions fighting North Korea in the early 1950s and Iraq in the early 1990s. This scholarship has focused on two basic questions: What makes multinational battlefield coalitions fight more effectively? What factors prevent multinational battlefield coalitions from fighting together more effectively?[1]

Today scholars and policy-makers increasingly recognise that this tradi-tional focus on multinational coalitions, battlefield coalitions comprised solely of states, is too narrow. Wars within states are at least as frequent as wars between states, and civil wars are often fought by coalitions of rebel groups. Further, many contemporary conflicts in countries such as Syria, Libya, and Ukraine involve hybrid coalitions, in which non-state actors such as rebel groups fight alongside third-party nation-state militaries for a common cause. Despite the frequency of rebel coalitions and the contem-porary phenomena of hybrid groups, there is limited scholarly discussion of rebel coalitions and almost no consideration of hybrid coalitions. This chapter thus begins to fill this lacuna in the literature to improve our un-derstanding of both. It does so through the lens of coalition command and how variation in the command structures adopted affects the operational effectiveness of the coalition. Building on conventional wisdom regarding multinational coalitions, I compose a conceptual framework to begin to understand the organisation and effectiveness of rebel and hybrid coalitions.

To begin, the chapter briefly reviews a core proposition from the study of multinational coalitions: unified coalition command positively affects oper-ational military effectiveness but, due to sovereignty concerns of states, is difficult to achieve. It then presents two caveats to the proposition that unified command always improves military effectiveness. First, hoped-for benefits are unlikely to be achieved if such formal structures are ignored and coalition members go their own way. The chapter thus presents the idea of 'effective unified command', to stress the importance of de facto unified

DOI: 10.4324/9781003399896-9

command, beyond de jure unified command. Second, unified command will not improve military effectiveness if the coalition embraces a military approach that requires greater member decision-making autonomy or pursues misguided force employment plans or military strategies.

The chapter then compares multinational and rebel coalitions regarding coalition command and its implications for operational military effectiveness. It first notes some similarities with respect to the relationship between unified command and military effectiveness as well as coalition member autonomy concerns that make implementation difficult. It then describes three differences between multinational and rebel coalitions. First, rebel coalition members experience unique risks with regard to sacrificing autonomy if they accede to unified command. Second, rebel coalition members have more variable kinds of authority relations with one another than do state coalition members. Third, unified command has different military effectiveness implications for rebel coalitions, as such collectives are less likely to engage in conventional warfare and must think carefully about decisions to target civilians.

Last, the chapter builds on the discussions of multinational and rebel coalitions to describe command structures and operational military effectiveness in hybrid coalitions. Command structures within hybrid coalitions vary significantly and, like in rebel coalitions, autonomy, and credible commitment concerns impede the formation of unified coalition command. Different from rebel coalitions, nation-state coalition members of hybrid coalitions have tools available with which they can reduce credible commitment concerns and encourage rebel groups to accept unified coalition command. There are also potentially high military operational military effectiveness benefits of unified command in a hybrid coalition, given the variety of comparative advantages of coalition members. The final section of the chapter offers policy recommendations for nation-states leading hybrid coalitions.

The reconceptualisation and exploration of battlefield coalitions offers several scholarly and policy payoffs. For scholars, it contributes to a growing trend of tearing down conceptual walls between types of conflict, rejecting assumptions that civil and interstate wars are different classes of phenomena that deserve distinct conceptual structures and policy approaches.[2] Relatedly, scholars are beginning to relax the assumption that interstate wars are only conventional and civil wars are only guerrilla.[3] Pushing these trends further, this chapter recognises that some conflicts are hybrid civil/interstate, with nation-states and rebel groups fighting alongside one another and employing a variety of military approaches. Scholars have begun to apply individual theories both to interstate and civil conflicts, and the proposed framework will facilitate further work along such lines.[4] Some have gone so far as to suggest abandoning the political science subfields of IR and comparative politics, replacing them with subfields of conflict, institutions, and political economy.[5]

For practitioners, it is critical to broaden understandings of coalition warfare. Certainly, there will be coalition warfare into the 21st century, involving coalitions of states, coalitions of rebel groups, and hybrid coalitions. Peacekeeping missions often involve multiple states, and sometimes states fighting with rebel groups. Counterinsurgency operations will continue to be multinational and will combat rebel or perhaps hybrid coalitions. NATO, of course, plans for multinational coalition operations, and US defence planning also examines coalition warfare to defend allies such as Taiwan, South Korea, and NATO members. States continue to work together in coalition operations in other ways, to combat maritime piracy, in counterproliferation missions, and others.[6]

6.1 Multinational Battlefield Coalitions

Battles between states often involve at least one coalition of belligerents.[7] Scholars and historians have long asked: How can coalitions fight better together? How can coalitions be more than the sum of their parts, or at least not less than the sum of their parts?

Common answers to these questions focus on coalition command, or the existence and nature of arrangements for the coordinated deployment and use of the military forces of coalition members.[8] Conventional wisdom regarding interstate coalition command emphasises the operational benefits of unified command, or a single individual directing a coalition.[9] Unified coalition command improves the operational military effectiveness of interstate coalitions by facilitating coordination among members, just as a unified commander of a country's various military branches helps that country's military fight better.[10] The military advantages of such coordination amongst belligerents are several and straightforward.[11] Coordination allows disparate members to efficiently combine their assets and execute in concert a single strategy, as compared with dividing assets to implement different strategies or a single strategy with poor coordination. Better coordination through unified coalition command can also facilitate sharing intelligence, joint military training in preparation for coalition warfare, the transfer of technology and doctrinal innovations, and improved logistical interoperability.[12] Relatedly, greater coordination within the coalition can exploit comparative advantages on the battlefield. In the North African theater in World War II, the establishment of a unified command facilitated exploiting comparative advantage, with German forces executing mobile attacks and Italian forces providing static defences.[13]

Evidence also suggests that unified command can help interstate coalitions win their wars. Daniel Morey described three types of interstate coalition leadership: the lack of a formal coalition leader, a joint coalition leadership composed of representatives of the coalition members, and a unified command comprising a single coalition representative. Examining the outcomes of all interstate wars since 1816, Morey found that joint command coalitions were more likely to win compared with coalitions without command, and unified command coalitions were more likely to win compared with joint command coalitions.[14]

The advantages of unified coalition command are demonstrated by examining the loss in military effectiveness when coalition command is not unified. South Korea, fearing significant casualties, resisted unified command during the Vietnam War, and as a consequence, its hundreds of thousands of troops deployed to South Vietnam stayed in their bases, limiting their ability to support aggressive operations undertaken by American troops.[15] Belgium was allied with Britain and France before World War I, but also resisted unified command and refused to allow the prewar deployment of Allied forces to its territory, an action that might have deterred Germany from attacking.[16] During that same war, in the East, Austria-Hungary and Germany in the early years of the war stumbled against Russia, eventually improving their operational military effectiveness when they formed a unified coalition command in the fall of 1916.[17] One of the reasons for the disastrous outcome in the 1993 Battle of Mogadishu was the lack of unified UN peacekeeping command over the various national forces there.[18]

There are, however, two caveats to the claim that unity of command improves operational military effectiveness. The first is that military effectiveness depends on a particular form of consolidated unified command. Specifically, though formally unified command structures may exist, a commander may lack substantial de facto authority to direct coalition military forces. This was initially the case when General Ferdinand Foch was given formal command of Allied forces in the West in March 1918, during World War I.[19] Rosella Cappella Zielinski and Ryan Grauer make a similar point, that unity of command (what they describe as U-form command structures) does not provide military effectiveness payoffs if coalition members do not accede to the authority of formal coalition command.[20] Conversely, there can be de facto unified command even when de jure there is not unified command. For example, in North Africa in World War II, Germany, and Italy had de jure joint or combined coalition command (meaning, formally, coalition decisions were supposed to be made by a committee composed of coalition member representatives rather than by a single person),[21] but Germany was the de facto coalition leader, enabling the coalition to fight effectively.[22] De facto unified command is necessary.

We label de facto unity of command as 'effective unity of command', observing that de jure unity of command may not result in effective coordination of coalition member forces. A coalition has more effective unity when there is a single actor who, in practice, has greater power to direct the employment of coalition military forces. In a unified command coalition, the less power the appointed commander has over the disposition of coalition forces, the lower the effective unity of command. In a joint or combined command where a commander has de jure authority, the coalition will be more effective when the appointed commander has greater actual ability to direct coalition forces. This approach of focusing on the effective unity of coalition command rather than just formal structures parallels approaches taken in the study of other international institutions. For example, in their study of international courts, Jeffrey Staton and Will Moore developed the concept of judicial power, combining autonomy and effectiveness into a single measure.[23] Scholars of international hierarchy have similarly recognised that power can be exerted formally or informally.[24]

A second caveat is that even effective unified command may not always boost military effectiveness. This is true for two reasons. First, for some forms of warfare, operational military effectiveness is improved if coalition members are given general instruction as to their combat tasks and control is limited, as members are given the necessary decision-making autonomy to execute these tasks as they see fit and to adapt to evolving combat conditions. In particular, in modern ground warfare, strategies which emphasise mobility and allow for combatants to exploit emerging battle opportunities are thought to be the most effective, and multinational coalitions executing such strategies may be more effective if some coalition authorities are more decentralised. That is, while some forms of unified command are helpful with crafting unified strategies and operational military goals, emphasis on unity can cut against operational military effectiveness if a unified command exercises too much control over members' execution of combat tasks, not providing them with the battlefield decision-making autonomy they might need to succeed. Morey recognised that unified coalition command, including as he measured it, might reduce military effectiveness because of reduction in autonomy of coalition member forces.[25]

For example, the UN ground campaign of the 1991 Gulf War involved a multinational coalition. The coalition's force employment strategy required the allocation of decision-making autonomy to the forces of individual coalition members, and the allocation of this authority as part of the broader force employment strategy helped ensure great success.[26] British and American ground forces sometimes fought side by side, and a certain degree of coalition member autonomy helped contribute to success in making a modern force employment strategy work, as at the February 27 Battle of Norfolk, one of the world's largest armor battles since 1945.[27] Further, counterinsurgency efforts can take place in a complex and evolving environment, and effectiveness in winning over local hearts and minds sometimes requires independent action by counterinsurgency forces. As the US Counterinsurgency Manual states, 'Tactical leaders must have the ability, within the commander's intent, to have freedom of action to work with a local society.'[28] Of course, providing coalition members with greater authority to act independently presumes that members have the capacity to take advantage of emerging opportunities effectively, and this may not always be the case if their forces lack the proper training or high-quality military leadership. It also presumes that coalition members are loyal, in the sense that, if directed more loosely, they will still take actions that serve the goals of the coalition.

The second reason that unified coalition command may not raise military effectiveness is that the proposition assumes that the unified coalition commander will make good decisions about deploying coalition forces. This may not necessarily be true, as coalition commanders may make suboptimal decisions because of biases from overlearning lessons of the past, cognitive bias, organisational culture or inertia, domestic political considerations, or other factors.[29] That is, giving more authority to a coalition command does not necessarily avoid mistaken judgment; it might instead make mistakes bigger.

Several historical episodes demonstrate that poor strategic choices can undermine the purported relationship between unified command and improved operational military effectiveness. Sometimes, unified command amplifies bad strategic decision making. In 1940, the fundamental flaw in the Anglo-French-Belgian defence plan was not insufficient coalition command unity, but failure to recognise that the main German offensive would come through the Ardennes.[30] During World War I, the Allied offensives on the Western Front failed primarily because they selected an inappropriate military strategy, not because the coalition command was insufficiently unified.[31] Anglo-American bombing of Germany in World War II failed to push Hitler to surrender not because of inadequate coalition command unity, but because of flaws in targeting strategy.[32] The coalition deployed to Iraq after the capture of Baghdad in 2003 fared poorly against the roiling insurgency, not because of insufficiently unified command, but because it was implementing the incorrect strategy, an error that when corrected in 2007–2008 quickly led to markedly better results.[33] Other times, unified command may suppress widespread adoption of ideas generated by individual coalition members who may understand how to implement military operations more effectively than the coalition commander. This was perhaps the environment for the US-led coalition in the Vietnam War, when Australian forces embraced different and arguably more effective counterinsurgency ideas as compared with American forces and maintained substantial independence within the multi-national coalition fighting against the Viet Cong insurgents.[34]

If it is at least widely believed that more unified coalition command often facilitates higher operational military effectiveness, what prevents coalitions from adopting such structures,[35] especially if battlefield losses begin to mount?[36] In short, sovereignty. Agreeing to a more unified command means giving another actor greater authority over the employment of a coalition's military forces. Delegating such authority might be costly for normative, political, and/or strategic reasons.[37]

A coalition member may oppose allowing a multinational command directed by a committee or by an officer from another nation to make decisions about the deployments and perhaps deaths of its own troops. This was an essential part of the 1994 American Presidential Decision Directive 25 (PDD-25), adopted after the disastrous 1993 Battle of Mogadishu. PDD-25 emphasised the American prerogative to keep US troops under US command, even in the context of multilateral peacekeeping operations.[38] Perhaps not unrelatedly, a coalition member may believe that the coalition command supports an inferior military strategy, as was the case regarding Australia's participation in the Vietnam War, discussed above. Differences in strategy may be driven by differences in national priorities, as in the early years of World War I, when joint Anglo-French planning for offensives in the West was hampered by a British preference for liberating Belgium, as compared with French desires to clear German forces out of France.[39] Further, diplomatic prerogatives may prevent a coalition member from agreeing to unified coalition command. Building an effective coalition command and effective inter-military coordination requires pre-war planning. However, some states may be formally neutral, and may fear that prewar coordination with other militaries may compromise their neutrality stance, potentially violating international law and treaty commitments, and perhaps attract aggression. In both 1914 and 1940, Belgium's insistence on maintaining strict neutrality led it to refuse to put its forces under a unified command with Britain or France.[40] Faced with a potentially unacceptable infringement on their autonomy, belligerents might be unwilling to accept unified coalition command.[41]

In summary, there are three central propositions about multinational coalitions. First, command structures for multinational coalitions can vary. Second, political considerations about member autonomy can prevent the formation of unified command structures. Third, unified command can improve operational military effectiveness, though not under all circumstances.

6.2 Rebel Coalitions

Rebel battlefield coalitions are, like multinational coalitions comprised solely of states, collections of belligerent actors fighting against a common enemy. The difference is that rebel coalitions contain non-state actors rather than nation-states. Indeed, in many civil wars, several rebel groups fight together.[42] As with multinational coalitions, in rebel coalitions there can be varying forms of command, including an absence of coalition command, joint coalition command composed of coalition members, and unified coalition command.[43] For example, during one phase of the Colombian Civil War several groups including FARC and M-19 were led by the Simon Bolivar Guerrilla Coordination Board.

The limited scholarship on rebel coalitions and rebel coalition command has often drawn theoretical ideas from scholarship on multinational coalitions.[44] For example, Fotini Christia draws on the oldest theory of (nation-state) coalition-joining behavior—balance of power theory[45]—to explain why rebel groups join and leave rebel coalitions.[46] Similarly, following the interstate coalition warfare debate on whether political ideology factors such as regime type affect decisions to enter and exit coalitions,[47] scholarship on civil war coalitions has also considered the potential role of ideology and regime type. Christia expresses skepticism that the ideology of rebel group affects coalition joining behavior, but other work found that factors such as ideology and ethnicity can affect rebel and terrorist coalition joining and exit behavior.[48]

Rebel and multinational coalitions do have important similarities. As with multinational coalitions, concern about the tide of combat can encourage rebel groups to strengthen coalition command and expand coordination, and accordingly more unified command can sometimes increase military effectiveness. Seden Akcinaroglu proposes that, when rebel groups work in concert, if one rebel group is badly battered, another group can launch an offensive, giving the damaged group time to retreat, regroup, and rebuild to fight again in the future. He provided quantitative empirical evidence in support of the proposition.[49] Others have made a related point that institutionalisation providing formal coordination is more likely when rebels fight using conventional tactics, as improved cooperation offers greater payoffs in those situations than when rebels use guerrilla tactics.[50] There is a related scholarship within the terrorism literature suggesting that deeper cooperation among terrorist groups can improve their effectiveness by facilitating the diffusion of innovations such as suicide terrorism, thereby increasing the lethality of their attacks and lengthening group survival times.[51]

The military advantages of more unified command within at least some rebel coalitions have been demonstrated. In the Ethiopian civil war, the EPLF and TPLF formed an alliance in reaction to the Ogaden settlement between the Ethiopian and Somali governments, which promised to allow the Ethiopian government to devote an additional 15,000–20,000 troops to fight in the civil war. The rebel coalition began to enjoy increased effectiveness as its coordination expanded, as when in February 1989 EPLF armor worked with TPLF troops to defeat Ethiopian government forces.[52] The Nationalist rebel coalition in the Spanish Civil War eventually achieved victory after the creation of a unified command under Francisco Franco. Similarly, Al Qaeda and ISIS enjoyed greater combat success in the Sahel region of Africa when they began to coordinate their operations there.[53]

Also as with multinational coalitions, there may be a difference between de facto and de jure military command in rebel coalitions. These may stem from a variety of factors, but an important driver is autonomy considerations. Rebel coalition members might hesitate to place their forces under a unified command for a variety of normative, political, and/or strategic reasons that often mirror the concerns articulated by states. Like nation-states, different rebel groups might have different policy goals, and might value independence of action for its own sake.

There are three important differences between multinational and rebel coalitions, however, that suggest the straightforward application of multinational coalition theoretical frameworks are inappropriate. First, because of the different implications of sacrificing autonomy, the willingness of rebel groups to accept unified coalition command differs compared with nation-states. For nation-states, the costs of sacrificing autonomy regard policy preferences in that, under unified command, a coalition member may be less able to pursue policies that best achieve its preferences. Sacrificing autonomy to a unified command will only rarely jeopardise the security of a multinational coalition member. Here, 5th century BCE Athens is perhaps the only prominent exception, as the city-state occasionally attacked its Delian League allies to keep them in line.[54] In the modern era, multinational coalition leaders turn on coalition members usually only after those members have themselves defected by seeking a separate peace, as when Germany turned on former allies Italy, Finland, Romania, and Hungary in World War II.

For rebel groups, the stakes around retention of autonomy within a coalition are greater than they are for nation-states, as the likelihood and consequences of a coalition command leader defecting both during and after a war are greater. During the war, defection from a rebel coalition might mean one rebel group attacking a fellow rebel group member or cutting a separate deal with the government, which would have the effect of exposing the abandoned rebel group to severe jeopardy. Unified command of a rebel coalition might heighten a coalition member's vulnerability if the coalition commander acquired inside information about coalition members that could jeopardise their security. Giving an outsider authority over a group's forces may also make it more difficult for the group to defend itself from threats.[55] These concerns are especially acute given the great temptation for rebel groups to attack each other during wartime, to exploit emerging opportunities, and to capture economic resources critical for funding their campaigns.[56]

The stakes of war outcomes are also higher for members of a rebel coalition than members of a multinational coalition. If a civil war ends in stalemate or rebel victory, then the post-war political order will result in allocations of political power to at least some members of the rebel coalition, including cabinet seats, positions in the post-war military, legislative seats, and recognition of new political parties, as well as subtler but equally important decisions such as what form a new constitution will take. Unified command may provide the lead rebel group within the coalition, especially strong leverage regarding the shaping of the postwar settlement and aggrandisement for substantial political power when the fighting ends, possibly facilitating the emergence of a postwar authoritarian regime led by that group.[57] Smaller members of a winning rebel coalition under unified command, however, may be extinguished in the postwar struggle for power. The Nationalist coalition during the Spanish Civil War contained a number of different groups, and Francisco Franco's eventual creation of a unified command of the group enabled him to become the dictatorial leader of Spain when the Nationalists won in 1939. Unified command of multinational coalitions, by contrast, does not present the same postwar opportunities and dangers to nation-state members (the unification of Prussia and the small German states to create the Second German Reich following the 1870–71 Franco-Prussian War, for example, was voluntary).

The stakes for rebel groups failing to form unified command can also be higher than for nation-states. Nation-states that lose interstate wars rarely suffer the most severe consequences of annexation, occupation, or foreign-imposed regime change.[58] Conversely, rebel groups face the prospect of being exterminated if they do not win their civil war. As a result, this may provide an incentive for rebel groups to be more willing than members of multinational coalitions to sacrifice autonomy for the improved operational military effectiveness advantages of unified coalition command.

The second way in which multinational coalitions differ from rebel coalitions is in their approach to collective authority relations. Political scientists have long described the systemic environments within which political actors operate as ranging from anarchy to hierarchy. Anarchy is an environment in which there are not recognised authority relationships of one actor over others, and hierarchy is an environment in which there are such recognised authority relationships. Anarchy and hierarchy are not pure types, and environments can range from higher anarchy to limited anarchy/limited hierarchy to higher hierarchy. Realists have long asserted that international relations are inevitably characterised by anarchy, and critics have replied that hierarchy exists within at least some areas of international relations, regardless of whether such authority relationships are stated explicitly in formal treaties or manifest formally in institutions.[59] Multinational coalitions tend to function in environments of limited anarchy. Coalition members are still sovereign, and retain juridical control of their territory and the ability to exit their coalitions. However, the coalition limits member independence, especially when effective unified command allows one coalition member some control over the disposition of coalition member military forces.[60]

Rebel coalitions are more varied in their degrees of anarchy and hierarchy. They can be relatively anarchic, like some multinational coalitions. In some civil wars, the frequent entering into and exiting from coalitions demonstrates higher levels of anarchy.[61] However, rebel coalitions can also be substantially more hierarchical than all multinational coalitions. Specifically, several rebel groups may agree to formal umbrella political and military leadership, or even unify into a single group. In India, for example, the Maoist Naxalite rebellion was composed of perhaps 40 different groups in the 1970s but, over time, experienced consolidation, including that of the two largest factions, the People's War Group, and the Maoist Communist Center, in 2004.[62] There has been extensive scholarship on the causes and consequences of variation of military and political command within a single group, and fragmentation within a group.[63] The more politically hierarchical a rebel coalition becomes, the easier it may be to form an effectively unified military command. For example, during the American Civil War, the Confederate rebels were in some sense a coalition of independent southern states common in their desire to safeguard the legality of slavery. However, the creation of a unified political leadership under Confederacy President Jefferson Davis afforded a greater degree of effective unified military command that, in combination with superiority in military leadership, was a critical contributor to Confederate military effectiveness.

A third difference between multinational and rebel coalitions concerns modes of warfare. Most interstate wars are characterised by conventional warfare, where unified command can offer important operational military effectiveness advantages, especially for implementing combined arms operations, modern force employment strategies, maneuver-based ground strategies, amphibious operations, and so forth.

However, the connections between unified command and operational military effectiveness are different for rebel coalitions. Though some rebel coalitions use conventional military tactics, many do not—or, more accurately, many rebel coalitions do not use conventional military tactics across the course of a civil war. This is the heart of Mao's theory of guerrilla warfare, that insurgents should employ guerrilla tactics in what he described as the second phase of revolution, only resorting to conventional tactics in the final phase when government forces became vulnerable.[64] For rebel coalitions employing guerrilla tactics like small raids on government forces and avoiding concentrating in mass, the coordination advantages of unified coalition command are reduced. A handful of rebel combatants from a single group can independently harass counterinsurgent forces, place land mines, engage in sniping, and so forth. Further, the simplicity of guerrilla tactics requires less coordination compared with conventional operations, the latter of which may require careful coordination of timing and placement of combatants across large battlespaces. Indeed, under some circumstances, more unified coalition command may undermine the military effectiveness of guerrilla operations. Insurgent and terrorist groups recognise that maintaining loose connections between combatants can improve security, as the absence of tight connections and frequent communications means that, if one combatant or cell is captured by government forces, the existence, location, or activities of other combatants or cells are not exposed.[65]

Unified command is not entirely counterproductive for rebel coalitions, however, and may provide some effectiveness advantages even when the rebels implement a guerrilla strategy. Specifically, a unified rebel coalition command may help the coalition best manage targeting civilians. For rebels, targeting civilians is a double-edged strategy. If targeting pushes civilians away from supporting rebels, then civilians might be more likely to provide critical intelligence to government forces and, in the longer term, be less likely to support a rebel government. However, rebels may also perceive advantages from at least selective violence against civilians, if such attacks permit the elimination or deterrence of collaborators.[66] One study found that, as rebel groups enjoyed more unified command, civilian targeting by rebels decreased. But, when there are foreign fighters present, more unified command increases civilian targeting. This suggests the possibility that, under certain conditions, rebels can have the best of both worlds: central command can direct foreign fighters to launch civilian attacks, achieving the benefits of civilian targeting, while avoiding the backlash costs because foreign fighters permit plausible deniability.[67]

6.3 Hybrid Coalitions

As noted, hybrid coalitions of rebel groups and nation-states have frequently fought wars in the past, and their appearance will likely be even more frequent in the future.[68] They are severely under-theorised in existing scholarship, however. We can improve our understanding of hybrid coalitions by comparing them to multinational and rebel coalitions, and assessing similarities and differences regarding coalition command and operational military effectiveness. First, as we detail below, similar to interstate and rebel commands, de jure command in hybrid coalitions does not always correlate with de facto command. Hybrid coalitions frequently lack formal agreements allocating authority to a unified commander, but sometimes in practice, there is de facto a unified commander.

Second, as with multinational and rebel coalitions, there is great diversity in command arrangements among hybrid coalitions. At one extreme, some hybrid coalitions have not only no formal coalition command structure, but also almost no communication or coordination among the coalition members. Operation Unified Protector in Libya in 2011 is an example of this type of coalition. NATO initially authorised airstrikes to protect civilians during the Libyan civil war, but those attacks evolved into de facto assistance to rebel groups seeking to overthrow the Qadaffi government despite there being no formal agreement or direct communication between the alliance and the rebels. NATO's Operation Deliberate Force in Bosnia in the 1990s similarly did not directly coordinate with Bosnian Muslim or Bosnian Croat militias, but NATO's attacks on Bosnian Serb forces had the effect of aiding the groups, described by one study as the 'de facto beneficiaries of NATO's air campaign.'[69]

Some hybrid coalitions have slightly more coordination. There is no formal command structure and the nation-state coalition members are not directing or coordinating military operations, but they are meeting rebel requests for aid and providing assistance. An example is the 1980s hybrid coalition in Afghanistan. The United States, Pakistan, the United Kingdom, Saudi Arabia, and China all provided military aid and training to the mujahideen groups fighting the Soviet military in Afghanistan, but these nation-state coalition members did not direct rebel group military operations.

Some hybrid coalitions have more coordination beyond the provision of aid, though still lack direct control or formal command structures. One example is US and UK forces fighting with the Northern Alliance rebels in Afghanistan in

Fall of 2001. American and British troops were deployed to Afghanistan, and there were direct contacts with Northern Alliances leaders, including a meeting between US General Tommy Franks and Northern Alliance leadership on October 30 to plan an attack on Mazar-e-Sharif. However, there was no formal command structure, and Northern Alliances groups sometimes pursued their own interests rather than those of Anglo-American forces.[70]

Finally, there are hybrid coalitions in which the nation-state actor is clearly the unquestioned de facto if not de jure unified coalition commander. During the Vietnam War, the North Vietnamese government had complete command of its coalition with the National Liberation Front rebels. Russia almost certainly has very firm command within its coalition that includes Ukrainian separatist rebels in the ongoing civil war in Ukraine.

Third, as with multinational and rebel coalitions, there may be political limits to the ability of a hybrid coalition to form a unified command. Nation-states may wish to keep their involvement secret,[71] which in turn may limit the extent of their involvement and ability to command rebel groups. The Pakistani government has publicly denied accusations of its affiliation with Islamist militant groups operating in India to reduce risks of direct conflict with its rival. Secrecy aside, as was the case in Libya in 2011, international diplomatic considerations may limit the mandate of nation-state involvement in civil wars, restricting the ability of nation-states to create unified coalition command. Even more commonly, some rebel groups may simply refuse to accept nation-state command of a coalition because their goals do not align with those of sovereign third-party actors.

Fourth, as with multinational and rebel coalitions, unified command of hybrid coalitions may offer operational military effectiveness advantages, and those advantages might under some circumstances be even greater for hybrid coalitions. Consider that a central effectiveness motivation for unified coalition command is coordination; coordination has especially high payoffs when there is heterogeneity of comparative advantages within the coalition, as command can then efficiently allocate military tasks based on particular capabilities. Within hybrid coalitions, there are often starkly different comparative advantages, as state actors usually bring advantages such as materiel, air power (including drone strikes and reconnaissance), and advanced technologies, while rebel groups usually bring in advantages of local knowledge and the trust of the population. Also, state actors may be politically constrained from contributing many troops, whereas rebel groups may be able to field more extensive contingents of fighters.

The relationships between hybrid coalition command structure, victimisation of civilians, and operational military effectiveness is complicated. The degree to which victimisation of civilians might aid or impede operational military effectiveness is itself complicated and likely contextual, as noted above. The participation of nation-states in a hybrid coalition can have a variety of effects on whether or not rebels kill civilians. In some circumstances, the nation-state may wish to reduce civilian targeting by rebel groups to reduce international diplomatic backlash or to help the insurgents win the hearts and minds of the population. Such a goal could be pursued by stressing the importance of not targeting civilians in rebel training programs, or threatening to withdraw from the coalition if rebels attack civilians. However, in other circumstances, the nation-state participant in the coalition may either be willing to turn a blind eye to rebel atrocities against civilians, as appeared to be the case when Cuba supported the People's Movement for the Liberation of Angola rebels in the Angolan Civil War, or the nation-state may see targeting at least some civilians as an effective military strategy, a view that Russia has embraced in the Syrian Civil War. There are also potential indirect effects, as when outside nation-states provide military aid to rebels; such assistance can incentivise rebels to recruit opportunistic individuals who may be more likely to victimise civilians.[72]

Fifth, as with multinational and rebel coalitions, unified command of hybrid coalitions might not be sufficient to improve operational military effectiveness appreciably, if the command attempts to implement a flawed plan for victory. Consider the 1961 hybrid coalition of the United States and Cuban rebels, attempting to overthrow the Fidel Castro regime. The United States government unified command of the coalition did not prevent the April Bay of Pigs invasion from failing because of flaws in the invasion plan, including insufficient American support of troops and materiel.[73]

Last, hybrid coalitions can face credible commitment problems similar to those that rebel coalitions face, as rebel groups within hybrid coalitions, might also be concerned about being attacked or abandoned by partners during the war or sidelined in postwar political arrangements. Rebel groups do not always trust nation-states to maintain their wartime commitments to stay in the war until victory. This fear may be especially acute if there are democratic nation-state members of hybrid coalitions, given democracies' casualty sensitivities, especially in conflicts that do not present direct threats to the national interest.[74] America's rebel allies in the Syrian Civil War, especially the Kurds, were deeply disappointed by the Trump administration's 2019 decision to exit the war.

One difference from rebel coalitions in this regard, however, is that nation-state coalition members in hybrid coalitions can take actions to alleviate these credible commitment concerns. Nation-state coalition members are often motivated to reduce credible commitment concerns, as rebel group partners reduce the need for their own troop contributions, maximise the chances for victory, and help to form broader and hence more popular post-war ruling coalitions. They can act on this motivation by committing to rebel group coalition members that postwar political institutions will represent all rebel groups, and/or will create democratic processes providing for fair political competition. The peacekeeping literature has demonstrated that rebel groups are attracted to credible promises of postwar power sharing, and belief in such commitments can make rebel groups more likely to accept a compromise settlement to a civil war.[75]

Nation-states can make these credible commitments to their rebel partners because they have less to gain from violating such promises. The stakes of the outcome of a conflict in which a hybrid coalition fights are different for nation-states than they are for rebel groups; if their favored rebel group or leader fails to attain power, then the nation-state might lose access to natural resources, or a smaller state might become a security liability rather than an asset, but these are relatively limited gains or losses. Conversely, if a rebel group fails to achieve political power in the postwar environment, it faces at best political irrelevance within its country and at worst the imprisonment or execution of its leaders. Nation-states also sometimes have more to gain from abiding by a commitment to inclusive postwar political institutions, as such commitments may increase the chances of transition to stability and even democracy, an outcome favored by many nation-states. Rebel groups, by contrast, may be more interested in acquiring political power after war's end rather than achieving democracy or stability.

The similarities and differences across the three types of coalitions are summarised in Table 6.1:

Table 6.1 Multinational, Rebel, and Hybrid Coalitions

	Multinational Coalitions	Rebel Coalitions	Hybrid Coalitions
Members	Nation-States	Rebel Groups	Nation- States and Rebel Groups
Can coalition have unified or non-unified command, which presents autonomy tradeoffs for members?	Yes	Yes	Yes
Does unified command present potential effectiveness advantages for conventional warfare?	Yes	Yes	Yes
Specialisation advantages to coordination for military effectiveness?	Moderate	Moderate	High
Does de jure unified command always mean de facto unified command?	No	No	No
Potential risks of unified command to member autonomy and survival?	Low	High	High
Ability of coalition command to alleviate credible commitment fears?	Low	Low	Moderate
Degree of anarchy within coalition?	Moderate	Can be Low, Moderate, or High	Can be Low, Moderate, or High
Frequency of Conventional Warfare?	Almost	Always	Sometimes

6.4 Conclusions

The chapter has demonstrated the conceptual benefits of framing multi-national, rebel, and hybrid coalitions within a common conceptual framework. Though the chapter has laid out some important similarities and differences across the three types of coalitions, more theoretical and empirical work is needed. For example, we need deeper exploration of the effect of command structures on operational military effectiveness in insurgency campaigns, better understanding of what conditions cause de jure command not to translate into de facto command, and others.

The chapter also generates policy recommendations, especially regarding how nation-states should manage their hybrid coalitions. Most simply, nation-state members of hybrid coalitions should recognise the operational military effectiveness advantages of unified command, coupling nation-state technology, and materiel with rebel group local knowledge and troop contributions. That said, nation-states should also recognise the complex relationships between command and military effectiveness. Unified command may provide fewer effectiveness advantages for some kinds of military operations, such as guerrilla attacks. Further, unified command will not help effectiveness if it is used to implement a flawed military strategy.

Nation-states also need to be sensitive to credible commitment concerns of rebel group coalition members. Victory may present very high risks for rebel group coalition members, as they may be left out of the ruling postwar coalition, and, worse, exposed to personal jeopardy if other groups turn on them. These credible commitment concerns might prevent rebel groups from accepting unified command. Nation-states should find ways to reduce these credible commitment concerns, using peacekeeping forces and postwar power-sharing to induce rebel groups to accept unified command. Linking these commitments to the involvement of international organisations such as the UN may be especially helpful.

Notes

1 See Rosella Cappella Zielinski and Ryan Grauer, 'A Century of Coalitions in Battle: Incidence, Composition, and Performance, 1900–2003', Chapter 2, this volume.
2 For a broad discussion of such phenomena see Dan Reiter, 'Should We Leave Behind the Subfield of International Relations?' *Annual Review of Political Science* 18 (2015) https://doi.org/10.1146/annurev-polisci-053013-041156.
3 Allan C. Stam III, *Win, Lose, or Draw: Domestic Politics and the Crucible of War* (Ann Arbor, MI: University of Michigan Press 1996); Stathis N. Kalyvas and Laia Balcells, 'International System and Technologies of Rebellion: How the End of the Cold War Shaped Internal Conflict', *American Political Science Review* 104/3 (August 2010) 415–429 https://doi.org/10.1017/S0003055410000286; Jason Lyall, *Divided Armies: Inequality and Battlefield Performance in Modern War* (Princeton: Princeton University Press2020); Stephen Biddle, *Nonstate Warfare: The Military Methods of Guerrillas, Warlords, and Militias* (Princeton: Princeton University Press 2021).

4 Dan Reiter, *How Wars End* (Princeton: Princeton University Press, 2009).
5 Dan Reiter, 'Should We Leave Behind the Subfield of International Relations?' Annual Review of Political Science 18 (2015) https://doi.org/10.1146/annurev-polisci-053013-041156.
6 See Sara Bjerg Moller, 'Learning from Losing: How Defeat Shapes Coalition Command Arrangements in Wartime', Chapter 7, this volume.
7 Cappella Zielinski and Grauer, 'A Century of Coalitions in Battle,' Chapter 2, this volume; Scott Wolford, *The Politics of Military Coalitions* (Cambridge: Cambridge University Press 2016).
8 Joshua C. Fjelstul and Dan Reiter, 'Explaining Incompleteness and Conditionality in Alliance Agreements', *International Interactions* 45/6 (2019) 976–1002 https://doi.org/10.1080/03050629.2019.1647838; Paul Poast, *Arguing About Alliances: The Art of Agreement in Military Pact Negotiations* (Ithaca: Cornell University Press 2019).
9 Anthony J. Rice, 'The Essence of Coalition Warfare', *Parameters* 27 (Spring 1997): 152–167; Moller, 'Learning from Losing', Chapter 7, this volume.
10 The focus in this chapter is on the operational level of military effectiveness, coordination within a single theater or campaign, but some also argue for the advantages of unified coalition command at the strategic level, or the coordination of coalition members across operations and theaters, as well as at the political level, or coordination across coalition members of economic production. Rosella Cappella Zielinski and Paul Poast, 'Supplying Allies: The Political Economy of Coalition Warfare', *Journal of Global Security Studies* 6/1 (March 2021), https://doi.org/10.1093/jogss/ogaa006. On operational, strategic, and political levels of military effectiveness, see Allan R. Millett and Williamson Murray, eds., *Military Effectiveness*, three vols (Cambridge: Cambridge University Press 2010). For further discussion of military effectiveness, see Dan Reiter, ed., *The Sword's Other Edge: Tradeoffs in the Pursuit of Military Effectiveness* (Cambridge: Cambridge University Press 2017).
11 Patricia A. Weitsman, *Waging War: Alliances, Coalitions, and Institutions of Interstate Violence* (Stanford: Stanford University Press 2014); Poast, *Arguing About Alliances*.
12 Nora Bensehel, 'International Alliances and Military Effectiveness: Fighting Alongside Allies and Partners', in Risa Brooks and Elizabeth Stanley, eds, *Creating Military Power: The Sources of Military Effectiveness* (Stanford: Stanford University Press 2007), 186–206.
13 Rosella Cappella Zielinski and Ryan Grauer, 'Organizing for Performance: Coalition Effectiveness on the Battlefield', *European Journal of International Relations* 26/4 (2020) 953–978, https://doi.org/10.1177/1354066120903369.
14 Daniel S. Morey, 'Centralized Command and Coalition Victory', *Conflict Management and Peace Science* 37/6 (2020) 716–734, https://doi.org/10.1177/0738894220934884.
15 Glenn Baek, 'A Perspective on Korea's Participation in the Vietnam War', Asian Institute for Policy Studies, Issue Brief no. 53, 2013; Neil Sheehan, *A Bright Shining Lie: John Paul Vann and America in Vietnam* (New York: Vintage 1989), 776.
16 Paul Poast and Dan Reiter, 'Tripwires are Not Enough: Forward Troop Deployments and the Prevention of War', *Texas National Security Review* 4/3 (Summer 2021), http://dx.doi.org/10.26153/tsw/13989.
17 Moller, 'Learning from Losing', Chapter 7, this volume.
18 Mark Bowden, *Black Hawk Down: A Story of Modern War* (New York: Grove 2010).
19 Rice, 'Essence of Coalition Warfare.'

20 Cappella Zielinski and Grauer, 'Organizing for Performance.'
21 Morey, 'Centralized Command' supplemental appendix classifies Italy-Germany in World War II as being a joint coalition.
22 Richard Carrier, 'Some Reflections on the Fighting Power of the Italian Army in North Africa, 1940–1943', *War in History* 22/4 (November 2015) 503–528, https://doi.org/10.1177/0968344514524395; Cappella and Zielinski and Grauer, 'Organizing for Performance.'
23 Jeffrey K. Staton and Will H. Moore, 'Judicial Power in Domestic and International Politics', *International Organization* 65/3 (Summer 2011) 553–87, https://doi.org/10.1017/S0020818311000130.
24 David A. Lake, *Hierarchy in International Relations* (Ithaca: Cornell University Press 2009).
25 Morey, 'Centralized Command', 724.
26 Stephen Biddle, *Military Power: Explaining Victory and Defeat in Modern Battle* (Princeton: Princeton University Press 2004).
27 https://history.army.mil/books/www/www8.htm.
28 *Insurgencies and Countering Insurgencies*, FM 3–24 (Washington: Department of the Army 2014), 6–3.
29 Jack Snyder, *The Ideology of the Offensive: Military Decision Making and the Disasters of 1914* (Ithaca: Cornell University Press 1984).
30 Ernest R. May, *Strange Victory: Hitler's Conquest of France* (New York: Hill and Wang 1999).
31 Biddle, *Military Power.*
32 Robert A. Pape, *Bombing to Win: Air Power and Coercion in War* (Ithaca: Cornell University Press 1996).
33 Thomas E. Ricks, *The Gamble: General Petraeus and the Untold Story of the Surge in Iraq* (New York: Penguin 2010).
34 Ian McNeill, 'The Australian Army and the Vietnam War', in *Australia's Vietnam War*, Jeff Doyle, Jeffrey Grey, and Peter Pierce, eds. (College Station, TX: Texas A&M University Press 2002), 16–54.
35 For a review of theories predicting to intra-coalition politics, see Thomas Stow Wilkins, 'Analysing Coalition Warfare from an Intra-Alliance Politics Perspective: The Normandy Campaign 1944', *Journal of Strategic Studies* 29/6 (December 2006) 1121–1150, https://doi.org/10.1080/01402390601016592. For discussion of whether or not coalitions involving democracies are more likely to experience deeper coordination including possibly unified command during war, see Dan Reiter and Allan C. Stam, *Democracies at War* (Princeton: Princeton University Press 2002), chapter 4. Power balances within the coalition can also matter, as when one member is more powerful that others the command structure is likely to reflect that powerful member's preferences. Marina E. Henke, *Constructing Allied Cooperation: Diplomacy, Payments, and Power in Multilateral Military Coalitions* (Ithaca: Cornell University Press 2019); Cappella Zielinski and Grauer, 'Organizing for Performance.'
36 Moller, 'Learning from Losing', Chapter 7, this volume.
37 Moller, 'Learning from Losing', Chapter 7, this volume. There is a more general argument that states sometimes refuse to join alliances in order to maintain autonomy. For a summary, see James D. Morrow, 'Alliances: Why Write Them Down?' *Annual Review of Political Science* 3 (2000) 63–83, https://doi.org/10.1146/annurev.polisci.3.1.63.
38 https://fas.org/irp/offdocs/pdd25.htm.
39 William Philpott, 'Britain and France Go to War: Anglo-French Relations on the Western Front 1914–1918', *War in History* 2/1 (March 1995) 43–64, https://doi.org/10.1177/096834459500200103.

40 Poast and Reiter, 'Tripwires'; May, *Strange Victory*; Dan Reiter, *Crucible of Beliefs: Learning, Alliances, and World Wars* (Ithaca: Cornell University Press 1996).
41 Moller, 'Learning from Losing', Chapter 7, this volume.
42 David E. Cunningham, *Barriers to Peace in Civil War* (Cambridge: Cambridge University Press 2009).
43 For data on variation in rebel command, see David E. Cunningham, Kristian Skrede Gleditsch, and Idean Saleyhan, 'Non-state Actors in Civil Wars: A New Dataset', *Conflict Management and Peace Science* 30/5 (2013) 516–531, https://doi.org/10.1177/0738894213499673.
44 See Barbara Elias, 'Why Rebels Rely on Terrorists: The persistence of the Taliban-al-Qaeda Battlefield Coalition in Afghanistan', Chapter 9, this volume.
45 Kenneth N. Waltz, *Theory of International Politics* (New York: Addison Wesley 1979).
46 Fotini Christia, *Alliance Formation in Civil Wars* (Cambridge: Cambridge University Press 2012). See also Costantino Pischedda, *Conflict Among Rebels: Why Insurgents Fight Each Other* (New York: Columbia University Press 2020).
47 Brian Lai and Dan Reiter, 'Democracy, Political Similarity, and International Alliances, 1816–1992', *Journal of Conflict Resolution* 44/2 (April 2000) 203–227, https://doi.org/10.1177/0022002700044002003; Reiter and Stam, *Democracies at War*.
48 Emily Kalah Gade, Mohammed M. Hafez, and Michael Gabbay, 'Fratricide in Rebel Movements: A Network Analysis of Syrian Militant Infighting', *Journal of Peace Research* 56/3 (2012): 321–335, https://doi.org/10.1177/0022343318806940; Victor H. Asal, Hyun Hee Park, R. Karl Rethemeyer, and Gary Ackerman, 'With Friends Like These ... Why Terrorist Organizations Ally', *Journal of Peace Research* 19/1 (2016) 1–30, https://doi.org/10.1080/10967494.2015.1027431.
49 Seden Akcinaroglu, 'Rebel Interdependencies and Civil War Outcomes', *Journal of Conflict Resolution* 56/5 (October 2012): 879–903, https://doi.org/10.1177/0022002712445741. See also 'Why Rebels Rely on Terrorists.'
50 Kristin Bakke, Kathleen Gallagher Cunningham, and Lee J. M. Seymour, 'A Plague of Initials: Fragmentation, Cohesion, and Infighting in Civil Wars', *Perspectives on Politics* 10/2 (June 2012) 270, https://doi.org/10.1017/S1537592712000667.
51 Michael C. Horowitz and Phillip Potter, 'Allying to Kill: Terrorist Intergroup Cooperation and the Consequences for Lethality', *Journal of Conflict Resolution* 58/2 (March 2014) 199–225, https://doi.org/10.1177/0022002712468726; Brian J. Phillips, 'Terrorist Group Cooperation and Longevity', *International Studies Quarterly* 58/2 (June 2014) 336–347, https://doi.org/10.1111/isqu.12073. A caveat is that these studies analyse data including some ties between terrorist groups that are not fighting the same adversary, such as the IRA and the ETA, meaning they are not coalitions as defined here.
52 John Young, *Peasant Revolution in Ethiopia: The Tigray People's Liberation Front, 1975–1991* (Cambridge: Cambridge University Press 1997), 157, 164.
53 Danielle Paquette and Joby Warrick, 'Isis and Al Qaeda Join Forces in West Africa,' *The Independent*, 23 February 2020. https://www.independent.co.uk/news/world/africa/isis-al-qaeda-terror-west-africa-mali-burkina-faso-niger-a9353126.html
54 Dan Reiter, 'Gulliver Unleashed? International Order, Restraint, and The Case of Ancient Athens', *International Studies Quarterly* 65/3 (September 2021) 582–593, https://doi.org/10.1093/isq/sqab061.

55 Pischedda, *Conflict Among Rebels*; Navin A. Bapat and Kanisha D. Bond, 'Alliances between Militant Groups', *British Journal of Political Science* 42/4 (October 2012) 793–824, https://doi.org/10.1017/S0007123412000075.

56 Hanne Fjelde and Desirée Nilsson, 'Rebels Against Rebels: Explaining Violence Among Rebel Groups', *Journal of Conflict Resolution* 56/4 (August 2012) 604–628, https://doi.org/10.1177/0022002712439496; Pischedda, *Conflict Among Rebels*.

57 Terrence Lyons, 'The Importance of Winning: Victorious Insurgent Groups and Authoritarian Politics', *Comparative Politics* 48/2 (January 2016) 167–184, https://doi.org/10.5129/001041516817037745.

58 Dan Reiter, *How Wars End* (Princeton: Princeton University Press 2009).

59 The classic realist statement of the inevitably of anarchy in international relations is Waltz, *Theory of International Politics*. For claims that there are areas of hierarchy in international relations, see Lake, *Hierarchy*; Staton and Moore, *Judicial Power*.

60 Lake, *Hierarchy*.

61 Christia, *Alliance Formation*.

62 Sameer Lalwani, 'India's Approach to Counterinsurgency and the Naxalite Problem', *CTC Sentinel* 4 (October 2011) 6.

63 Lindsay Heger, Danielle Jung, and Wendy H. Wong, 'Organizing for Resistance: How Group Structure Impacts the Character of Violence', *Terrorism and Political Violence* 24/5 (2012) 746–768, https://doi.org/10.1080/09546553.2011.642908; Paul Staniland, *Networks of Rebellion: Explaining Insurgent Cohesion and Collapse* (Ithaca: Cornell University Press 2014); Austin C. Doctor, 'A Motion of No Confidence: Leadership and Rebel Fragmentation', *Journal of Global Security Studies* 5/4 (October 2020) 598–616, https://doi.org/10.1093/jogss/ogz060.

64 Andrew F. Krepinevich Jr., *The Army and Vietnam* (Baltimore: Johns Hopkins University Press 1986).

65 Jacob M. Shapiro, *The Terrorist's Dilemma: Managing Violent Covert Organizations* (Princeton: Princeton University Press 2013).

66 Stathis N. Kalyvas, *The Logic of Violence in Civil War* (Cambridge: Cambridge University Press 2006).

67 Austin C. Doctor and John D. Willingham, 'Foreign Fighters, Rebel Command Structure, and Targeting Civilians in Civil War', *Terrorism and Political Violence* 34/6 (2022) 1125–1143, https://doi.org/10.1080/09546553.2020.1763320.

68 Cappella Zielinski and Grauer, 'A Century of Coalitions in Battle', Chapter 2, this volume.

69 Anthony M. Schinella, *Bombs Without Boots: The Limits of Airpower* (RAND: Washington 2019), 18.

70 Don Chipman, 'Air Power and the Battle for Mazar e Sharif', *Air Power History* 50/1 (Spring 2003) 34–45; Walter L. Perry and David Kassing, *Toppling the Taliban: Air-Ground Operations in Afghanistan, October 2001-June 2002* (Washington: RAND 2015), https://doi.org/10.7249/RR381.

71 Austin Carson, *Secret Wars: Covert Conflict in International Politics* (Princeton: Princeton University Press 2018).

72 Jeremy Weinstein, *Inside Rebellion: The Politics of Insurgent Violence* (Cambridge: Cambridge University Press 2006).

73 Jim Rasenberger, *The Brilliant Disaster: JFK, Castro, and America's Doomed Invasion of Cuba's Bay of Pigs* (New York: Scribner 2011).

74 Reiter and Stam, *Democracies at War*.

75 Barbara F. Walter, *Committing to Peace* (Princeton: Princeton University Press 2002).

Bibliography

Akcinaroglu, Seden. 'Rebel Interdependencies and Civil War Outcomes.' *Journal of Conflict Resolution* 56/5 (October 2012): 879–903. 10.1177/0022002 712445741

Asal, Victor H., Hyun Hee Park, R. Karl Rethemeyer, and Gary Ackerman. 'With Friends Like These … Why Terrorist Organizations Ally.' *Journal of Peace Research* 19/1 (2016): 1–30. 10.1080/10967494.2015.1027431

Baek, Glenn. 'A Perspective on Korea's Participation in the Vietnam War.' *Asian Institute for Policy Studies* (2013), Issue Brief no. 53.

Bakke, Kristin Kathleen Gallagher Cunningham, and Lee J. M. Seymour. 'A Plague of Initials: Fragmentation, Cohesion, and Infighting in Civil Wars.' *Perspectives on Politics* 10/2 (June 2012): 265–283. 10.1017/S1537592712000667

Bapat, Navin A. and Kanisha D. Bond. 'Alliances between Militant Groups.' *British Journal of Political Science* 42/4 (October 2012): 793–824. 10.1017/S0007123412 000075

Bensehel, Nora. 'International Alliances and Military Effectiveness: Fighting Alongside Allies and Partners.' In *Creating Military Power: The Sources of Military Effectiveness*, edited by Risa Brooks and Elizabeth Stanley, 186–206. Stanford: Stanford University Press, 2007.

Biddle, Stephen. *Military Power: Explaining Victory and Defeat in Modern Battle.* Princeton: Princeton University Press, 2004.

Biddle, Stephen. *Nonstate Warfare: The Military Methods of Guerrillas, Warlords, and Militias.* Princeton: Princeton University Press, 2021.

Bowden, Mark. *Black Hawk Down: A Story of Modern War.* New York: Grove, 2010.

Cappella Zielinski, Rosella and Ryan Grauer. 'Organizing for Performance: Coalition Effectiveness on the Battlefield.' *European Journal of International Relations* 26/4 (2020): 953–978. 10.1177/1354066120903369

Cappella, Zielinski and Paul Poast. 'Supplying Allies: The Political Economy of Coalition Warfare.' *Journal of Global Security Studies* 6/1 (March 2021). 10.1093/jogss/ogaa006

Carrier, Richard. 'Some Reflections on the Fighting Power of the Italian Army in North Africa, 1940–1943.' *War in History* 22/4 (November 2015): 503–528. 10.1177/0968344514524395

Carson, Austin. *Secret Wars: Covert Conflict in International Politics.* Princeton: Princeton University Press, 2018.

Chipman, Don. 'Air Power and the Battle for Mazar e Sharif.' *Air Power History* 50/1 (Spring 2003): 34–45.

Christia, Fotini. *Alliance Formation in Civil Wars.* Cambridge: Cambridge University Press, 2012.

Cunningham, David E. *Barriers to Peace in Civil War.* Cambridge: Cambridge University Press, 2009.

Cunningham, David E., Kristian Skrede Gleditsch, and Idean Saleyhan. 'Non-state Actors in Civil Wars: A New Dataset.' *Conflict Management and Peace Science* 30/5 (2013): 516–531. 10.1177/0738894213499673

Doctor, Austin C. 'A Motion of No Confidence: Leadership and Rebel Fragmentation.' *Journal of Global Security Studies* 5/4 (October 2020): 598–616. 10.1093/jogss/ogz060.

Doctor, Austin C. and John D. Willingham. 'Foreign Fighters, Rebel Command Structure, and Targeting Civilians in Civil War.' *Terrorism and Political Violence* 34/6 (2022): 1125–1143. 10.1080/09546553.2020.1763320

Fjelde, Hanne and Desirée Nilsson. 'Rebels Against Rebels: Explaining Violence Among Rebel Groups'. *Journal of Conflict Resolution* 56/4 (August 2012): 604–628. 10.1177/0022002712439496

Fjelstul, Joshua C. and Dan Reiter. 'Explaining Incompleteness and Conditionality in Alliance Agreements.' *International Interactions* 45/6 (2019): 976–1002. 10.1080/03050629.2019.1647838

Gade, Emily Kalah, Mohammed M. Hafez, and Michael Gabbay. 'Fratricide in Rebel Movements: A Network Analysis of Syrian Militant Infighting.' *Journal of Peace Research* 56/3 (2012): 321–335. 10.1177/0022343318806940

Heger, Lindsay, Danielle Jung, and Wendy H. Wong. 'Organizing for Resistance: How Group Structure Impacts the Character of Violence.' *Terrorism and Political Violence* 24/5 (2012): 746–768. 10.1080/09546553.2011.642908

Henke, Marina E. *Constructing Allied Cooperation: Diplomacy, Payments, and Power in Multilateral Military Coalitions.* Ithaca: Cornell University Press, 2019.

Horowitz, Michael C. and Phillip Potter. 'Allying to Kill: Terrorist Intergroup Cooperation and the Consequences for Lethality.' *Journal of Conflict Resolution* 58/2 (March 2014): 199–225. 10.1177/0022002712468726

Insurgences and Countering Insurgencies, FM 3-24. 2014. Washington: Department of the Army.

Kalyvas, Stathis N. *The Logic of Violence in Civil War.* Cambridge: Cambridge University Press, 2006.

Kalyvas, Stathis N. and Laia Balcells. 'International System and Technologies of Rebellion: How the End of the Cold War Shaped Internal Conflict.' *American Political Science Review* 104/3 (August 2010): 415–429. 10.1017/S0003055410000286

Krepinevich, Andrew F. Jr. *The Army and Vietnam.* Baltimore: Johns Hopkins University Press, 1986.

Lai, Brian and Dan Reiter. 2000. 'Democracy, Political Similarity, and International Alliances, 1816–1992'. *Journal of Conflict Resolution* 44/2 (April 2000): 203–227. 10.1177/0022002700044002003

Lake, David A. *Hierarchy in International Relations.* Ithaca: Cornell University Press, 2009.

Lalwani, Sameer. 'India's Approach to Counterinsurgency and the Naxalite Problem.' *CTC Sentinel* 4 (October 2011): 5–9.

Lyall, Jason. *Divided Armies: Inequality and Battlefield Performance in Modern War.* Princeton: Princeton University Press, 2020.

Lyons, Terrence. 'The Importance of Winning: Victorious Insurgent Groups and Authoritarian Politics.' *Comparative Politics* 48/2 (January 2016): 167–184. 10.5129/001041516817037745

May, Ernest R. *Strange Victory: Hitler's Conquest of France.* New York: Hill and Wang, 1999.

McNeill, Ian. 'The Australian Army and the Vietnam War.' In *Australia's Vietnam War*, edited by Jeff Doyle, Jeffrey Grey, and Peter Pierce, 16–54. College Station, TX: Texas A&M University Press, 2002.

Millett, Allan R. and Williamson Murray, eds. *Military Effectiveness*, three vols. Cambridge: Cambridge University Press, 2010.

Morey, Daniel S. 'Centralized Command and Coalition Victory.' *Conflict Management and Peace Science* 37/6 (2020): 716–734. 10.1177/0738894220934884

Morrow, James D. 'Alliances: Why Write Them Down?.' *Annual Review of Political Science* 3 (2000): 63–83. 10.1146/annurev.polisci.3.1.63

Pape, Robert A. *Bombing to Win: Air Power and Coercion in War*. Ithaca: Cornell University Press, 1996.

Paquette, Danielle and Joby Warrick. 'Isis and Al Qaeda Join Forces in West Africa.' *The Independent*, (23 February 2020).

Perry, Walter L. and David Kassing. *Toppling the Taliban: Air-Ground Operations in Afghanistan, October 2001-June 2002*. Washington: RAND, 2015. 10.7249/RR381

Phillips, Brian J. 'Terrorist Group Cooperation and Longevity.' *International Studies Quarterly* 58/2 (June 2014): 336–347. 10.1111/isqu.12073.

Philpott, William. 'Britain and France Go to War: Anglo-French Relations on the Western Front 1914–1918.' *War in History* 2/1 (March 1995): 43–64. 10.1177/096834459500200103

Pischedda, Costantino. *Conflict Among Rebels: Why Insurgents Fight Each Other*. New York: Columbia University Press, 2020.

Poast, Paul. *Arguing About Alliances: The Art of Agreement in Military Pact Negotiations*. Ithaca: Cornell University Press, 2019.

Poast, Paul and Dan Reiter. 'Tripwires are Not Enough: Forward Troop Deployments and the Prevention of War.' *Texas National Security Review* 4/3 (Summer 2021). 10.26153/tsw/13989

Rasenberger, Jim. *The Brilliant Disaster: JFK, Castro, and America's Doomed Invasion of Cuba's Bay of Pigs*. New York: Scribner, 2011.

Reiter, Dan. *Crucible of Beliefs: Learning, Alliances, and World Wars*. Ithaca: Cornell University Press, 1996.

Reiter, Dan. *How Wars End*. Princeton: Princeton University Press, 2009.

Reiter, Dan. 'Should We Leave Behind the Subfield of International Relations?.' *Annual Review of Political Science* 18 (2015). 10.1146/annurev-polisci-053013-041156

Reiter, Dan. 'Gulliver Unleashed? International Order, Restraint, and The Case of Ancient Athens.' *International Studies Quarterly* 65/3 (September 2021): 582–593. 10.1093/isq/sqab061

Reiter, Dan, ed. *The Sword's Other Edge: Tradeoffs in the Pursuit of Military Effectiveness*. Cambridge: Cambridge University Press, 2017.

Reiter, Dan and Allan C. Stam. *Democracies at War*. Princeton: Princeton University Press, 2002.

Rice, Anthony J. 'The Essence of Coalition Warfare.' *Parameters* 27 (Spring 1997): 152–167.

Ricks, Thomas E. *The Gamble: General Petraeus and the Untold Story of the Surge in Iraq*. New York: Penguin, 2010.

Schinella, Anthony M. *Bombs Without Boots: The Limits of Airpower*. Washington: RAND, 2019.

Shapiro, Jacob M. *The Terrorist's Dilemma: Managing Violent Covert Organizations*. Princeton: Princeton University Press, 2013.

Sheehan, Neil. *A Bright Shining Lie: John Paul Vann and America in Vietnam.* New York: Vintage, 1989.

Snyder, Jack. *The Ideology of the Offensive: Military Decision Making and the Disasters of 1914.* Ithaca: Cornell University Press, 1984.

Stam, Allan C. III. *Win, Lose, or Draw: Domestic Politics and the Crucible of War.* Ann Arbor, MI: University of Michigan Press, 1996.

Staniland, Paul. *Networks of Rebellion: Explaining Insurgent Cohesion and Collapse.* Ithaca: Cornell University Press, 2014.

Staton, Jeffrey K. and Will H. Moore. 'Judicial Power in Domestic and International Politics.' *International Organization* 65/3 (Summer 2011): 553–587. 10.1017/S0020818311000130.

Walter, Barbara F. *Committing to Peace.* Princeton: Princeton University Press, 2002.

Waltz, Kenneth N. *Theory of International Politics.* New York: Addison Wesley, 1979.

Weinstein, Jeremy. *Inside Rebellion: The Politics of Insurgent Violence.* Cambridge: Cambridge University Press, 2006.

Weitsman, Patricia A. *Waging War: Alliances, Coalitions, and Institutions of Interstate Violence.* Stanford: Stanford University Press, 2014.

Wilkins, Thomas Stow. 'Analysing Coalition Warfare from an Intra-Alliance Politics Perspective: The Normandy Campaign 1944.' *Journal of Strategic Studies* 29/6 (December 2006): 1121–1150. 10.1080/01402390601016592

Wolford, Scott. *The Politics of Military Coalitions.* Cambridge: Cambridge University Press, 2016.

Young, John. *Peasant Revolution in Ethiopia: The Tigray People's Liberation Front, 1975–1991.* Cambridge: Cambridge University Press, 1997.

7 Learning from Losing

How Defeat Shapes Coalition Dynamics in Wartime

Sara Bjerg Moller

When and why do battlefield coalition partners adapt in wartime? Specifically, under what conditions do coalition partners refashion their command arrangements? Existing scholarship on military coalitions emphasises factors leading to their formation and performance, with less attention paid to the actual operation and management of these wartime fighting arrangements.[1] Much of what makes coalitional wars different from fighting alone, however, occurs in the interim stage – between the onset of fighting and its outcome – in the shared management of the battlespace. As with everything war touches, coalitions rarely emerge from combat unaltered. Even victorious coalitions that emerge with their memberships intact rarely resemble their original selves by war's end, owing to the organisational, doctrinal, and technological transformations embraced by respective armed forces in combat. Yet for all the advances made in recent years in understanding how militaries change in wartime, we still lack an appreciation of when and why coalitions adapt.[2]

Building on insights from organisational research and bargaining theories of war, this chapter introduces a two-step model of coalitional military adaptation, defined here as: a major change that is collectively adopted by all members of a coalition in a given operational theater. Starting from the premise that fighting reveals previously hidden information not only about the enemy's capabilities but also those of coalition partners, I argue that battlefield losses drive leaders to pursue adaptation. However, because (1) war is a dynamic environment, (2) the learning process is different for different militaries, and (3) coalition adaptation requires agreement from all members, the convergence process in coalitions takes longer than in national militaries. Additionally, because states prise independence of action above all else, both military and civilian leaders are often reluctant to embrace adaptations that risk limiting their control.[3] Consequently, adaptation in coalitions is far more difficult and multi-directional than the linear process described by traditional models. Specifically, I argue that the process unfolds in a two-step process: first, coalition militaries must decide they need to adapt collectively to confront their common adversary, which is a choice made at the strategic level of warfighting; second, they must collectively

DOI: 10.4324/9781003399896-10

implement, institutionalise, and routinise the changes at the operational level of warfighting. Dissonance between the coalition partners at the strategic level often creates turbulence and can inhibit operational-level coordination.

The rest of this chapter proceeds as follows: the first section reviews existing arguments about military adaptation in war. In section two, I use insights from the organisational learning and military adaptation literatures to draw inferences about the processes underpinning coalition adaptation, focusing on one organisational feature in particular: command and control arrangements. I then illustrate my argument with a case study of the Austro-German coalition on the Eastern Front in World War I. The final section discusses the implications of my findings and provides suggestions for future research on coalition adaptation.

7.1 Learning in War

That militaries and states learn from their mistakes is a common theme in international relations scholarship. The notion that decision-makers learn from past experiences is especially prominent in strategic studies, where it has been used to explain everything from alliance formation to how states perceive threats and behave in crises to the timing of wartime settlements.[4] Building on this core insight of the behavioral revolution – that organisations undergo structural transformation following changes to their external environment – strategic studies scholars developed new theories of military adaptation.[5]

Some of the earliest adopters of the experiential approach to organisational learning emerged from the bargaining theory of war, a body of scholarship that holds battlefield developments reveal previously hidden information about the enemy's capabilities causing actors to update their priors.[6] According to this school of thought, fighting occurs because leaders on both sides believe they will be successful. During fighting, however, private information about each belligerent's resolve and capabilities is revealed, giving rise to revised expectations about the likely outcome. As more information becomes available, expectations converge, leading to settlement offers.[7] Unlike in the classic bargaining model, the battlespace in coalition wars is often shared by more than two parties. Following the bargaining school's logic, campaigns featuring two or more states fighting alongside each other should reveal battlefield information not only about one's enemy but also one's coalition partners.

The military adaption literature, a branch of study concerned with how military praxis changes over time, also draws heavily on organisational theories of learning. Adaptation represents the final phase of learning—the stage where, following the recognition that something is not working as envisioned, change is enacted. As with the bargaining literature, this literature emphasises learning from battlefield experiences.[8] While different explanations have been put forth as to how this learning process unfolds, as

well as competing arguments about whether the impetus for change originates with civilian or military leaders, common to most explanations is the
idea that militaries adapt to their operational environment, with changes
typically following unsatisfactory outcomes in combat performance.[9]

Just as work on bargaining in war is limited in its applicability to coalitional adaptation by its assumption that only two actors are playing the
game, however, the utility of much of the existing literature on military
adaptation during wartime is restricted by two assumptions: (1) adaptation
is unidirectional and always beneficial to the fighting forces; and (2) the
learning process is linear.[10] These assumptions are of questionable value, as
empirical evidence supporting the first assumption is at best mixed and it is
far from clear that, once changed, new patterns of behaviour will prevail and
remain in place unchallenged until the end of hostilities.[11] Adaptation is by
its nature fluid; a dynamic process that ebbs and flows, often reversing course
within an organisation's lifespan.

7.2 A Theory of Military Adaptation in Coalitions

The shortcomings in existing scholarship on military adaptation are odd
given that, as discussed, it draws so heavily on insights from the field of
organisational studies. Conceptually, organisations are simply 'coalitions of
conflicting interests.'[12] Accordingly, although military coalitions differ from
stand-alone national military organisations in that they are, by definition, ad
hoc and temporary, many of the same insights used to explain drivers of
adaptation in national militaries may still apply to the coalitional experience.
However, it also follows that, given their multinational makeup, the learning
process in coalitions is unlikely to perfectly resemble the experiences of
national militaries. To determine the extent to which existing understandings
of military adaptation illuminate and obscure coalitional adaptation, I derive
testable hypotheses to develop a deductive model of coalitional adaptation
and apply them to the specific issue of change in collective command and
control arrangements. In the next section, I test my hypotheses using a case
study of the Austro-German military coalition on the Eastern Front in World
War I.

7.2.1 *Defining Coalitional Adaptation*

Before we can understand how coalition adaptation occurs, it is necessary to
define the term. Coalitional military adaptation is defined here as: a major
change that is collectively adopted by all the members of the coalition in a
given operational theater.[13] Put simply, coalition adaptation consists of a
major adjustment in operational praxis by coalition militaries sharing the
same battlespace. This definition embodies two of the key elements present
in existing definitions of military adaptation: first, in that it focuses on how
militaries function in the field, and, second, in that the change in question

must be of sufficient magnitude to transform how the organisation (in this case, the entire coalition) operates.[14]

A crucial distinction does exist between coalition adaptation and single-actor adaptation, however: the former always entails a two-step interaction. First, individual militaries must jointly decide at the strategic level that they need to adapt to confront their adversary more effectively. Then, they must implement, institutionalise, and routinise the changes at the operational level. While a similar two-step process may occur within some single-actor militaries as they adapt to combat requirements, the process is complicated in coalitional settings by the presence of multiple, sovereign actors. As a result, to be collectively adopted, the new practice must first be accepted by each partner at the strategic level before the new tactic, technique, technology, or shared feature can be institutionalised by the individual organisational units. Failure to navigate the political negotiation and bargaining in the first step is likely to preclude success in the second step.

7.2.2 Adapting from Fear

Like existing explanations of military adaptation, my model starts from the premise that learning occurs in the aftermath of a negative outcome.[15] What constitutes a sufficiently consequential negative outcome, though? According to Farrell, the trigger for organisational adaptation happens when militaries 'appear to be losing' and must entertain 'prospective defeat.' Meanwhile, Murray suggests the impetus comes from 'great losses in men and national treasure.'[16]

Following Farrell and others, I argue that the catalyst for adaptation is the fear of apparent defeat, where defeat here refers to either substantial losses in men or territory over the course of the campaign.[17] Two provisional hypotheses about coalitional adaptation follow from this premise. First, the larger the losses in either life or territory, the greater the pressure to undergo adaptation. Second, recurrent battlefield defeats, especially when accumulated in swift succession, are likely to exert more pressure for change, owing to the logic of cumulative effects.

Thinking about coalitional adaptation rather than single military adaptation forces consideration of additional factors shaping the relationship between fear and change, however. Since no two nations are alike, it follows that every national military belonging to a coalition will have a different threshold for pain. Factors that may make a country more, or less, sensitive to battlefield losses include population size, wealth, geography, ideology, and war aims.[18] Losses in battle may be unevenly felt for other reasons still, such as different skill levels in national militaries or the enemy's decision to target one sector of the battlefield over another. Consequently, coalition members may arrive at the point of departure – the recognition that adaptation is required – in stages. Thus, the process of coalition adaptation is only likely to begin once all members share a common strategic assessment of their collective prospects on the battlefield (a process that may be hastened through intra-coalitional negotiation and bargaining).[19]

7.2.3 Collective Action

As noted, the realisation that adaptation is required is only the first step in the process. Having acknowledged that a change in standard operating procedures is needed, coalition partners must agree on which changes to implement. Scholarship on collective action and coalitions suggest that negotiation on this issue is likely to be a lengthy and arduous process.[20] Scholars have long recognised that adaptation in any military organisation is laborious, and thus, often incremental.[21] If this is true in the case of a single military organisation, it follows that the obstacles to change are compounded when multiple military organisations attempt adaptation simultaneously. The fear of prospective defeat may generate urgency but, owing to the rigid nature of the organisations involved, reaching consensus on a new course of action is likely to be a protracted endeavor. The time needed to implement any agreed-upon changes only draws out the process further. Coalition adaptation will thus take longer to germinate than the unilateral adaptation of the individual militaries that make up the coalition, and is likely to be preceded by each of the forces tweaking their own tactics, techniques, and organisational practices internally first since both stages of adaptation—strategic and operational—are easier to accomplish singly than collectively.

7.2.4 Civilian Influences

Of course, military leaders are not the only actors invested in adaptation in coalitional wars. Like their military counterparts, civilian leaders are reluctant to relinquish control over decision making to foreign parties.[22] Civilian resistance arises from worries that surrendering control over decision making will hinder the government's ability to issue strategic directives, leaving leaders unable to exert their influence on the political direction of the war. These concerns arise primarily at the strategic level, but have implications for the process and success of operational-level adaptation.

7.2.5 Coalition Adaptation in Practice

The drivers of coalitional adaptation are many and varied, while the process of change is complex and slow. How should we expect it to play out during conflicts? Following other adaptation scholars, I apply my argument about adaptation to perhaps the most consequential coalitional structure—their command and control arrangements.[23] Considered the 'eternal function' of military organisations, command and control is the essence of coalition warfare.[24] Command and control refers to the exercise of authority and direction of forces in battle. At the operational level, it includes all stages of battle, from planning and preparation to execution and assessment. Because of its centrality to coalition military operations, members' sovereignty and national prestige concerns complicate the structuring and delegation of this authority.[25] Consequently, most coalition partners opt to keep command of their own forces, rather than surrender it to foreigners. Not all coalition members retain such firm command of their own forces, however, and many alter their views on the matter during fighting.

Previous scholarship on command and control structures has established that these arrangements often take varying shapes and forms.[26] When it comes to coalitional command structures, military doctrine typically recognises three kinds of arrangements.[27] Following existing practice, I classify coalitional command arrangements according to one of three types (independent; combined; or unified), depending on their degree of centralisation.[28] The least centralised type of coalitional command arrangement is independent command, which is defined by separate (usually, national) chains of command.[29] Unlike independent commands, which do not involve the sharing of operational authority, combined commands involve the bifurcation of control over forces. Dividing operational authority between nations – either according to geography, service branches, or some other feature – is often seen as an effective way to deal with political sensitivities while guaranteeing some semblance of military coordination.[30] The final type of coalition command structure is unified command, an arrangement in which a single commander exercises operational authority over all forces. Of the three ideal types identified here, this command comes closest to embodying the military principle of unity of command, an axiom stating that, whenever possible no member of a military organisation should report to more than one superior on any task or function.[31]

Within coalitions, standard practice is to retain national control of forces, owing to the principle of unity of command. While military practitioners have long recognised the benefits that accrue from having a single decision-maker at the helm of a military organisation, as well as the defects which can occur when more than one person is in charge and the communication lines become muddled, they are slow to shed control. Because vesting all decision-making authority over an organisational unit in one person not only reduces uncertainty and lowers transaction costs but also strengthens the monitoring and enforcement mechanisms of informational exchange, scholars argue that unity of command in coalitions can strengthen cohesion and help with the aggregation of combat power.[32] Thus, one might expect that military officers enthusiastically embrace the principle of unity of command in coalition warfare and seek to approximate it wherever possible.[33] Yet, empirically, we know that coalition wars rarely start out this way, indicating something must both inhibit unity of command and, ultimately, trigger its selection.

Differences across national civil-military traditions can account for hesitance at the strategic level of adaptation, and the fear of apparent defeat can explain willingness to change at the operational level, albeit at different rates across the coalition. Military officers are notorious for being status-conscious creatures. Most military cultures revolve around the privileges and power that come from holding a command. Even within national militaries, officers often resist efforts to reassign forces from their command. In coalitions, resistance to changes in command arrangements is even greater, owing to the fact that it often involves the transfer of control over national forces to a foreigner.[34] The strength of such cultures varies across militaries, however. Taken together, this suggests that adaptation in coalitions is likely to be even more complex than that of single-actor adaptation: the pull of their training leads military officers to both search for ways to strengthen unity of command following unsatisfactory battle outcomes and, to preserve honor and national prestige, oppose any effort to assign forces from their command to foreigners.

Civilian leaders' similar reluctance to surrender control adds yet another layer of complication. Civilian leaders worry that, if they cede control over their own forces, coalition partners might, if faced with a choice, hesitate to assign resources and soldiers to safeguard their nation's territory (or portion of the front) if it means leaving their own territory vulnerable to enemy attack. Similar concerns exist regarding other strategic considerations in wartime. The delegation of military authority to a foreign commander, for example, may make it harder for civilian leaders to influence other important operational matters, such as when to sue for peace or halt fighting altogether. Civilian leaders will therefore also think twice before assigning control of military forces to coalition partners for fear that doing so will significantly curtail their ability to direct the war's course.

Although they come at it from slightly different perspectives, both civilian and military leaders thus have powerful incentives to oppose changes to coalition command arrangements that entail a sacrifice of national authorities, such as combined and unified command structures. Consequently, we should only expect to see the adoption of these types of command structures once both the military *and* civilian leadership have overcome their respective objections, which is most likely to occur in the face of apparent defeat. (I make no claims about whether this attitudinal shift occurs within the civilian leadership or military leadership first, since, as with most things in war, it is context specific.[35]) For these reasons, the process is likely to be iterative, slow, and fragile.

When combined with the earlier insights about crisis-induced adaptation, we get the following three testable propositions: first, adaptation in coalitional command structures – especially changes that entail the surrender of military authority by one party to another, such as the adoption of combined and unified command arrangements – will only occur when there is strategic-level agreement among the coalition partners on the need for the change, which is especially likely following substantial losses in battle. Second, because both civilians and military officials have reasons to oppose the transfer of such authorities, the process is likely to involve substantial infighting among coalition partners. Friction along the strategic dimension of adaptation inevitably creates turbulence in operational adaptation. Finally, following these propositions, we should only expect to see movement toward coalition arrangements featuring greater unity of command following a convergence of expectations about prospective defeat by *both* civilian and military leaders. Put differently, operational adaptation in coalitions occurs only when friction along the strategic adaptation dimension is minimised.

7.3 Testing the Argument

As an initial test of my argument, I examine the case of the Austro-German coalition on the Eastern Front in World War I. Although the First World War features prominently in the military adaptation literature, existing scholarship focuses largely on the experiences of belligerents on the Western Front.[36] In theoretical terms, the Austro-German coalition represents a hard case for my argument.[37] If ever there was a wartime partnership that should have been well- equipped to manage military adaptation, it is this one. The parties' common language, similar organisational cultures, and prewar civil-military exchanges all point in the same direction and suggest that achieving agreement on command arrangements should have been relatively straight-forward for these two longtime allies.[38] The case also provides within-case variation owing to the nature of fighting on the Eastern Front, which, unlike the Western Front, featured far more fluid movements of battle. Unlike the Entente, who took until 1918 to achieve unity of command, the Central Powers underwent command reforms sooner but struggled to maintain them.

7.3.1 The Austro-German Coalition on the Eastern Front

At the outset of the war, German operations in East Prussia were managed from Berlin, by the German Supreme Army Command (Oberste Heeresleitung, hereafter OHL). Austrian operations were overseen by the Austro-Hungarian High Command (Armee Oberkommando, hereafter AOK) located in Teschen. Despite pre-war plans calling for the German Eighth Army and the Austro-Hungarian armies to launch an attack across the Narew River in a planned pincer movement, the German Chief of Staff Helmuth von Moltke informed his Austrian counterpart Franz Conrad von Hötzendorf on 3 August 1914 that he would instead assume the defensive. Unbeknownst to Moltke, Conrad had already ordered two of his three Army groups (comprising 21 of Austro-Hungary's then 48 infantry divisions) to the Serbian front.[39] As a result, the Austro-Hungarian Third Army suffered a major defeat at the hands of the Russian Eighth Army in Galicia. After capturing the strategically important city of Lemberg in early September, the Russians surrounded Przemysl Fortress, trapping more than 100,000 Habsburg soldiers inside.[40] While the Germans held the Russians back in East Prussia with fewer than 40,000 casualties, the Austro-Hungarians suffered casualties of more than 300,000 men in the Galician campaign.[41]

OHL responded to the Austrian setback by forming the Ninth Army from elements of the German Eighth Army in preparation for a combined offensive with the Austrians in October. While the Ninth Army, under the command of Paul von Hindenburg, concentrated its forces northeast of Cracow, the Austro-Hungarian First Army assembled south of the Vistula River. Unable to agree on either the operation's timing or objective, Hindenburg and Conrad wasted days trying to get the other to accede to their own campaign plans, allowing the Russian Ninth Army to slip away. Mistaking the Russian strategic withdrawal for a retreat, both commanders decided to press the advance, albeit in different directions. While Hindenburg dispatched General August von Mackensen's three corps toward Warsaw, Conrad ordered his forces south across the San River toward Lemberg. As the Ninth and First armies advanced in separate directions, a gap opened between the two. By 11 October, both men had become aware of the folly of their uncoordinated advances. Alarmed at the danger awaiting them from the Russian forces now assembling in the gap which they had let form between their two armies, both ordered a retreat.[42] The Central Powers' first experiment with a combined offensive was a resounding failure, resulting in a combined loss of approximately 90,000 men.[43]

Following their failure at the Battle of the Vistula River, the two allies resumed independent operations along their respective sections of the front. In early November, Germany established a separate command – known as Oberkommando Ost (hereafter, Ober Ost) – at Posen under Hindenburg and Erich Ludendorff to manage operations in East Prussia. Seeking to drive a wedge between the Russian First Army in East Prussia and the Russian Second Army in Galicia, Hindenburg ordered the Ninth Army under Mackensen's command to link up with the Eighth in a southeast thrust toward Lodz. To ensure enough freedom of movement for the coming operation, Ober Ost asked Austrian forces located near Cracow to move south. Hindenburg also asked that Conrad send his Second Army north of the Vistula to protect Mackensen's flank in the upcoming offensive. Acceding to both requests, Conrad allowed the Second Army to be temporarily placed under the command of German General Remus von Woyrsch, provided he in turn came under the command of AOK, as part of an early experiment in combined command.[44]

Barred from participating in the attack on Lodz but unable to sit by and watch, the Austrian chief of staff ordered his Fourth Army to attack near Cracow on the night of 15 November. Ober Ost immediately cabled its disapproval, asking Conrad to wait until Mackensen's offensive had begun to show results before staging his own operation. Ignoring the request, Conrad ordered Woyrsch's Army Detachment and the First Army to join Fourth Army's attack. Facing stubborn Russian resistance, Conrad's attack faltered. At around the same time further north, the enemy came close to encircling one of Mackensen's corps, forcing a German retreat.[45]

On New Year's Day 1915, the two chiefs met to plan their next move. They agreed the German Eighth and Tenth armies would attack the Russian Tenth Army in East Prussia while Conrad launched a parallel attack in the Carpathian Mountains aimed at liberating the besieged fortress at Przemysl.[46] Short of troops following the devastating losses of the opening months of the war, Conrad requested German reinforcements. As a condition of their deployment, Hindenburg insisted that they be placed in a new *armeegruppe* under German command, rather than in the Austro-Hungarian Third Army, as Conrad wished. After first objecting, Conrad agreed and the resulting operational-level adaptation – a new army group, named Südarmee – was assigned to General Alexander von Linsingen.[47] Together with the Third Army, the two armies had a combined force of 175,000. In late January, Südarmee attacked along a 160 km front in temperatures of minus 30 degrees Celsius. Considered the Stalingrad of the First World War, the 1915 Carpathian winter campaign fought between January and March cost the Central Powers nearly 800,000 men.[48]

In April, Conrad traveled to Berlin for discussions with Ober Ost Commander Erich von Falkenhayn, who argued that Germany be allowed to oversee the next offensive. After initially refusing, Conrad acceded to Berlin's demand and the two chiefs of staff set about planning what would become the Battle of Gorlice-Tarnow. Launched in early May 1915, the Central Powers' summer offensive was a resounding success. Led by Mackensen, the German Eleventh Army and Austrian Fourth Army attacked together from the upper Vistula to the Lupka Pass, along a front of more than 100 km. By mid-June, the combined Austro-German force had liberated Lemberg. By the time the campaign drew to a close that September, the Central Powers had captured nearly half-a-million Russian soldiers and advanced more than 300 km.[49] The breakthrough at Gorlice- Tarnow was an impressive victory in strategic terms, as well. From then on, the war on the Eastern Front would be waged entirely on Russian territory. By driving the Russians from East Galicia and inflicting such heavy losses on them in the process, the Central Powers had effectively knocked the enemy out of action for a year.

By May 1916, both headquarters received intelligence indicating the Russians were planning a springtime offensive of their own. On 4 June, the Russians – led by General Aleksei Brusilov – attacked along the entire southeastern front from the Pripet River to Czernowitz. Outnumbered by more than 130,000 men at the critical center of the front, more than 200,000 German and Austro-Hungarian soldiers were taken prisoner as the Russians broke through in Bukovina.[50] While the Austrian and German chiefs of staffs debated command arrangements, their staffs rushed reinforcements from other theaters to Galicia. To facilitate better coordination, the two head-quarters agreed to appoint German General Hans von Seeckt Chief of Staff to the Austro-Hungarian Seventh Army and place Linsingen's Südarmee under OHL. With these moves, the Germans were now in control of the front between Czernowitz and Tarnopol, while the AOK controlled the front between Tarnopol and Brest.[51] A counteroffensive launched in June by both militaries failed. By July, the Russian breach extended 140 km and was 60 km deep in some places.[52] On 27 July, Conrad and Supreme Commander of the Austro- Hungarian Army Archduke Friedrich met with Falkenhayn, Hindenburg, Ludendorff, and the German Kaiser at Pless to discuss how to proceed. Two days later, the leaders agreed that Hindenburg would take control of the front stretching north from Brest all the way to the Baltic Coast in the closest approximation to date of unified command. In addition to the armies already under his command, Hindenburg was given the Austro-Hungarian Second Army, albeit with the proviso that he not commit those forces to battle without first attaining AOK's approval.[53] In the wake of continued Russian gains and by now entirely dependent on German man-power, the Austrian military leadership abandoned its remaining objections to the creation of a unified command on the Eastern Front under Hindenburg, enabling the two emperors (who supported such a move) to sign off on the agreement. On 6 September 1916, Berlin formally announced the creation of a 'United Supreme Command.'[54]

7.3.2 Explaining Austro-German Command Adaptation

The experience of the Austro-German coalition on the Eastern Front largely conforms to my theoretical expectations. In each instance, the move toward a more centralised command arrangement was precipitated by a defeat, despite the Germans having informed the Austrians prior to the war that any direct cooperation between the two allies would commence 'only after the first mutual victories.'[55] After launching independent military operations on the Eastern Front at the start of the war, the Germans grew alarmed once they learned of the danger to their forces prompted by Conrad's decision to divert his armies to the Serbian front, and sought greater coordination with AOK.[56] Initially, both powers believed it was simply a question of more German troops: while the Austrians thought the Germans should bear more of the burden of stopping the Russian steamroller than the 18 divisions they had thus far committed, the Germans – believing their soldiers qualitatively superior to those of the Austrians – thought their forces could halt the numerically superior enemy's advance on their own.[57] Although the Germans dispatched the newly created Ninth Army to the southeastern front in September 1914 to help their beleaguered allies, Falkenhayn and Hindenburg did everything in their power at the strategic level to ensure that German forces remained operationally detached from those of the Austrians. Once the danger from the Russian trap laid along the Warsaw-Krasnik line became apparent, the Kaiser telegrammed the emperor asking that the First Army be placed under Hindenburg's command. Although the Austrian emperor was inclined to consider it, Conrad quickly put a stop to the idea.[58] Thus, as the first year of the war drew to a close, the first step in the two-stage coalition adaptation process – consensus as the strategic dimension – was still incomplete.

The two allies derived different lessons from the failure of their first combined operation during the autumn 1914 Galician campaign. While the German High Command responded by establishing a separate command headquarters for the east, Conrad requested that Franz Joseph ask the Kaiser to assign all German forces in Galicia to the Austrian supreme commander, Archduke Friedrich.[59] However, OHL continued to insist the solution lay in separate operations, rather than in more centralised command arrangements. Convinced their success in the upcoming offensive depended on keeping the Austrians as far away as possible, Ober Ost requested the Austrians vacate the operational sector before the German assault on Lodz.[60] Thus, it was not until after the disastrous Carpathian winter offensive (December 1914 – March 1915) that the German leadership seriously began to contemplate changes to coalition command arrangements. Already in January 1915, the Kaiser proposed that the Austrian Third Army be placed under a 'unified command' led by Hindenburg. But Falkenhayn, who doubted whether the liberation of Przemysl was worth 'one drop of German blood,' still opposed the idea.[61] The Austrians, who preferred to determine the command arrangements on a case-by-case operational basis, agreed with Falkenhayn. As a result, the two headquarters instead decided to share control of Südarmee, in an arrangement loosely resembling a combined command.[62] Thus, by early 1915, each partner had approached the other with the suggestion of establishing a unified command on the eastern front. Yet, because each partner insisted command be given to one of their own officers, these proposals went nowhere. Despite the enormous losses suffered during the Carpathian offensives, neither Germany nor Austria was as yet prepared to surrender control of their forces to the other.

Unable to agree on command arrangements, the two chiefs instead focused on improving the liaison activities between their respective headquarters. Prior to winter 1915, German liaison officer August von Cramon recalled, relations between AOK and OHL had been 'fairly casual,' with the Austrians only informing the Germans of their decisions after the fact. After the disastrous Carpathian winter campaign, however, the two headquarters began using their liaison officers as official intermediaries.[63] The two staffs worked closely together to plan the successful Gorlice-Tarnow campaign. In an arrangement that marked the closest approximation of unity of command in the war up until that point, an army group comprising the German Eleventh Army and the Austro-Hungary Fourth Army was formed, with Mackensen in command.[64] Yet following the campaign's successful conclusion Ober Ost and AOK resumed independent operations again.[65] In November, Mackensen's Army Group, now operating in the Balkan Theater, secured a major victory. When OHL refused Conrad's request that the combined Austro-German force press the advance, the Austrian chief of staff ordered his forces to do so on their own, ending the brief experiment in unified command.[66] Enraged by Conrad's behavior, the German chief of staff warned him that such actions were contrary to the need for close coordination between the two militaries 'upon which alone hinges the fate of our dynasties and peoples.'[67] Conrad replied that it was Falkenhayn, by withdrawing German divisions from the Serbian front without first informing the Austrians, who had broken the spirit of the arrangement and then formally removed the Third Army from Mackensen's control on 20 December 1915.[68]

Consequently, when Brusilov launched his offensive the following summer, the command arrangements of the two militaries were as fractured as ever. After refusing Conrad's proposal that the front be divided into three sectors and Linsingen's Army Group placed under the control of AOK in early June, Falkenhayn instead suggested Mackensen be given control of the entire front from the Pripet to the Romanian border. After Conrad refused, it took a further three days to agree to Seeckt's appointment as chief of staff to the commander of the Austrian Seventh Army and restore some semblance of coordination between the two militaries.[69] Throughout July, as the Russians continued their advance, the Germans and Austrians traded proposals and counterproposals. On 23 July, the two chiefs agreed to assign Hindenburg control of the entire 500 km front between the Pripet and Dniester, but the plan was scrapped by Hindenburg, who thought it a demotion. With the prospect of defeat at the hands of Brusilov growing each day, Archduke Friedrich and Conrad traveled to OHL the following week, where, in the presence of the kaiser, Hindenburg was given control of operations on the Eastern Front.[70]

In contrast to the linear adaptation process often portrayed in existing models, adaptation in Austro-German command arrangements occurred in fits and spurts, often ebbing and flowing depending on battlefield developments.[71] While major losses like those incurred during the Carpathian winter campaign typically spurred more centralisation, the opposite was also true with the two allies often going their separate ways following battlefield successes. Moreover, major changes to the coalition command arrangements followed only after earlier attempts by the coalition members to modify their own command arrangements, such as Germany's decision to stand up Ober Ost in November 1914. Likewise, movement toward more centralised coalition arrangements occurred piecemeal, and only after previous experiments involving shared command arrangements like Südarmee failed to achieve the desired results, causing all parties to accept the strategic need for adaptation.

Additionally, there is substantial support for the proposition that at each stage the shift toward more centralised command arrangements was contentious, requiring the backing of both countries' civil-military establishments before it could proceed. While the Kaiser was an early proponent of establishing a German-led unified command in the east, his own general staff initially opposed the idea. On this, the Austrians concurred with their German counterparts. When the Kaiser suggested that the Austrian First Army come under the German Ninth in November 1914, the Austrian chief of staff threatened to resign in protest. Conrad refused, he wrote, to let

Habsburg soldiers service under a 'foreign, even if allied, power.'[72] But, unlike Falkenhayn, Conrad was not entirely opposed to the idea of a more unified command arrangement, having previously raised the idea himself. The devil lay in the details since each ally wished for the other's forces to be placed under its control. In contrast to the Kaiser and his chief of staff, the Austrian emperor was prepared 'where possible' to entertain the idea by late 1914 but, having already delegated supreme command of his armies to Archduke Friedrich, was advised against it.[73] By August 1916, an important obstacle was removed when Hindenburg replaced Falkenhayn as chief of staff. Although Conrad never warmed to the idea of a unified command under Germany, the twin disasters of the Brusilov Offensive and Romania's entry into the war in the summer of 1916 reduced his stature in the eyes of the emperor, allowing Franz Joseph to push through the reforms.[74]

7.4 Conclusion

Despite the emergence of a burgeoning research agenda on battlefield coalitions in recent years, we still know comparatively little about why these fighting arrangements look the way they do, and even less about the factors that cause them to undergo organisational changes during wartime. I argue that the process unfolds in a two-step fashion, first along the strategic dimension, and only afterward at the crucial operational level. As with other organisations, battlefield coalitions have multiple design choices from which to choose. As sovereign nations, their members typically prefer to operate in a self-contained manner. As battlefield developments reveal information about both the enemy and coalition partners' capabilities, uncertainty regarding the prospects of victory and defeat fluctuates, spurring changes first in national military behavior, then collective reforms. Consistent with existing accounts on organisational and battlefield learning, my findings suggest the greater the uncertainty, the greater the organisational reforms.[75] However, as the Austro-German case indicates, even when militaries converge on the same answer, adaptation takes time because of coalitional politics. Accordingly, adaptation failures may arise not only because decision-makers are too slow to learn or learn the wrong lessons, but also because they refuse to implement what they have learned.[76] In short, while defeat incentivises adaptation, coalition politics often determines it.

My argument has important implications for both the study of military adaptation and battlefield coalitions. First, it suggests that coalition adaption is likely to be more cumbersome the larger the battlefield coalition given the number of decisionmakers (and hence veto players) involved. These findings are particularly salient given today's trend toward ever-larger coalitions (formed to, among other things, signal international legitimacy) and imply that larger coalitions come with disadvantages, as well. The experience of the Austro-German coalition on the Eastern Front in World War I also point to several fruitful avenues for future research. Along with additional case studies to ascertain whether the experiences of the Austro-German coalition are generalisable across time and space, future research should explore whether drivers beyond defeat spur adaptation in coalitions. Questions such as the precise threshold of losses needed to induce collective reforms as well as the relationship between victory and reversion to the status quo ante also merit further investigation, as do considerations about the degree to which factors like regime type, alliance institutionalisation, and intra-coalition balance of power dynamics affect the adaptation timeline. Although not examined in detail here, it is worth noting that the latter appears to be correlated with command structure changes in the Austro-German case: as German troop contributions on the Eastern Front grew in 1915–1916, so too did OHL's ability to impose its will on AOK. This suggests that asymmetric coalitions in which one partner is stronger than another may undergo adaptation faster than those where coalition partners are more evenly matched.[77] Future research might also investigate the conditions under which reforms stick (even the in the wake of victories) and which reforms revert back. Lastly, scholars may wish to study whether the adaptation process described here looks the same for coalition features other than command and control arrangements. Determining which elements of military adaptation are applicable (and which are not) to battlefield coalitions is equally important for policymakers, given the frequency with which states are choosing to wage war through coalitions.[78] In short, the continued prevalence of coalitional warfare coupled with our as yet still limited understanding of internal dynamics underpinning these fighting organisations suggests that the study of coalitions will remain a promising research area for years to come.

Notes

1 Patricia Weitsman, 'Alliance Cohesion and Coalition Warfare,' *Security Studies* 12/3 (Spring 2003), 79, https://doi.org/10.1080/09636410390443062
2 The focus here is on battlefield coalitions, configurations which arise when units from more than one political community operate in a shared battlespace. On the distinction between alliances, wartime coalitions, and battlefield coalitions, see Rosella Cappella Zielinski and Ryan Grauer, 'Understanding Battlefield Coalitions,' Chapter 1, this volume, and Patricia Weitsman, 'Alliances,' in *Encyclopedia of Power*, Keith Dowding, ed. (Thousand Oaks, California: Sage, 2011), 14–17. On military adaptation, see: Williamson Murray, *Military Adaptation in War: With Fear of Change* (Cambridge: Cambridge University Press 2011), 2; Theo Farrell, Frans Osinga, and James Russell, eds., *Military Adaptation in Afghanistan* (Stanford, CA: Stanford University Press 2013). Following Murray, I use the term innovation for peacetime changes and adaptation for wartime changes.
3 David Auerswald and Stephen Saideman, *NATO in Afghanistan: Fighting Together, Fighting Alone* (Princeton: Princeton University Press 2014).
4 Dan Reiter, 'Learning, Realism, and Alliances: The Weight of the Shadow of the Past,' *World Politics* 46/4 (Jul., 1994), 490–526, https://doi.org/10.2307/2950716; Alex Weisiger, 'Learning from the Battlefield: Information, Domestic Politics, and

Interstate War Duration,' *International Organization* 70/2 (Spring, 2016), 347–375, https://doi.org/10.1017/S0020818316000059.

5 Alfred Chandler, *Strategy and Structure: Chapters in the History of the Industrial Enterprise* (Cambridge, MA: MIT Press 1962); Richard Cyert and James March, *A Behavior Theory of the Firm* (Englewood Cliffs, NJ: Prentice Hall 1963). James March and Johan Olsen, 'The Uncertainty of the Past: Organizational Learning Under Ambiguity,' *European Journal of Political Research* 3 (1975), 147–171, https://doi.org/10.1111/j.1475-6765.1975.tb00521.x; James Wilson, 'Innovation in Organizations: Notes Toward a Theory,' in James Thompson, editor, *Approaches to Organizational Design* (University of Pittsburgh Press 1966), 193–218.

6 Geoffrey Blainey, *The Causes of War* (New York: Free Press 1988); James Fearon, 'Rationalist Explanations for War,' *International Organization* 49/3 (1995), 379–414, https://doi.org/10.1017/S0020818300033324.

7 Weisiger, 'Learning from the Battlefield'; See also Eric Min, 'Speaking with One Voice: Coalitions and Wartime Diplomacy,' Chapter 10, this volume.

8 Adam Grissom, 'The Future of Military Innovation Studies,' *Journal of Strategic Studies* 29/5 (2006), 920–4, https://doi.org/10.1080/01402390600901067; Theo Farrell, 'Improving in War: Military Adaptation and the British in Helmand Province, Afghanistan, 2006–2009,' *Journal of Strategic Studies* 33/4 (2010), 567–594, https://doi.org/10.1080/01402390.2010.489712.

9 Murray, Military Adaptation in War, 6; Farrell, 'Improving in War,' 571; David Barno and Nora Bensahel, *Adaptation Under Fire: How Militaries Change in Wartime* (New York: Oxford University Press 2020), 3. For the argument that defeat in wartime is not necessary for innovation, see Stephen Rosen, 'New Ways of Understanding War: Understanding Military Innovation,' *International Security* 13/1 (Summer 1988), 135, https://doi.org/10.2307/2538898.

10 Grissom, 'The Future of Military Innovation Studies,' 907; Farrell, 'Improving in War,' 569; and Murray, *Military Adaptation in War*, 28, 98, Barno and Bensahel, *Adaptation Under Fire*, 9.

11 James March, *The Ambiguities of Experience* (Ithaca: Cornell University Press 2010), 2; Royston Greenwood and C.R. Hinings, 'Organizational Design Types, Tracks and the Dynamics of Strategic Change,' *Organizational Studies* 9/3 (1988), 293–316, https://doi.org/10.1177/017084068800900301; Bo Hedberg, 'How organizations learn and unlearn,' in *Handbook of Organizational Design*, ed. Paul C. Nystrom and William H. Starbuck (Oxford: Oxford University Press 1981).

12 March, *The Ambiguities of Experience*, 6.

13 I thus exclude adaption that involves changes to only individual militaries that may be part of a coalition.

14 This definition departs from existing conceptualisations of military adaptation in one important way: it is agnostic on the issue of whether the adaptation in question improves the performance of the organisation. Grissom, 'The Future of Military Innovation Studies,' 906–7; Barno and Bensahel, *Adaptation Under Fire*, 9, 22; and Farrell, 'Improving in War,' 567–9.

15 Scott Gartner, *Strategic Assessment in War* (New Haven: Yale University Press, 1997); Weisiger, 'Learning from the Battlefield,' 350–1.

16 Barno and Bensahel likewise argue that 'disruptive shock(s) on the battlefield' led to doctrinal adaptation in Iraq and Afghanistan. Farrell, 'Improving in War,' 510; Murray, *Military Adaptation in War*, 6; Barno and Bensahel, *Adaptation Under Fire*, 3.

17 In contrast, Min, 'Speaking with One Voice,' Chapter 10 of this volume, argues that that battlefield losses can prompt coalition members to seek a negotiated

exit. Empirically, coalition defection remains quite rare, with most cases occurring when countries are fighting alone on a front. Alex Weisiger, 'Exiting the Coalition: When Do States Abandon Coalition Partners during War?' *International Studies Quarterly* 60/4 (December 2016), 753–765, doi: 10.1093/isq/sqw029.

18 Another variable of interest is regime type. A large research agenda argues that democracies exhibit lower levels of tolerance for casualties than non-democracies. Hugh Smith, 'What Costs Will Democracies Bear? A Review of Popular Theories of Casualty Aversion,' *Armed Forces and Society* 32/4 (2005), 487–512, https://doi.org/10.1177/0095327X05031004.

19 While losses incurred by a coalition partner in another theater of war are likely to also be viewed with alarm, defeats occurring in a shared battlespace will cause fellow coalition members greater anxiety because the consequences for their own situation are more immediate.

20 Mancur Olson, *The Logic of Collective Action: Public Goods and the Theory of Groups* (Cambridge, MA: Harvard University Press 1965).

21 Grissom, 'The Future of Military Innovation Studies.'

22 Some states also worry today's friend could become tomorrow's enemy. Nora Bensahel, 'The Coalition Paradox: The Politics of Military Cooperation,' (Ph.D. diss., Stanford 1999); Kenneth N. Waltz, 'Structural Realism after the Cold War,' *International Security* 25 (Summer 2000): 10, https://www.jstor.org/stable/2626772.

23 Rosen, 'New Ways of Understanding War,' 140.

24 Martin Van Creveld, *Command in War* (Cambridge: Harvard University Press 1985), 9; Kenneth Allard, *Command, Control, and the Common Defense* (Washington, D.C.: National Defense University 1996),18; Anthony J. Rice, 'Command and Control: The Essence of Coalition Warfare,' *Parameters* (Spring 1997), 152–67.

25 See Sara Bjerg Moller, Fighting Friends: Institutional Cooperation and Military Effectiveness in Multinational war,' (PhD diss., Columbia University 2016), 88–89; Daniel Morey, 'Military Coalitions and the Outcome of Interstate Wars,' *Foreign Policy Analysis* 12/4 (October, 2016), 536, https://doi.org/10.1111/fpa.12083.

26 Grauer identifies four ideal-type organisational forms based on differentiation and centralisation, while Cappella Zielinski and Grauer, following Chandler, distinguish between two broad types: unitary form (U-form) and multidivisional (M-form) command structures. Ryan Grauer, *Commanding Military Power* (Cambridge: Cambridge University Press 2016); Rosella Cappella Zielinski and Ryan Grauer, 'Organizing for performance: coalition effectiveness on the battlefield,' *European Journal of International Relations* 26/4 (December 2020), 953–978, https://doi.org/10.1177/13540661209033.

27 U.S. doctrine identifies three types of multinational command arrangements, called integrated; lead nation; and parallel; while NATO uses the terms fully integrated; lead nation; and framework nation. Joint Chiefs of Staff, Joint Publication 1: Doctrine for the Armed Forces of the United States, II-22; Joint Chiefs of Staff, Joint Publication 3–16: Multinational Operations, II-4, II-15; North Atlantic Treaty Organization, AJP-01 (D): Allied Joint Doctrine, 3–4, 3–5.

28 Morey also adopts a triptych typology but calls the second category 'joint command.' Because 'joint' doctrinally refers to inter-service coordination, not cross-national coordination, I use the term 'combined.' Daniel Morey, 'Centralized Command and Coalition Victory,' *Conflict Management and Peace Science* 37/6 (2020) 721–22, https://doi.org/10.1177/0738894220934884

29 Independent command structures do not preclude operational cooperation between militaries, though often it will make it more difficult. See Moller, 'Fighting Friends,' 73.

30 During the Lopez War (1864–70), Brazil, Argentina, and Uruguay agreed to rotate command based on geography. Command of all operations undertaken within Argentine territory or along the Argentinian-Paraguay border was assigned to the Argentinean commander, while command of all other operations was given to the Brazilian commander. Chris Leuchars, *To the Bitter End: Paraguay and the War of the Triple Alliance* (Westport: Greenwood Press 2002), 44, 164, 178; Thomas Whigham, *The Paraguayan War: Causes, and Early Conduct*, vol. 1 (Lincoln: University of Nebraska Press 2002), 357–9.

31 Rice, 'Command and Control: The Essence of Coalition Warfare,' 156.

32 Patricia Weitsman, *Waging War: Alliances, Coalitions, and Institutions of Interstate Violence* (Stanford, CA: Stanford University Press 2014), 18, 190; Moller, 'Fighting Friends,' 83–85, 115; Cappella Zielinski and Grauer, 'Organizing for Performance,' 954; Weitsman, 'Alliance Cohesion and Coalition Warfare,' 104; Morey, 'Centralized Command and Coalition Victory,' 730; Kelly A. Grieco, 'Fighting and Learning in the Great War: Four Lessons in Coalition Warfare,' *Parameters* 48/3 (Autumn 2018), 33.

33 Perfect unity of command is never fully achieved in coalitions since some military authorities are never relinquished. Moller, 'Fighting Friends,' 70–79.

34 Richard M. Leighton, 'Allied Unity of Command in the Second World War: A Study in Regional Military Organization,' *Political Science Quarterly* 67/3 (September 1952), 399–425, https://doi.org/10.2307/2145165.

35 All adaptation in war in at least one sense begins on the battlefield since that is where the impetus for improved performance originates.

36 Stephen Biddle, *Military Power: Explaining Victory and Defeat in Modern Battle* (Princeton: Princeton University Press 2006); Grieco, 'Fighting and Learning in the Great War,'; Michael Hunzeker, 'Perfecting War: The Organizational Sources of Doctrinal Optimization' (PhD diss., Princeton University Press 2013).

37 Jack Levy, 'Case Studies: Types, Designs, and Logics of Inference,' *Conflict Management and Peace Science* 25/1 (Feb., 2008), 1–18.

38 While the Austro-Hungarian armies were multiethnic as well as multilingual, the language of command (Kommandosprache) was German. Richard Bassett, *For God and Kaiser: The Imperial Austrian Army* (New Haven: Yale University Press 2016), 367.

39 Holger Herwig, *The First World War: Germany and Austro-Hungary, 1914–1918* (New York: Oxford University Press, 1997), 53.

40 Hew Strachan, *The First World War: To Arms*, vol.1 (Oxford: Oxford University Press, 2001), 350–7; Herwig, *The First World War*, 89; Ronald Louis Ernhath, 'The Tragic Alliance: Austro-German Military Cooperation, 1871–1918.' PhD diss., Columbia University (1970), 143, 148; Norman Stone, *The Eastern Front, 1914–1917* (London: Hodder and Stoughton, 1975), 88.

41 Among the dead was Conrad's son, Herbert, who fell at the Battle of Ravaruska. Russian losses are estimated at 290,000. Herwig, *The First World War*, 92, 94–95; Stone, *The Eastern Front*, 90–1; Strachan, *The First World War: To Arms* 354–6; Michael Clodfelter, *Warfare and Armed Conflicts: A Statistical Encyclopedia of Casualty and Other Figures, 1494–2007*, 3rd ed., (Jefferson, NC: McFarland, 2008), 436–7.

42 Without informing Conrad, Hindenburg ordered the Ninth Army to retreat all the way to Thorn, near the Silesian border. The Austrians were thus forced to pull their own forces back to the Dunjac-Biala line, the very line along which they had launched their offensive. Edmund Glaise-Horstenau, ed., *Österreich-Ungarns*

Letzter Krieg, 1914–1918 [*Austria-Hungary's Last War, 1914–1918*], trans. Stan Hanna, (Vienna: Publisher of Military Science Releases, 1930), vol. 1, 534–8, 548. (Hereafter, *OULK*); Prit Buttar, *Collision of Empires; The War on the Eastern Front in 1914* (Oxford: Osprey Group 2014), 349–353.

43 The Russians by contrast lost 70,000. Buttar, *Collision of Empires*, 355.

44 Despite the unusual arrangement, the offensive was to be an entirely German-managed affair, with Woyrsch's forces tasked with protecting Mackensen's southern flank. *OULK*, vol. 1, 562. Ernharth, 'The Tragic Alliance,' 164–5; Buttar, *Collision of Empires*, 224.

45 Strachan, *The First World War: To Arms*, 370–2; Stone, *The Eastern Front*, 105–6; *OULK*, vol. 1, 582–4, 603; Herwig, *The First World War*, 109; Winston Churchill, *The Unknown War: The Eastern Front* (New York: Charles Scribner's Sons, 1931), 264.

46 Ernharth, 'The Tragic Alliance,' 168, 172; Herwig, *The First World War*, 111; Richard DiNardo, *Breakthrough: The Gorlice-Tarnow Campaign, 1915* (Santa Barbara, California: Praeger, 2010), 21–2.

47 Ernharth, 'The Tragic Alliance,' 179–80; *OULK*, vol. 2, 117–119.

48 DiNardo, *Breakthrough*, 25; Herwig, *The First World War*, 137; Graydon Tunstall, *Blood on the Snow: The Carpathian Winter War of 1915* (Lawrence: University Press of Kansas, 2020), 3, 99.

49 *OULK*, vol 2, 508–559, 660–1; Dinardo, *Breakthrough*, 83, 99, 132–3.

50 *OULK* vol. 4, 466, 482, 494; Herwig, *The First World War*, 209; Erich von Falkenhayn, *General Headquarters 1914–1916 and its Critical Decisions* (London: Hutchinson & Co. 1919), 249.

51 OULK vol. 4, 503–4.

52 Nicholas Golovin, 'Brusilov's Offfensive: The Galician Battle of 1916,' *The Slavonic and East European Review* 13 (Apr. 1935): 585–6.

53 OULK, vol. 5, 132–3; Herwig, *The First World War*, 215; Falkenhayn, *General Headquarters*, 250–52, 272; August von Cramon, *Unser Österreich-Ungarischer Bundesgenosse im Weltkriege* (Berlin: E.S. Mittler 1920), 66–70; Hermann Cron, *Geschichte des deutschen Heeres im Welkrieg 1914–1918* (Berlin: Siegismund 1937), 55.

54 Although the kaiser was assigned the title of 'Supreme Commander,' Hindenburg and Ludendorff were in charge. By September 1916, the only Austro-Hungarian unit not directly under German control on the Eastern Front was Archduke Karl's Army Group. However, here too, Berlin managed to exert its influence by insisting that Seeckt serves as Karl's chief of staff. Herwig, *The First World War*, 215.

55 Ernharth, 'The Tragic Alliance,' 47.

56 Falkenhayn, *General Headquarters*, 20

57 Clodfelter, *Warfare and Armed Conflicts*, 437.

58 Ernharth, 'The Tragic Alliance,' 160; OULK vol 1, 477.

59 *OULK*, vol 1, 53, Buttar, *Collision of Empires*, 211.

60 Hindenburg and Ludendorff were not alone. Falkenhayn also supported giving the Ninth Army 'greater liberty of movement' from the Austrians. Falkenhayn, *General Headquarters*, 27; *OULK*, vol 1, 552–3.

61 Falkenhayn, *General Headquarters*, 53–55; Ernharth, 'The Tragic Alliance,' 179–80; *OULK*, vol 2, 117.

62 During the course of the campaign, Linsingen refused to follow the agreed-upon chain of command and report to AOK, reporting instead to the OHL in Posen. Tunstall, *Blood on the Snow*, 193; *OULK*, vol. 2., 131, 163–5.

63 Cramon, *Unser Österreich-Ungarischer Bundesgenosse im Weltkriege*, 10.

64 The Austro-Hungarian Second Army was later added to Mackensen's command, as well. The command arrangements for Gorlice-Tarnow did not achieve perfect

unity of command, however. Although Mackensen was put in charge of the entire operation, at Conrad's insistence Falkenhayn agreed to have him subordinated to the AOK provided the Austrian's first consulted OHL before issuing any orders. In practice, however, the AOK functioned as little more than a transmission belt. Herwig, *The First World War*, 141–6; Gordon Craig, 'The World War I Alliance of the Central Powers in Retrospect: The Military Cohesion of the Alliance,' *Journal of Military History* 37 (Sep., 1965), 342; DiNardo, *Breakthrough*, 42, 97; OULK vol 2. 327.

65 While Conrad attacked in the direction of the Lutsk-Rovno line, Hindenburg ordered his armies to advance north of the Pripet Marches. Stone, *The Eastern Front*, 190–1.
66 Cramon, *Unser Österreich-Ungarischer Bundesgenosse im Weltkriege*, 61–66, Alfred Krauss, *Die Ursachen Unserer Niederlage: Erinnerungen Un Urteile Aus Dem Weltkrieg* (Munich: J.F. Lehmanns Verlag, 1921), 181–5.
67 *OULK*, vol 3., pp. 602–3.
68 *OULK*, vol 3, 323, *OULK*, vol 4, 205–6, Herwig, *The First World War*, 159.
69 *OULK*, vol 4, 503–4.
70 *OULK*, vol 5, 132–3; Herwig, *The First World War*, 214; Falkenhayn, *General Headquarters*, 250–2, 272; Cramon, *Unser Österreich-Ungarischer Bundesgenosse im Weltkriege*, 66–70.
71 Nor was the relationship entirely one-way. During the fall of 1914 operations in southwest Poland, a German corps was placed under the command of AOK. *OULK*, vol. 1, 403, 454, 474.
72 Herwig, *The First World War*, 108; Strachan, *The First World War: To Arms*, vol.1, 372; Ernharth, 'The Tragic Alliance,' 163.
73 *OULK*, vol. 1, 53; Buttar, *Collision of Empires*, 21; Ernharth, 'The Tragic Alliance,' 160; *OULK*, vol. 1, 477.
74 When Franz Joseph died two months later his successor, Emperor Charles I, insisted that the command be dissolved so that he could assume personal command of his armies. Because of the effect the emperor's death had on coalition command arrangements, a German officer described the death of Franz Joseph as the 'most serious' event of the war short of the Battle of the Marne. Cramon, *Unser Österreich-Ungarischer Bundesgenosse im Weltkriege*, 86.
75 Weisiger, 'Learning from the Battlefield.'
76 James March, 'Footnotes to Organizational Change,' *Administrative Science Quarterly* 26/4 (1981); 564–66, https://doi.org/10.2307/2392340.
77 Glenn Snyder, *Alliance Politics* (Ithaca: Cornell University Press, 1997).
78 Rosella Cappella Zielinski and Ryan Grauer, 'A Century of Coalitions in Battle: Incidence, Composition, and Performance, 1900–2003,' Chapter 2, this volume.

Bibliography

Allard, Kenneth. *Command, Control, and the Common Defense*. Washington, D.C.: National Defense University, 1996.
Auerswald, David and Stephen Saideman. *NATO in Afghanistan: Fighting Together, Fighting Alone*. Princeton, NY: Princeton University Press, 2014.
Barno, David and Nora Bensahel. *Adaptation Under Fire: How Militaries Change in Wartime*. New York: Oxford University Press, 2020.
Bassett, Richard. *For God and Kaiser: The Imperial Austrian Army*. New Haven: Yale University Press, 2016.
Bensahel, Nora. 'The Coalition Paradox: The Politics of Military Cooperation.' (PhD diss., Stanford University, 1999).

Biddle, Stephen. *Military Power: Explaining Victory and Defeat in Modern Battle.* Princeton, NJ: Princeton University Press, 2006.

Blainey, Geoffrey. *The Causes of War.* New York, NY: Free Press, 1988.

Buttar, Prit. *Collision of Empires: The War on the Eastern Front in 1914.* Oxford: Osprey Group, 2014.

Cappella Zielinski, Rosella and Ryan Grauer. 'Organizing for performance: coalition effectiveness on the battlefield.' *European Journal of International Relations* 26/4 (December 2020): 953–978. 10.1177/13540661209033

Chandler, Alfred. *Strategy and Structure: Chapters in the History of the Industrial Enterprise.* Cambridge, MA: MIT Press, 1962.

Churchill, Winston. *The Unknown War: The Eastern Front.* New York, NY: Charles Scribner's Sons, 1931.

Clodfelter, Michael. *Warfare and Armed Conflicts: A Statistical Encyclopedia of Casualty and Other Figures*, 1494–2007, 3rd ed. Jefferson, NC: McFarland, 2008.

Craig, Gordon. 'The World War I Alliance of the Central Powers in Retrospect: The Military Cohesion of the Alliance.' *Journal of Military History* 37 (September 1965): 336–344.

Cramon, August von. *Unser Österreich-Ungarischer Bundesgenosse im Weltkriege.* Berlin: E.S. Mittler, 1920.

Cron, Hermann. *Geschichte des deutschen Heeres im Welkrieg 1914–1918.* Berlin: Siegismund, 1937.

Cyert, Richard and James March. *A Behavior Theory of the Firm.* Englewood Cliffs, NJ: Prentice Hall, 1963.

DiNardo, Richard. *Breakthrough: The Gorlice-Tarnow Campaign, 1915.* Santa Barbara, CA: Praeger, 2010.

Ernharth, Ronald Louis. 'The Tragic Alliance: Austro-German Military Cooperation, 1871–1918' (PhD diss., Columbia University, 1970).

Falkenhayn, Erich von. *General Headquarters 1914–1916 and its Critical Decisions.* London: Hutchinson & Co, 1919.

Farrell, Theo. 'Improving in War: Military Adaptation and the British in Helmand Province, Afghanistan, 2006–2009.' *Journal of Strategic Studies* 33/4 (2010): 567–594. 10.1080/01402390.2010.489712

Farrell, Theo, Frans Osinga, and James Russell, eds. *Military Adaptation in Afghanistan.* Stanford, CA: Stanford University Press, 2013.

Fearon, James D. 'Rationalist Explanations for War.' *International Organization* 49/3 (Summer, 1995): 379–414. 10.1017/S0020818300033324

Gartner, Scott. *Strategic Assessment in War.* New Haven, CT: Yale University Press, 1997.

Glaise-Horstenau, Edmund, ed. *Österreich-Ungarns Letzter Krieg, 1914–1918* [Austria-Hungary's Last War, 1914–1918, 7 vols.] Translated by Stan Hanna. Vienna: Publisher of Military Science Releases, 1930.

Golovin, Nicholas. 'Brusilov's Offensive: The Galician Battle of 1916.' *The Slavonic and East European Review* 13 (April 1935): 571–596.

Greenwood, Royston and C.R. Hinings. 'Organizational Design Types, Tracks and the Dynamics of Strategic Change.' *Organizational Studies* 9/3 (1988): 293–316. 10.1177/017084068800900301

Grauer, Ryan. *Commanding Military Power.* Cambridge: Cambridge University Press, 2016.

Grieco, Kelly. 'Fighting and Learning in the Great War: Four Lessons in Coalition Warfare,' *Parameters* 48/3 (Autumn 2018): 27–36.

Grissom, Adam. 'The Future of Military Innovation Studies.' *Journal of Strategic Studies* 29/5 (2006): 919–930. 10.1080/01402390600901067

Hedberg, Bo. 'How organizations learn and unlearn.' In Vol. 1 of *Handbook of Organizational Design*, edited by Paul C. Nystrom and William H. Starbuck, 3–27. Oxford: Oxford University Press, 1981.

Herwig, Holger. *The First World War: Germany and Austro-Hungary*, 1914–1918. New York, NY: Oxford University Press, 1997.

Hunzeker, Michael. 'Perfecting War: The Organizational Sources of Doctrinal Optimization.' (PhD diss., Princeton University, 2013).

Krauss, Alfred. *Die Ursachen Unserer Niederlage: Erinnerungen Un Urteile Aus Dem Weltkrieg*. Munich: J.F. Lehmanns Verlag, 1921.

Leighton, Richard M. 'Allied Unity of Command in the Second World War: A Study in Regional Military Organization.' *Political Science Quarterly* 67/3 (September, 1952): 399–425. 10.2307/2145165.

Leuchars, Chris. *To the Bitter End: Paraguay and the War of the Triple Alliance*. Westport, CT: Greenwood Press, 2002.

Levy, Jack S. 'Case Studies: Types, Designs, and Logics of Inference.' *Conflict Management and Peace Science* 25/1 (Feb., 2008): 1–18.

March, James. 'Footnotes to Organizational Change,' *Administrative Science Quarterly* 26/4 (1981), 563-55. 10.2307/2392340.

March, James. *The Ambiguities of Experience*. Ithaca, NY: Cornell University Press, 2010.

March, James and Johan Olsen. 'The Uncertainty of the Past: Organizational Learning Under Ambiguity.' *European Journal of Political Research* 3 (1975): 147–171. 10.1111/j.1475-6765.1975.tb00521.x.

Moller, Sara Bjerg. 'Fighting Friends: Institutional Cooperation and Military Effectiveness in Multinational War.' (PhD diss., Columbia University, 2016).

Morey, Daniel. 'Centralized Command and Coalition Victory.' *Conflict Management and Peace Science* 37/6 (2020): 716–734. 10.1177/0738894220934884

Morey, Daniel. 'Military Coalitions and the Outcome of Interstate Wars.' *Foreign Policy Analysis* 12, no. 4 (2006): 533–551. 10.1111/fpa.12083

Murray, Williamson. *Military Adaptation in War: With Fear of Change*. Cambridge: Cambridge University Press, 2011.

Olson, Mancur. *The Logic of Collective Action: Public Goods and the Theory of Groups*. Cambridge: Harvard University Press, 1965.

Rosen, Stephen. 'New Ways of War: Understanding Military Innovation.' *International Security* 13/1 (Summer 1988): 134–168. 10.2307/2538898.

Reiter, Dan. 'Learning, Realism, and Alliances: The Weight of the Shadow of the Past.' *World Politics* 46/4 (Jul, 1994): 490–526. 10.2307/2950716

Rice, Anthony. 'Command and Control: The Essence of Coalition Warfare.' *Parameters* 27 (Spring 1997): 152–167.

Smith, Hugh. 'What Costs Will Democracies Bear? A Review of Casualty Aversion.' *Armed Forces & Society* 31/4 (2005): 487–512. 10.1177/0095327X05031004

Snyder, Glenn. *Alliance Politics*. Ithaca, NY: Cornell University Press, 1997.

Stone, Norman. *The Eastern Front, 1914–1917*. London: Hodder and Stoughton, 1975.

Strachan, Hew. *The First World War: To Arms*, vol.1. Oxford: Oxford University Press, 2001.

Tunstall, Graydon. *Blood on the Snow: The Carpathian Winter War of 1915*. Lawrence, KS: University Press of Kansas, 2010.

Van Creveld, Martin. *Command in War*. Cambridge: Harvard University Press, 1985.

Waltz, Kenneth N. 'Structural Realism after the Cold War.' *International Security* 25/ 1 (Summer, 2000): 5–41. https://www.jstor.org/stable/2626772.

Weisiger, Alex. 'Learning from the Battlefield: Information, Domestic Politics, and Interstate War Duration.' *International Organization* 70/2 (Mar., 2016): 347–375. 10.1017/S0020818316000059

Weisiger, Alex. 'Exiting the Coalition: When Do States Abandon Coalition Partners during War?' *International Studies Quarterly* 60/4 (December 2016): 753–765. 10.1093/isq/sqw029

Weitsman, Patricia. 'Alliance Cohesion and Coalitional Warfare.' *Security Studies* 12/3 (Spring 2003): 79–113. 10.1080/09636410390443062

Weitsman, Patricia. 'Alliances.' In *Encyclopedia of Power*, edited by Keith Dowding, 14–17. Thousand Oaks: Sage, 2011.

Weitsman, Patricia. *Waging War: Alliances, Coalitions, and Institutions of Interstate Violence*. Stanford, CA: Stanford University Press, 2014.

Whigham, Thomas. *The Paraguayan War: Causes, and Early Conduct*, vol. 1. Lincoln: University of Nebraska Press, 2002.

Wilson, James. 'Innovation in Organizations: Notes Toward a Theory.' In *Approaches to Organizational Design*, edited by James D. Thompson, 193–218. Pittsburgh, PA: University of Pittsburgh Press, 1966.

Part IV
Execution

8 Regime Type, War Aims, and Coalition Member Effort in Combat

Rosella Cappella Zielinski, Ryan Grauer, and Alastair Smith

At the outset of World War I, the French and British quickly formed a battlefield coalition to combat the German offensive. The new collective did not operate smoothly, however. In the run-up to the Battle of the Marne, the British commander, Sir John French, feared being forced to shoulder the weight of the fight and repeatedly refused to conform to French commander Joseph Joffre's plans for arresting the German onslaught.[1] It was not until Joffre made concrete moves to assure the British that their forces would be protected by French troops during the battle that the latter acceded to the operation.[2] Similarly, as they were battling the British in North Africa in December 1941, German commander Erwin Rommel suggested that the Panzer forces and Italian units under his command retreat through Cyrenaica to escape building pressure and prepare for a counterstroke. The Italian High Command refused; with 20,000 Italian colonials and an array of industries located in the region, the Italian General Gastone Gambara 'feared political difficulties for the Duce if Cyrenaica were lost.'[3] Rommel nevertheless issued his order and the Italians, offended by their partner's disregard for their political aims, soon retaliated. Rommel writes, '[Chief of the Comando Supremo Ugo] Cavallero took his revenge by holding back part of the Italian Corps in the Mersa el Brega area and part in Agedabia, so that it was more or less removed from my command.'[4]

These cases illustrate a common phenomenon: states form battlefield coalitions to combat a shared foe but are frequently frustrated by their partners' underwhelming efforts while fighting. This chapter analyses battlefield coalition members' under-provision of combat effort and how their choices are shaped by domestic political institutions. We argue that, while under-provision of effort is a near-ubiquitous phenomenon, the root causes and manifestation of such behaviour differ across political regimes. Using selectorate theory, we contend regime type conditions both the composition of forces belligerents generate and their war aims, which then shape battlefield coalition activities within wars. Democratic states like Britain in 1914, with capital-heavy forces and war aims that tend to emphasise the provision of international collective goods, often attempt to shirk and free-ride on their partners' efforts during the fighting. Non-democratic states like

DOI: 10.4324/9781003399896-12

Italy in 1941 tend to rely on labour-heavy forces and emphasise private ends in their war aims; consequently, they are often reluctant to form coalitions and, when fighting jointly, push for operations that benefit them more than the collective. When democratic and non-democratic states fight together, their divergent types of forces and war aims create conflicting pressures in which the former prefer to free-ride on the latter, but the latter prioritise their private ends over collective combat goals, rendering free-riding untenable.

Battlefield coalitions often overcome attempts at and consequences of effort under-provision, but not all do. Unpacking the different logics of effort under-provision accordingly offers important caveats to the body of work on democratic wartime cooperation, efficiency, and effectiveness.[5] It also contributes to a growing body of work assessing the efficiency and effectiveness of non-democracies in combat.[6] More broadly, our argument sheds new light on dynamics surrounding coalition formation, duration, and termination as well as signals of strength and resolve sent by coalitions to current adversaries, potential joiners, and possible future foes.[7]

We develop our argument in four sections. First, drawing on selectorate theory insights, we articulate the intuitive logic underpinning our argument that domestic political institutions shape force composition and war aims, and those factors, in turn, condition coalition members' effort provision on the battlefield. Second, we formalise the essential components of our argument and derive testable implications. Third, we assess our argument in three case studies drawn from the World Wars: (1) the United Kingdom, France, and Belgium in combat operations on the Western Front in May and June 1940, (2) Germany and Austria-Hungary in combat operations on the Eastern Front between August 1914 and June 1915, and (3) the United Kingdom, the United States, and China in combat operations in Burma between 1942 and 1945. Finally, we conclude with implications for future research.

8.1 War Aims and Battlefield Coalition Member Effort

Battlefield coalitions are groupings of officers, troops, and materiel provided by multiple distinct political actors, brought together for the purpose of fighting a common foe in a discrete combat action and battlespace. They are distinct from wartime coalitions, which feature groups of forces that wage war against the same foe but may not deploy troops to fight side-by-side in combat operations, and alliances, which are formal peacetime agreements to cooperate militarily in future conflicts.[8] Under-provision of effort by battlefield coalition members, or the phenomenon of belligerents fighting less hard when in partnership with others than they would on their own, including to the point that combat outcomes can be jeopardised, is both common and puzzling. Given the stakes involved in warfighting, why do actors who seemingly feel the need to fight alongside others on the battlefield in order to defeat their common foe then frequently behave in ways that make it less likely that they will succeed? To the extent scholars have investigated this question, they have emphasised two dynamics. Initially, most pointed to the general challenges of harnessing member contributions and effort in the pursuit of collective goods, focusing especially on free-riding and shirking.[9] More recently, others have highlighted the practical challenges of coordinating efforts in war, including problems such as incompatible strategic goals, weak institutionalisation, inefficient command and control, and insufficient resource management capabilities.[10]

Existing scholarship thus offers some insight but does not fully answer the question for at least three reasons. First, much focuses on effort-provision either in terms of budgetary allocations to alliance structures created during peacetime (especially the North Atlantic Treaty Organisation, or NATO) or during multilateral peacekeeping operations.[11] There is little reason to believe that pathologies driving states' behaviour in such circumstances are the same as those that cause under-provision of combat effort in ad hoc groupings fighting interstate wars. Second, with a few exceptions, existing scholarship investigates under-provision of effort at the war level.[12] The reasons why partners fighting a war together who then take the further step of deploying forces side-by-side in battle and still under-provide combat effort require further explanation. Finally, scholars have identified drivers of considerable friction, inefficiency, and ineffectiveness in democracies' wartime cooperative efforts.[13] Given the myriad differences in the wartime behaviour of different regime types, however, it is not obvious that members of democratic, non-democratic, and mixed-regime coalitions should under-provide combat effort for the same reasons.[14] Indeed, different drivers of effort under-provision highlighted in the Entente and Axis vignettes in the introduction to this chapter suggest that assuming similarity likely obscures more than it explains.

Our approach to explaining coalition members' under-provision of effort in combat builds on, but is distinct from, existing scholarship. Specifically, we argue that domestic political institutions shape the conduct of coalition member so the battlefield as well as in wars. The influence of such institutions obtains whether or not coalition partners concluded a peacetime alliance.

We begin from the selectorate theory insight that the number of supporters a leader needs to retain power drives policy choices. When a leader's winning coalition is large – that is, when she leads a democracy and depends on securing the support of many eligible voters to remain in power – she is incentivised to both reduce collective costs and pursue collective goods that offer benefits to a wide range of citizens. By contrast, when a leader's winning coalition is small – when she is the leader of a non-democracy – she is incentivised to reduce costs for and pursue specific, private goods that can be used to satiate the narrower set of people within the state that facilitate her retention of power.[15] These tendencies obtain in foreign as well as domestic policymaking, including in the determination of war aims.[16]

War aims are overarching goals that belligerents hope to achieve through the use of military force. Some war aims are common to all belligerents – namely, survival in the current conflict and, ideally, success. We argue that belligerents also possess additional war aims that are conditioned by their domestic political governance institutions. Following selectorate theory logic, democratic leaders are incentivised to use relatively few well-trained and well-equipped forces to not only survive and win, but also to secure international collective goods that flow from victory, like the restoration of peace and stability in the region (e.g., the United States' intention to defeat and occupy Japan during World War II).[17] Doing so minimises the cost to the population in terms of casualties while maximising the benefit that any individual citizen may enjoy from victory and the likelihood that such benefits are sustained for an extended period of time. Non-democratic leaders are incentivised to not only survive and win, but also to preserve capital and use larger numbers of minimally equipped troops to pursue private goods – or spoils – that redound to the benefit of the elites within the state, like the occupation of territory and extraction of resources (e.g., the Soviet Union's intention to dominate Eastern Europe during World War II). To be sure, states at war often pursue both international collective goods and spoils, and can gain both through a single outcome: the US not only sought the collective good of pacifying Japan, but also gained the spoil of basing access after World War II. However, the dominant themes of democratic states' additional war aims should emphasise international collective goods while those of non-democratic states should emphasise spoils.

While not perfectly determinative, belligerents' war aims beyond mere survival and success condition how they use their variously composed forces in combat, particularly when fighting with partners in battlefield coalitions. Democracies, as rational actors pursuing international collective goods war aims while seeking to avoid raising costs for their citizens, will attempt to push manpower and materiel combat expenditures onto their partners. Accordingly, they are likely to seek many battlefield coalition partners and under-provide effort in combat by shirking and free-riding on others' efforts. The more successful democracies are in the first task, the more opportunities there will be for them to pass off costs in the second.[18]

Non-democracies' pursuit of spoils has two important implications for battlefield coalitions. First, non-democracies fighting on the same side are likely to have divergent aims: though they are combatting the same foe, they are also in competition for private goods. Accordingly, they will generally avoid forming large coalitions, as fewer co-belligerents means fewer divergent interests to reconcile as well as fewer competitors for conquered territory and plundered resources. Second, when they do form battlefield coalitions, non-democracies have few incentives to shirk or free-ride, as such methods of under-provision can have deleterious effects on their ability to capture spoils. They are likely, however, to cheat by using their manpower-heavy forces to pursue their individual objectives at the expense of collective interests, especially when the two diverge.

Democracies' and non-democracies' additional war aims are thus likely to incentivise under-provision of effort, but that under-provision will manifest in different ways. Democracies will under-provide by attempting to free-ride and shirk while non-democracies will cheat by diverting effort to secure private, rather than collective, ends.[19] When battlefield coalitions of forces fielded by a single regime type form, under-provision should manifest primarily as either free-riding and shirking or cheating, and intra-coalitional conflict should centre on resolving problems created by that behaviour. When democracies and non-democracies fight together and form mixed-regime battlefield coalitions, their divergent types of forces and additional war aims are likely to create conflict of a different sort. Rational democratic attempts at free-riding and shirking while in pursuit of international collective goods only succeed if partners also seek those same international collective goods. Rational non-democratic cheating in pursuit of private goods works well if partners seek international collective goods rather than spoils. Mixed-regime battlefield coalitions are thus likely to be riven by considerable tension and conflict over members' respective obligations and where their efforts should be made.

To be clear, nothing in our theory about different regime types' dominant additional war aims and the effect those goals have on their effort provision speaks to the likelihood of battlefield success or failure; battlefield coalitions can and have succeeded and failed with and without free-riding, shirking, and cheating. Combat is a risky enterprise that depends upon the balance of quantitative and qualitative resources, strategic choices, and a host of other intangible factors.[20]

8.2 Model of Battlefield Coalitions

To generate testable implications from our theory, we formalise our intuitive logic. The game is about the forces, war aims, and combat effort of two battlefield coalition partners, A and B. We subscript all notation by A and B, as appropriate.

We use a simple selectorate framework to model domestic politics in each combat participant.[21] Leaders seek to stay in office and, accordingly, make decisions that maximise the welfare of their supporters. The policies that maximise supporters' welfare differ depending on leaders' domestic winning coalition size (W), or the number of supporters they need to retain power. If leader A has R_A resources and a domestic winning coalition of size W_A, she can provide each supporter with $\frac{R_A}{W_A}$ private rewards. When W is small, a leader can provide each supporter with many private goods. However, as W grows, each supporter's share of private goods declines. As a result, leaders beholden to large numbers of supporters seek to provide public goods that reward all in society, including her supporters.

War affects available resources and rewards. Making military effort in combat is costly. In building and committing military forces to battle, leaders must use, and face trade-offs between, labour and capital. At one extreme, leaders might employ many poorly trained and equipped troops in battle. At the other, leaders might use a smaller force of well-trained and -equipped soldiers. The labour-intensive approach is relatively cheap financially but likely to be costly in casualties; the capital-intensive approach is financially expensive, but likely to keep casualties low.

Military effort can thus be understood in terms of two inputs, labour (l), and capital (k). The first input can be thought of as the number of soldiers, while the second input refers to the tanks, artillery, planes, and other materiel used by the troops. If a leader spends capital, then those resources can no longer be used to reward members of her winning coalition. For A, the relative labour and capital costs are λ_A and c_A, and the total labour and capital costs are $\lambda_A l_A^2 + c_A k_A^2$. The squared terms indicate increasing marginal costs from additional effort.[22] We assume a simple Cobb-Douglas production function: if member A uses l_A labour and k_A capital, then its military effort is $m_A = k_A^\alpha l_A^{1-\alpha}$, where $\alpha = \frac{1}{2}$.

Military effort is used, however, to gain new resources that might be distributed as rewards for supporters. There are two basic types of resources to be gained through making military effort, which we refer to as war aims. One is defeat of the enemy in the field and subsequent stabilisation of the region, with the value of victory being v for everyone in A and B. The other is spoils, r, such as territory and pillage. Military capacity can be dedicated to the pursuit of v, r, or both. Let $s_A \in [0, 1]$ refer to A's war aims focus, or the proportion of A's military effort dedicated to the capture of spoils.

Leaders' war aims (s) and effort (m) determine the likely rewards to be gained from victory and spoils. The reward of v from victory benefits every person in state A, including, importantly, the leader's domestic winning coalition. If A and B deploy m_A and m_B military efforts with war aims of s_A and s_B, respectively, then their cumulative military effort dedicated to attaining victory is $(1 - s_A)m_A + (1 - s_B)m_B$ and the probability they defeat the common enemy – that is, secure v – is $F((1 - s_A)m_A + (1 - s_B)m_B)$, where $F: \mathbb{R}^+ \to [0, 1]$ is a smooth, increasing, and concave function. Supporters in W_A also stand to accrue private gains, or spoils. In particular, if A expects to gain $rG(m_A s_A)$ resources, where r measures the abundance of spoils and $G: \mathbb{R}^+ \to \mathbb{R}^+$ is a smooth, increasing, and concave function, then each supporter in W_A can expect to gain a $\frac{1}{W_A}$ share of these spoils.[23]

These gains are offset to some degree by costs. All people in A suffer from the loss of their fellow citizens – what we have termed labour costs, $\lambda_A l_A^2$. Additionally, any private gains are offset by A's financial costs of conflict. As a leader spends capital on the war effort, she reduces the resources available for private rewards. If the capital cost of effort is $c_A k_A^2$, then each supporter sees a $\frac{c_A k_A^2}{W_A}$ reduction in private rewards.

The creation of a battlefield coalition is a simultaneous choice by the leaders in A and B as to their war aims (s), the level of effort to make (m), how to generate that effort in terms of labour and capital inputs (l and k). Note that A and B are in a different relationship with respect to victory and spoils. Victory is an international collective good that benefits both partners, and possibly the larger international community. In contrast, spoils are a private good that benefits only the belligerent that captures them. In this context, leaders make choices that maximise the expected welfare of their essential supporters. Given the strategy profile (l_A, k_A, s_A, l_B, k_B, s_B), the payoff to a domestic winning coalition supporter in A is

$$U_A(l_A, k_A, s_A, l_B, k_B, s_B) = \underbrace{vF(m_A(1 - s_A) + m_B(1 - s_B)) - \lambda_A l_A^2}_{\text{National rewards and costs}}$$
$$+ \underbrace{\frac{rG(s_A m_A)}{W_A} - \frac{c_A k_A^2}{W_A}}_{\text{Private rewards and costs}} \tag{8.1}$$

where $m_i = k_i^\alpha l_i^{1-\alpha}$.

8.2.1 Model Analysis

For tractability, we break the analysis into two steps. We first establish how political institutions affect the composition of military forces to be employed. We then embed these results within our primary analysis: war aims and military efforts within battlefield coalitions.

8.2.1.1 Composition of Forces

Political institutions determine the capital/labour ratio used in producing military effort. As W increases, leaders increasingly emphasise capital over labour.

Proposition 1: *For leader A, the (political) cost-minimising ratio of capital and labour is*

$$\frac{k_A}{l_A} = \sqrt{\frac{W_A \lambda_A}{c_A}} \tag{8.2}$$

and the cost of producing m_A military effort is

$$cost_A(m_A) = 2m_A^2 \sqrt{\frac{c_A \lambda_A}{W_A}} \tag{8.3}$$

All proofs are in the online appendix.[24]

We refer to $MC_A = \sqrt{\frac{c_A \lambda_A}{W_A}}$ as the military cost to member A.

Implication 1: (Force Composition): *The ratio of capital to labour* $\left(\dfrac{k_A}{l_A}\right)$ *increases in wealth and domestic winning coalition size and decreases in population.*

8.2.1.2 War Aims and Effort

Given the results in Proposition 1, it is convenient to rewrite the leaders' objective functions in terms of war aims (s) and effort (m): for $i = A, B$,

$$U_i(m_A, s_A, m_B, s_B) = \underbrace{vF(m_A(1 - s_A) + m_B(1 - s_B))}_{\text{Collective good of victory}}$$
$$+ \underbrace{\frac{rG(s_i m_i)}{W_i}}_{\text{Private Spoils}} - \underbrace{2m_i^2 \sqrt{\frac{c_i \lambda_i}{W_i}}}_{\text{Cost of fighting}} \tag{8.4}$$

Consider a game in which A and B simultaneously choose war aims and effort.

Proposition 2: *There exists a pure strategy Nash Equilibrium profile* (m_A, s_A, m_B, s_B) *such that* $m_A, m_B > 0$ *that solves*

$$\frac{\partial U_i(m_A, s_A, m_B, s_B)}{\partial m_i} = v(1 - s_A)F'(m_A(1 - s_A) + m_B(1 - s_B))$$
$$+ \frac{s_i rG'(s_i m_i)}{W_i} - 4m_i\sqrt{\frac{c_i \lambda_i}{W_i}} = 0 \tag{8.5}$$

for $i = A, B$. *For interior cases,* $s_A, s_B \in (0, 1)$,

$$\frac{\partial U_i(m_A, s_A, m_B, s_B)}{\partial s_i} = -vm_i F'(m_A(1 - s_A) + m_B(1 - s_B))$$
$$+ \frac{m_i rG'(s_i m_i)}{W_i} = 0 \tag{8.6}$$

for $i = A, B$.

Corollary 1: *For the interior equilibrium* $s_A, s_B \in (0, 1)$,

$$\frac{G'(m_A s_A)}{G'(m_B s_B)} = \frac{W_A}{W_B} \tag{8.7}$$

and

$$\frac{m_A}{m_B} = \sqrt{\frac{c_B \, \lambda_B}{c_A \, \lambda_A}} \sqrt{\frac{W_A}{W_B}} \qquad (8.8)$$

Implication 2: (International Collective Goods War Aims): *Wealthy, populous, and more democratic battlefield coalition members make greater effort in pursuit of victory in combat than less wealthy, less populous, and less democratic battlefield coalition members.*

From equation (8.8), the ratio of efforts depends upon the ratio of force production costs. Battlefield coalition members with lower capital costs (small c), lower labour costs (small λ), and larger domestic winning coalitions (large W) make greater efforts to achieve victory on the battlefield than coalition members with higher military costs in capital and labour and smaller domestic winning coalitions.

Implication 3: (Private Goods War Aims): *Less democratic battlefield coalition members expend more resources to capture private spoils in combat than more democratic battlefield coalition members.*

From equation (8.7), the ratio of domestic winning coalition sizes determines the ratio of marginal returns for spoils. Less democratic battlefield coalition members focus relatively more on spoils than more democratic battlefield coalition members and expend greater resources in the search for such resources: if $W_A > W_B$, then $m_B \, s_B > m_A \, s_A$. Non-democratic forces can fight harder than democratic militaries; they just make a smaller effort to secure victory.

Implication 4: (Under-Provision of Effort toward Victory): *Battlefield coalition members under-provide effort to achieve victory in combat.*

Victory, and the regional or global security and stability that stems from it, is an international collective good and, as such, is under-provided. This is necessarily true when members pursue spoils in conjunction with or instead of victory. However, suppose there are no spoils ($r = 0$) and A and B incur the same costs ($MC_A = MC_B$). In such a setting, A and B's equilibrium efforts solve $vF'(2m_A^*) = 4\, MC_A m_A^*$. A and B could improve their welfare if they could commit to an effort of $\hat{m}$ where $\hat{m}$ solves $v2F'(2\hat{m}) = 4\, MC_A \hat{m}$. This cooperative level of effort improves the welfare of both leaders relative to the Nash equilibrium efforts but is larger than the equilibrium level and thus unattractive.

8.3 Assessing Battlefield Coalition Member Effort in Combat

To assess the validity of our claim, we evaluate how co-belligerents' domestic institutions condition their force composition, war aims beyond their baseline objectives, and combat effort in three case studies: the democratic battlefield coalition of the United Kingdom, France, and Belgium (the Western Allies) fighting Germany in May and June 1940; the non-democratic battlefield coalition of Germany and Austria-Hungary (the Central Powers) fighting Russia between August 1914 and June 1915; and the mixed-regime battlefield coalition of the United Kingdom, the United States, and China (the Pacific Allies) fighting Japan in Burma between 1942 and 1945. The case study findings are presented here; for more detailed treatments of each, consult our online appendix.[25]

These cases maximise our analytical leverage. While there is no existing systematic theory of under-provision of combat effort in battlefield coalitions against which we can compare our expectations, the cases hold constant variables that could plausibly have consistent effects on belligerent war aims and effort, including the scale and stakes of the conflict, nature of fighting, military technology used, and formal peacetime commitments regarding wartime behaviour.[26] In all three cases, the stakes were exceptionally high, with coalition partners fighting powerful adversaries and the territory of at least one of the partners was threatened. Additionally, as all combat occurred on highly mobile battlefields with similar weaponry, there is little reason to think variation in combat circumstances or military technology impacted co-belligerent provision of effort. Finally, in all cases, peacetime coordination and commitments to fight alongside one another were weak. In the Western Allies case, the members had fought together during World War I, but their coordination dwindled during the Interwar Period as the Belgians withdrew from military cooperation and declared neutrality, the British retreated back across the English Channel, and the French developed a national military strategy anchored around the Maginot Line.[27] In the Central Powers case, while Germany and Austria-Hungary had signed the Dual Alliance in 1879, the period prior to the outbreak of World War I up through the early months of the fighting was one of poor coordination between general staffs, with the two forming no joint command, coordinated strategy, or common doctrine for combined operations.[28] In the Pacific Allies case, the United States' retreat from engagement in European strategic affairs during the Interwar Period precluded any close military Anglo-American cooperation until naval staff talks between the two countries were authorised in October 1940; British and American military relations with China only began to take serious form at the end of 1941.[29]

These cases thus allow us to isolate and trace the impact of our postulated variables on belligerents' provision of combat effort when fighting as part of a battlefield coalition, with confidence that any variation observed is attributable to the causes our theory identifies.[30] Accordingly, if our selectorate-theory-based claim is correct, we should observe: (1) democracies and wealthier countries fielding forces marked by higher capital-to-labour ratios than non-democracies and larger states; (2) democracies' war aims beyond survival and success articulating visions of broader regional or even global stability and security; (3) non-democracies' war aims beyond survival and success articulating earnest preferences for spoils to be gained from the conflict; and (4) democracies and non-democracies both under-providing combat effort, with the former tending to shirk and free-ride and the latter tending to cheat and redirect effort toward the capture of private ends.

8.3.1 The Western Allies

When the German *Wehrmacht* struck west in May 1940, it battled Dutch, Belgian, British, and French forces. While the Dutch fought largely on their own, the others waged a coalitional effort to block the assault. The three democracies, or Western Allies, were ultimately unable to halt the German advance and France fell in June 1940. In their defeat, they invested in capital and labour at rates that largely fit our theory's expectations, emphasised the centrality of their collective goods war aims, and frequently sought to minimise their own combat burdens.

Regime Type and Force Composition: Implication 1 holds states' capital-to-labour investment ratios increase with wealth and democracy and decrease with population. The Western Allies were all democratic; the United Kingdom (UK) was the wealthiest, followed by Belgium and then France.[31] In population, exclusive of dominions and colonies, the UK was the largest, with 47.5 million people, followed by France with 42 million and Belgium with 8.4 million.[32] Our theory therefore anticipates Belgium's capital-to-labour investment ratio would be the highest, followed by the UK's and then France's.

The British and French capital-to-labour ratios align with our expectations while the Belgian ratio does not. In 1940, the UK invested 32.7% of its pre-war GDP in military expenditures on a force comprising roughly 5% of its population while France invested 286% of its pre-war GDP in military expenditures on the nearly 15% of its population serving in uniform.[33] These capital-to-labour ratios are consistent with our expectations in a situation in which one democratic co-belligerent is significantly richer, but slightly larger, than the other. They are also notable in comparison to the Soviet Union's capital-to-labour ratio; when the Germans invaded in 1941, the Soviets were investing approximately 17.9% of their pre-war GDP in their military, which encompassed roughly 4.3% of their population.[34] Belgium is the outlier. Though richer and much smaller than France, Belgium invested only 6.8% of its pre-war GDP in military expenditures on a military comprising 7% of its population in 1940.[35] This mismatch between our theory and the historical record is understandable, however, since Belgium pursued a strategy of neutrality during the interwar period and only began increasing investment in military capabilities shortly before war broke out.

Regime Type and War Aims: Per Implication 2, the Western Allies adopted war aims emphasising the collective good of victory. The UK was the clearest in this respect. Both Belgium and France saw defeating the

German invasion as an existential threat in a way the UK did not, but their concern for survival was supplemented by the international collective good belief that Nazism had to be defeated to secure peace in Europe.

Britain's collective goods war aims were proclaimed by Prime Minister Winston Churchill in a famous speech to the House of Commons three days after the German attack. He noted British policy was to 'wage war, by sea, land, and air, with all our might and with all the strength that God can give us' in order to achieve 'victory—victory at all costs, victory in spite of all terror; victory, however long and hard the road may be; for without victory there is no survival.'[36] In these words, Churchill articulated the position taken by a host of leading Conservative, Labour, and Liberal politicians as well as the majority of the British population – a position informed in large part by the memories of the costs and devastation of the great War, fear of hostile control of the Low Countries, and recognition of the gravity of the situation after German Chancellor Adolf Hitler violated the Munich Agreement.[37]

The French and Belgians, facing a more direct threat from the *Wehrmacht*, held comparable views. French Prime Minister Edouard Daladier noted, 'The ambitions of Napoleon were far inferior to the present aims of the German Reich,' and the Belgians, despite clinging to their policy of neutrality, laboured assiduously to block the rise of Nazi sympathisers in their domestic political system.[38] The domestic political debates over the threat of Germany were heated in both countries until the onset of German aggression against Poland in 1939 when, like in Britain, elite and public opinion converged on the need to combat the rising menace for the good of peace and stability in Europe.[39]

War Aims and Combat Effort: Per Implications 1 and 4, we expect the Western Allies to both face significant political costs to deploying large amounts of force and seek to minimise their individual combat costs. Indeed, this is what the historical record reveals.

Turning to the question of the political costs of deploying military force, the Western Allies consistently sought to bring additional partners on board, which would lessen their own contributions and costs. The coalition's core – Britain and France – pushed both Belgium and the Netherlands to abandon their neutrality policies and cooperate in the preparation of defences against potential German aggression.[40] They also sought to conclude a tripartite agreement with the Soviet Union to arrest German aggression, though that effort failed when the Western Allies demurred in response to Stalin's demands for territorial concessions in Eastern Europe.[41] Finally, Britain and France sought to bring the United States in on their side, renewing their appeals as late as 15 June 1940, when France was nearing total defeat.[42]

The Western Allies also attempted to minimise their combat costs via the structuring of their deployed forces. Specifically, the democracies each sought to sacrifice capital rather than labour. This is clear when comparing their forces to those of non-democratic Germany. The Western Allies together fielded 141 divisions, 4,164 tanks, 4,910 bombers and fighters, and 13,318 pieces of heavy artillery against Germany's 135 divisions, 2,439 tanks, 3,369 bombers and fighters, and 7,378 pieces of heavy artillery in May 1940 – a far more capital-heavy force. Considering the mustered resources in proportion, both Britain and France fielded more tanks, bombers and fighters, and pieces of heavy artillery per division than Germany. Even Belgium employed more pieces of heavy artillery per division than Germany.[43]

Finally, the Western Allies' tendencies to shift combat costs are seen in their shirking and free-riding during the fighting. Some of the under-provision of effort was formally enshrined in the national directions given to field commanders. For example, on the ground, the British sharply circum-scribed the authority of the combined command structure to direct their forces; Lord Gort, commander of the British Expeditionary Force, was to appeal to London for revised instructions if any command issued by his French superiors appeared to imperil the British forces.[44] Belgium adopted similar constraints and, as a result, tense negotiations and delayed commu-nications led to both the British and the Belgians taking unilateral decisions deviating from French-planned defensive operations to avoid anticipated heavy costs.[45] In the air, the UK and France disagreed on the best use of the former's bombers and, after consequent miscommunications led to British bombers flying without French fighter support and suffering grievous losses, the former pulled back in its conduct of sorties.[46] As the Germans advanced, the Belgians and the British both withdrew from the fighting – decisions that, while strategically defensive, especially given British concerns elsewhere, were perceived by the French as the height of cravenness.[47] France's then-Prime Minister Paul Reynaud underscored the French view on its partners' shirking after Belgium withdrew from the war on 28 May, declaring 'that [Belgian King] Leopold had laid down arms "without warning General Blanchard, without a thought or word for the French or British troops who went to the aid of his country in response to his agonized appeal",,' and 'rated it a "deed without precedent in history".'[48]

The Western Allies thus conform to our theoretical expectations in several ways. Comprised of democracies, the battlefield coalition featured co-belligerents with capital-heavy forces. Further, the partners articulated col-lective goods war aims designed to satisfy both national and international security demands. Finally, to lower the costs of deploying and using forces in pursuit of a collective good, the co-belligerents sought to increase the number of partners committed to a certain level of effort provision while minimising their individual exposure in the fight with the Germans.

8.3.2 The Central Powers

Between August 1914 and June 1915, Germany and Austria-Hungary fought a series of battles against Russia. These non-democracies – the Central

Powers – initially struggled to fight together but did improve over time. As with the Western Allies, their capital-to-labour ratios align with our expectations; the Central Powers also pursued private war aims and sought to avoid fighting for the collective good as often as possible.

Regime Type and Force Composition: Implication 1 anticipates capital-to-labour military investment ratios increasing with wealth and democracy and decreasing with population. While Germany and Austria-Hungary were both non-democratic, the former was significantly richer; Germany was also more populous, with 65 million people to Austria-Hungary's 50.6 million.[49] Our theory accordingly anticipates Germany's capital-to-labour ratio would be higher than Austria-Hungary's.

The Central Powers' capital-to-labour ratios match these expectations. In 1914, Germany invested 9.8% of its pre-war GDP in its military and it increased that investment to 27.3% in 1915. Its military grew from 1.3% of its population in 1914 to 5.9% in 1915.[50] Austria-Hungary spent 13.5% of its pre-war GDP on military expenditures in 1914 and 25.9% in 1915; its military grew from 1.7% of its population to 7.5% over the same period.[51] Germany, the wealthier partner, had a capital-to-labour ratio slightly lower than Austria-Hungary's when the war began, but it grew to be 35% larger by 1915.

Regime Type and War Aims: Implication 3 holds that non-democratic battlefield coalition partners emphasise private aims, or spoils. Both Germany and Austria-Hungary evinced this behaviour; their individual pursuit of private goods dominated their combined emphasis on the collective good of defeating Russia until such an approach proved untenable.

Germany's aims in World War I were not clearly articulated like Britain's in World War II. As late as March 1916, the German Chancellor Theobald von Bethmann-Hollweg noted he was 'anxious for the moment when it would be possible to define concrete aims.'[52] In practice, Germany behaved as if its goals were to prevent encirclement by Britain and France to the west and Russia to the east and, crucially, establish the private good of economic hegemony in Europe.[53] Within that set of aims, the danger to the west was prioritised. Helmuth von Moltke, Chief of the German General Staff, made the point clear in 1913, telling Franz Conrad von Hötzendorf, Chief of the Austro-Hungarian General Staff, that 'Austria's fate will not be definitively decided along the Bug but rather along the Seine.'[54] These aims were determined solely by the political elite, and driven by their specific interests rather than the preferences of the German population, in large part due to the German system of government: the Kaiser, backed by his Chancellor and the Supreme Army Command, held virtually all foreign and military policy decision-making power. The Reichstag, Bundesrat, and other governmental bodies were under the functional control of the Chancellor, who submitted to the final authority of Kaiser Wilhelm III.[55] Naturally, there was little public debate of German war aims before the war, and none that influenced the views of the decision makers.[56]

In Austria-Hungary, war aims were similarly focused on the interests of the political elites. The Austro-Hungarian Emperor, who, if anything, held a tighter grip on foreign and military policymaking than his German counterpart, was concerned with dangers to the south.[57] Serbia's gains in the 1912 and 1913 Balkans wars doubled its territory and increased its population by 1.5 million. Its growing strength, combined with nascent Yugo-Slav unity in the region, posed a significant threat to Austro-Hungarian political and economic domination of the region.[58] Indeed, the Austro-Hungarians made the point plain to the Germans, with Emperor Franz Joseph explicitly telling the German ambassador in July 1914 that he hoped his allies 'appreciated the dangers which threatened the Monarchy by reason of the neighborhood of Serbia.'[59] There was no check on this imperial focus; the legislature held no compulsive power over the Emperor and had been dissolved in favour of rule by decree in March 1914.[60]

Crucially, though the Central Powers intended to fight jointly against Russia, their parochial war aims dominated planning. Germany focused on the west and Austria-Hungary focused on the south; despite significant overlap in the ethnic composition of the ruling elite, the powerful actors within the countries exhibited relatively little interest in making common cause for the purpose of advancing a pan-German project.[61]

War Aims and Combat Effort: Per Implications 1 and 4, we expect that the non-democratic Central Powers to under-provide effort in pursuit of victory and instead work toward the achievement of their individual private war aims. The Central Powers behaved as we anticipate, until their combined under-provision jeopardised their survival; at that point, the battlefield coalition partners worked together to beat back the threat to their most essential, and shared, war aims.

In line with our expectations, the Central Powers did not work well together or seek to expand their coalition. Despite having signed the Dual Alliance in 1879, Germany and Austria-Hungary struggled to coordinate with one another. Personal rivalries and professional distrust on both sides precluded the German and Austro-Hungarian general staffs establishing a joint command, aligning strategies, and developing common doctrine for combined operations in the period before World War I.[62] To the extent other states were considered for membership in the coalition, Romania and Turkey were viewed as useful to have in the war generally, but not as serious contributors with whom the two core actors would often or meaningfully share battlespace.[63]

With respect to their force deployments, the Central Powers' war aims again pushed them away from cooperation on their collective interested in defeating Russia and toward application of effort to their own private interests. Both Germany and Austria-Hungary recognised Russia as a key danger, and frequently discussed the need for each to deploy forces sufficient to stem any attack from the east. Neither side confirmed it would do so, and both sides assumed the other would launch a major offensive against Russia at the outset of war while they struck at who they saw as their principal enemy.[64] Accordingly, Moltke sent the bulk of the German forces to the Western Front in 1914, where they were poorly positioned to either combat Russia or send assistance to the Austro-Hungarians, if needed.[65] Conrad, for his part, sent a significant portion of his force south toward Serbia.[66] As late as December 1914, there was still 'little to no coordination between Habsburg and Hohenzollern staffs. [... Each partner] conducted its own campaign with little regard for the other.'[67]

The private aims and resultant initial deployments ended in near-disaster for the Central Powers in the 1914 Battle of Galicia. The Austro-Hungarian battle plan assumed German support to compensate for its own weaknesses resulting from deployments against Serbia. When Germany could not provide support, the Russians defeated the Austro-Hungarians.[68] Suddenly, the Central Powers and their private interests were exposed. As Moltke's successor as Chief of the German General Staff, Erich von Falkenhayn, wrote, 'the loss of the frontier territories would have rendered the continuation of the war impossible after a comparatively short time. The industrial and agricultural districts of the East were as important as the industrial districts on both banks of the Rhine. Neither the exclusion of the one nor the other was practicable for Germany and her allies.'[69]

The significantly divergent German and Austro-Hungarian war aims then converged; both suddenly had a private interest in protecting their eastern territory and adjusted their troop deployments and efforts at coordination accordingly.[70] When the Central Powers again attempted a combined offensive at Gorlice-Tarnow in 1915, their pursuit of private aims dovetailed with the collective good of defeating the Russians and they did not underprovide effort. The Russians could not withstand the concerted attack.

Like the Western Allies, the Central Powers conform to our theoretical expectations. The non-democratic partners' capital-to-labour ratios and war aims manifested as we anticipate, with the latter emphasising private goods until such a focus could not be sustained. Conrad underscored out point when expressing his frustration with the Germans early in the war, declaring, 'Kaiser Wilhelm II cared more for his East Prussian stud farms and hunting grounds than he did for the common cause of the Central Powers!'[71]

8.3.3 The Pacific Allies

The Japanese invaded Burma in January 1942, ejecting the British, Chinese, and American forces present in the country. It took three and a half years for Pacific Allies to recapture the country and re-establish the land bridge needed to supply China. As they worked to regain Burma, the Pacific Allies exhibited the divergent capital-to-labour ratios and war aims as well as the resultant internal discord our theory implies.

Regime Type and Force Composition: The United States (US) and UK, beyond being democratic, were considerably wealthier than their battlefield coalition partner.[72] China, however, was much larger, with 412 million people; the US and the UK numbered roughly 131 million and 47 million, respectively.[73] Given the Pacific Allies' variable regime type, wealth, and size, Implication 1 anticipates that the UK and US would have higher capital-to-labour ratios than China.

Our expectations are borne out in the historical record. When the Burma Campaign began in 1942, the UK's military expenditures were 38% of its pre-war GDP and its military comprised 8.5% of its population. The US, having just entered the war, spent 26% of its pre-war GDP on its military, which was 3% of its population. By 1945, the US was outspending the UK significantly, investing the equivalent of 81.7% of its pre-war GDP in the military, which by then encompassed 9% of Americans; the UK's defence spending was 43.5% of its pre-war GDP and its military numbered 10.7% of its population.[74] By contrast, China's military expenditures in 1942 were a mere 0.1% of its pre-war GDP and only 0.5% in 1945; its military comprised 0.7% of its population in 1942 and 1.2% of its population in 1945.[75] The wealthier democracies' capital-to-labour ratios were vastly greater than that of non-democratic China.

Regime Type and War Aims: Our theoretical expectation regarding the Pacific Allies' different types of war aims also fits with the historical record. In line with Implication 2, the democracies tended to privilege the collective good of defeating Japan while, per Implication 3, the non-democratic Chinese invested much more effort in gaining spoils.

The US and UK were mostly united in their pursuit of a collective good – the ejection of Japan from Burma, re-establishment of the over-land supply line to China, and victory in the Pacific Theatre more generally.[76] The democracies' war aims were not perfectly aligned, however. While the better-resourced Americans were single-minded about reopening the Burma Road, supplying the Chinese, and keeping China in the coalition against the Japanese, the British were also intent on re-establishing their imperial domain over Burma before advancing into Malaya and retaking Singapore.[77] These differences derived from the specific American and British domestic winning coalitions; the US public was emphatic about the need to defeat the Japanese, who had attacked Pearl Harbor, while the UK public, and Churchill himself, were strongly inclined toward the reclamation of the Empire.[78]

The Anglo-American differences in war aims were minor compared to their combined divergence with the Chinese. China's leader, Generalissimo Chiang Kai-shek, had two foci: defeat of the Japanese and victory over Mao Tse-tung's Communist Part in an ongoing civil war.[79] Chiang made his aim clear to the US and UK, issuing 'Three Demands,' which had to be satisfied to ensure Chinese participation in the coalition: at least three American divisions to assist the Chinese forces in the region, a significant American air force presence, and at least 5,000 tons per month of supplies shipped to China.[80] He increased his demands for aid in December 1943, growing so audacious in his requests that the US threatened to cut off all aid to China.[81] The co-belligerents' divergent war aims and China's pursuit of spoils – to be acquired from its partners as much as its adversaries – undermined the coalition's overall cohesion.

War Aims and Combat Effort: The implications of our theory highlighting the different capital-to-labour ratios and war aims likely to attend different regime types collectively suggest that mixed-regime coalitions are likely to be riven with friction and conflict over the coordination of contributions and efforts made by each partner. Further complicating that inherent friction, implication 4 suggests all members of a battlefield coalition are likely to attempt to under-provide combat effort in pursuit of victory. The Pacific Allies' experience and efforts in Burma reflect this expectation.

First, there was considerable friction in the coordination of coalition efforts.[82] In the Anglo-American relationship, competition for leadership dominated the early days of the war, with fighting between British and American generals growing so intense that a special command conference was held to resolve the dispute in November 1943.[83] Meanwhile, China continually worked to minimise its commitment of forces in combat and quickly departed the theatre in early 1945, when the Burma Road had been partially reopened but the Japanese had not yet been defeated in the country.[84]

Second, consistent with Implications 3 and 4, China privileged its own interests over the collective good and the democratic coalition partners sought to free-ride on their partners' efforts. The British took advantage of American and Chinese efforts during the fighting to reopen the Burma Road by launching multiple independent offensives to retake the Burmese capital of Rangoon – an objective unnecessary for the defeat of the Japanese but required for the re-establishment of the Empire.[85] The Americans, who controlled Allied forces assigned to the northern Combat Area Command, took advantage of Britain's large in-country force by pushing attached British units extremely hard during the battles of Mogaung and Myitkyina, exhausting and wrecking them in the process.[86] The Chinese, as noted, sought to minimise their commitment of forces to combat in Burma and withdrew as soon as they could.

The Pacific Allies' experience in the Burma Campaign demonstrates that our theory sheds light not only on the composition and combat effort of battlefield coalitions comprised of forces drawn from like regimes, but also on mixed regime type groupings. The partners' divergence in force composition and war aims in mixed battlefield coalitions cause them to both struggle to cohere and under-provide effort in combat operations.

8.4 Conclusion

Belligerents form battlefield coalitions to increase their chances of success and lower their costs in war. Yet, due to rampant under-provision of effort by virtually all members of such groups, they are rarely satisfied with the performance of their partners. Our theory and model dissects the pathologies of battlefield coalitions and shows that, while dissatisfaction with partners is commonplace, the underlying basis for discontentment differs according to the domestic political institutions of the coalition members. Battlefield coalitions of democratic nations are plagued by free-ridging and shirking while non-democratic groupings suffer from the redirection of effort toward private goods and mixed collectives struggle with different regime types under-providing for different reasons. Evidence from the World Wars strongly supports our claim.

Our findings bring nuance to conventional wisdom about battlefield coalitions in two ways. First, they suggest fighting with others is not necessarily more effective or efficient than fighting alone, and pre-war efforts to safeguard against likely collective action pathologies is crucial to ensure that combat effort – a central component of burden sharing – is contributed at the rates necessary. Second, we amend the 'democracies fight better' narrative by highlighting how democracies have distinct pathologies that affect how they fight with partners. They may be able to overcome the pathologies in many cases – and scholarship on the performance of battlefield coalitions comprised solely of forces fielded by democracies suggests that they often do[87] – but the tendency to shirt and free-ride does impose a tax on collective martial capabilities.

Our argument and findings suggest avenues for future research. We have demonstrated the utility of exploring the dynamics of coalition behaviour at the sub-war level for understanding the effort made by members in combat. Taking account of such variance could improve understandings of coalition duration and termination, methods of ensuring equitable burden-sharing, and possibly improving – in combination with other drivers – combat performance. Additionally, while our focus has been on state members of battlefield coalitions, such groupings are increasingly incorporating non-state actors. Investigation into the systematic drivers of the war aims and combat efforts of those actors promises to be fruitful.

Notes

1 Sir Frederick Maurice, *Lessons Of Allied Co-Operation; Naval, Military, And Air, 1914–1918* (London: Oxford University Press, 1942), 10.

2 Elizabeth Greenhalgh, *The French Army and the First World War* (Cambridge: Cambridge University Press, 2014), 27; Roy A. Prete, *Strategy and Command: The Anglo-French Coalition on the Western Front, 1914* (Montréal Québec: McGill-Queen's University Press, 2009), 109–14.

3 Erwin Rommel and Sir Basil H. Liddell Hart, *The Rommel Papers* (New York: Da Capo Press, 1953), 175.

4 Rommel and Liddell Hart, 182.

5 David A. Lake, 'Powerful Pacifists: Democratic States and War,' *American Political Science Review* 86, no. 1 (March 1992): 24–37, https://doi.org/10.23 07/1964013; Dan Reiter and Allan C. Stam, *Democracies at War* (Princeton: Princeton University Press, 2002); Michael C. Desch, *Power and Military*

Effectiveness: The Fallacy of Democratic Triumphalism (Baltimore: Johns Hopkins University Press, 2008).

6 Risa Brooks, 'An Autocracy at War: Explaining Egypt's Military Effectiveness, 1967 and 1973,' *Security Studies* 15, no. 3 (September 2006): 396–430, https://doi.org/10.1080/09636410601028321; Caitlin Talmadge, *The Dictator's Army: Battlefield Effectiveness in Authoritarian Regimes* (Ithaca, NY: Cornell University Press, 2015).

7 Brett Ashley Leeds, 'Alliance Reliability in Times of War: Explaining State Decisions to Violate Treaties,' *International Organization* 57, no. 4 (September 2003): 801–27, https://doi.org/10.1017/S0020818303574057; Kristopher W. Ramsay, 'Settling It on the Field: Battlefield Events and War Termination,' *Journal of Conflict Resolution* 52, no. 6 (December 1, 2008): 850–79, https://doi.org/10.1177/0022002708324593; Stephen B. Long, 'A Winning Proposition? States' Military Effectiveness and the Reliability of Their Allies,' *International Politics* 52, no. 3 (2015): 335–48, https://doi.org/10.1057/ip.2015.8; Alex Weisiger, 'Exiting the Coalition: When Do States Abandon Coalition Partners during War?,' *International Studies Quarterly* 60, no. 4 (December 2016): 753–65, https://doi.org/10.1093/isq/sqw029.

8 Rosella Cappella Zielinski and Ryan Grauer, 'Understanding Battlefield Coalitions,' Chapter 1, this volume.

9 Mancur Olson, *The Logic of Collective Action* (Cambridge, MA: Harvard University Press, 1971); Mancur Olson and Richard Zeckhauser, 'An Economic Theory of Alliances,' *The Review of Economics and Statistics* 48, no. 3 (1966): 266–79, https://doi.org/10.2307/1927082; Mancur Olson and Richard Zeckhauser, 'Collective Goods, Comparative Advantage, and Alliance Efficiency,' in *Issues of Defense Economics*, ed. Roland N. McKean (New York: NBER, 1967), 25–63; John A. C. Conybeare and Todd Sandler, 'The Triple Entente and the Triple Alliance 1880–1914: A Collective Goods Approach,' *The American Political Science Review* 84, no. 4 (December 1990): 1197–1206, https://doi.org/10.2307/1963259; Andrew Bennett, Joseph Lepgold, and Danny Unger, 'Burden-Sharing in the Persian Gulf War,' *International Organization* 48, no. 1 (Winter 1994): 39–75, https://doi.org/10.1017/S0020818300000813.

10 Patricia Weitsman, *Dangerous Alliances: Proponents of Peace, Weapons of War* (Stanford, CA: Stanford University Press, 2003); Thomas Stow Wilkins, 'Analysing Coalition Warfare from an Intra-Alliance Politics Perspective: The Normandy Campaign 1944,' *Journal of Strategic Studies* 29, no. 6 (December 2006): 1121–50, https://doi.org/10.1080/01402390601016592; Sara Moller, 'Fighting Friends: Institutional Cooperation and Military Effectiveness in Multinational War' (PhD thesis, New York, Columbia University, 2016); Rosella Cappella Zielinski and Ryan Grauer, 'Organizing for Performance: Coalition Effectiveness on the Battlefield,' *European Journal of International Relations* 26, no. 4 (December 2020): 953–78, https://doi.org/10.1177/1354066120903369.

11 See, for example, Hirofumi Shimizu and Todd Sandler, 'Peacekeeping and Burden-Sharing, 1994–2000,' *Journal of Peace Research* 39, no. 6 (November 2002): 651–68, https://doi.org/10.1177/0022343302039006001; Songying Fang and Kristopher W. Ramsay, 'Outside Options and Burden Sharing in Nonbinding Alliances,' *Political Research Quarterly* 63, no. 1 (March 2010): 188–202, https://doi.org/10.1177/1065912908327528; Jordan Becker and Edmund Malesky, 'The Continent or the "Grand Large"? Strategic Culture and Operational Burden-Sharing in NATO,' *International Studies Quarterly* 61, no. 1 (March 2017): 163–80, https://doi.org/10.1093/isq/sqw039.

12 For two exceptions, see Cappella Zielinski and Grauer, 'Organizing for Performance'; Patrick A. Mello, 'National Restrictions in Multinational Military

Operations: A Conceptual Framework,' *Contemporary Security Policy* 40, no. 1 (January 2019): 38–55, https://doi.org/10.1080/13523260.2018.1503438.

13 Erik Gartzke and Kristian Skrede Gleditsch, 'Why Democracies May Actually Be Less Reliable Allies,' *American Journal of Political Science* 48, no. 4 (October 2004): 775–95, https://doi.org/10.1111/j.0092-5853.2004.00101.x; Atsushi Tago, 'When Are Democratic Friends Unreliable? The Unilateral Withdrawal of Troops from the "Coalition of the Willing",' *Journal of Peace Research* 46, no. 2 (March 2009): 219–34, https://doi.org/10.1177/0022343308100716; Patrick A. Mello, 'Paths Towards Coalition Defection: Democracies and Withdrawal from the Iraq War,' *European Journal of International Security* 5, no. 1 (February 2020): 45–76, https://doi.org/10.1017/eis.2019.10.

14 For discussions of democracies' behaviour in coalitions, see, for example, Ajin Choi, 'The Power of Democratic Cooperation,' *International Security* 28, no. 1 (Summer 2003): 142–53, https://www.jstor.org/stable/4137578; Ajin Choi, 'Fighting to the Finish: Democracy and Commitment in Coalition War,' *Security Studies* 21, no. 4 (October 2012): 624–53, https://doi.org/10.1080/09636412 .2012.734232; Benjamin A. T. Graham, Erik Gartzke, and Christopher J. Fariss, 'The Bar Fight Theory of International Conflict: Regime Type, Coalition Size, and Victory,' *Political Science Research and Methods* 5, no. 4 (October 2017): 613–39, https://doi.org/10.1017/psrm.2015.52.

15 Bruce Bueno de Mesquita et al., *The Logic of Political Survival* (Cambridge, MA: MIT Press, 2003); Jessica L. P. Weeks, *Dictators at War and Peace* (Ithaca, NY: Cornell University Press, 2014).

16 James D. Morrow et al., 'Selection Institutions and War Aims,' *Economics of Governance* 7, no. 1 (January 2006): 31–52, https://doi.org/10.1007/s10101-005-0108-z. This expectation aligns with the common assumption that belligerents fighting with partners in war seek to secure their rationally defined individual political and strategic objectives at the lowest possible cost. Sarah E. Kreps, *Coalitions of Convenience: United States Military Interventions after the Cold War* (New York: Oxford University Press, 2011); Scott Wolford, *The Politics of Military Coalitions* (New York: Cambridge University Press, 2015).

17 These 'international collective goods' are non-rival and non-excludable for both actors within the state and throughout the international system.

18 William H. Riker, *The Theory of Political Coalitions* (New Haven, CT: Yale University Press, 1962); Patricia Weitsman, *Waging War: Alliances, Coalitions, and Institutions of Interstate Violence* (Stanford, CA: Stanford University Press, 2013).

19 Note that we make no claim about the relative degree of under-provision likely to occur. Such considerations are reserved for future research.

20 Ryan Grauer, *Commanding Military Power: Organizing for Victory and Defeat on the Battlefield* (New York: Cambridge University Press, 2016), 9–18.

21 Bueno de Mesquita et al., *The Logic of Political Survival*.

22 Alternatively, we could have modelled diminishing marginal productivity by setting $m_A = k_A^\alpha l_A^\beta$, $\alpha + \beta < 1$, and had constant costs.

23 To focus on the substantively interesting cases, we assume as $x \to 0$, $F'(x) > 0$ and $G'(x) > 0$.

24 Available at http://www.ryangrauer.com.

25 Available at: http://www.ryangrauer.com.

26 Weitsman, *Dangerous Alliances*; Nora Bensahel, 'International Alliances and Military Effectiveness: Fighting Alongside Allies and Partners,' in *Creating Military Power: The Sources of Military Effectiveness*, ed. Risa A. Brooks and Elizabeth A. Stanley (Stanford, CA: Stanford University Press, 2007), 186–206; Evan N. Resnick, 'Hang Together or Hang Separately? Evaluating Rival Theories

of Wartime Alliance Cohesion,' *Security Studies* 22, no. 4 (October 2013): 672–706, https://doi.org/10.1080/09636412.2013.844520.

27 N.H. Gibbs, *Grand Strategy*, vol. 1: Rearmament Policy (London: Her Majesty's Stationery Office, 1976), 491–529, 605–88; Alistair Horne, *To Lose a Battle: France 1940* (New York: Penguin, 2007), 66–85.

28 Gordon A. Craig, 'The World War I Alliance of the Central Powers in Retrospect: The Military Cohesion of the Alliance,' *The Journal of Modern History* 37, no. 3 (September 1965): 337–9, https://www.jstor.org/stable/1875406; Holger H. Herwig, *The First World War: Germany and Austria-Hungary 1914–1918* (New York: Bloomsbury, 2014), 51–3, 136.

29 William Slim, *Defeat into Victory* (London: Pan Books, 2009), 21–3.

30 Hein Goemans and William Spaniel, 'Multimethod Research: A Case for Formal Theory,' *Security Studies* 25, no. 1 (January 2016): 25–33, https://doi.org/10.1080/09636412.2016.1134176; Peter Lorentzen, M Taylor Fravel, and Jack Paine, 'Qualitative Investigation of Theoretical Models: The Value of Process Tracing,' *Journal of Theoretical Politics* 29, no. 3 (July 2017): 467–91, https://doi.org/10.1177/0951629816664420. Some may argue that, because our cases are drawn from multifront wars the wars were multifront in nature and, as a consequence, belligerents' concern for events elsewhere in the world may have conditioned the effort they made in the battles we study. However, nearly all of the belligerents we examine were fighting on multiple fronts, meaning that such concerns, if not necessarily equally felt, were effectively omnipresent. While it is possible that the uneven distribution of belligerents' concern for events elsewhere could affect our inferences, we believe that danger is outweighed by the benefit of holding constant factors that other scholars have suggested could plausibly shape war aims and combat effort.

31 The UK's GDP per capita at the outset of the war was $5,983; Belgium's was $4,730 and France's was $4424. These and all other monetary figures in this chapter are denoted in 1990 international dollars. Mark Harrison, 'The Economics of World War II: An Overview,' in *The Economics of World War II: Six Great Powers in International Comparison*, ed. Mark Harrison (Cambridge: Cambridge University Press, 1998), 3, 7.

32 Harrison, 3, 7.

33 British military expenditures were $92.9 billion and its military was 2.3 million Britons. French military expenditures were $53.3 billion and its military was 7 million people. J. David Singer and Melvin Small, *The Wages of War, 1816–1965: A Statistical Handbook* (New York: John Wiley & Sons, Inc., 1972); Harrison, 'The Economics of World War II,' 3, 14.

34 Soviet GDP per capita at the outset of the war was $2,150 and its population was 167 million; Soviet military expenditures in 1941 were approximately $64.3 billion and its military numbered approximately 7.1 million. Harrison, 'The Economics of World War II,' 7, 14.

35 Belgian military expenditures were $2.7 billion and its military was 600,000 strong. Singer and Small, *The Wages of War, 1816–1965*; Harrison, 'The Economics of World War II,' 7; Michael Clodfelter, *Warfare and Armed Conflicts*, 3rd ed. (Jefferson, NC: McFarland & Company, 2007), 467.

36 Winston Churchill, *The Second World War*, vol. 2: Their Finest Hour (London: Cassell, 1949), 24.

37 Horne, *To Lose a Battle*, 97, 138; Ernest R. May, *Strange Victory: Hitler's Conquest of France* (New York: Hill and Wang, 2000), 165; Victor Rothwell, *War Aims in the Second World War: The War Aims of the Major Belligerents, 1939–1945* (Edinburgh: Edinburgh University Press, 2005), 65–6.

38 May, *Strange Victory*, 196, 302.

39 Horne, *To Lose a Battle*, 220; May, *Strange Victory*, 115–17, 153–212, 301–2. The Western Allies' discussion of partitioning and annexing portions of Germany, or pursuing spoils, did not occur until later in the war. Rothwell, *War Aims in the Second World War*, 72.

40 J.R.M. Butler, *Grand Strategy*, vol. 2: September 1939-June 1941 (London: Her Majesty's Stationery Office, 1957), 157–65; Gibbs, *Grand Strategy*, 1: 616–22.

41 Note that Soviet demands for a guarantee of a territorial buffer zone align with our expectation, outlined in Implication 3, that non-democracies are likely to prioritise spoils in their war aims. Gibbs, *Grand Strategy*, 1: 719–65; May, *Strange Victory*, 197–8.

42 Butler, *Grand Strategy*, 2: 20–1, 200–1; Churchill, *The Second World War*, 1949, 2: 116–17, 135–6; May, *Strange Victory*, 446.

43 May, *Strange Victory*, 477–8.

44 Butler, *Grand Strategy*, 2: 152; Horne, *To Lose a Battle*, 165.

45 Butler, *Grand Strategy*, 2: 189–90; Churchill, *The Second World War*, 1949, 2: 74–6; Horne, *To Lose a Battle*, 293, 437–9, 568–608.

46 Butler, *Grand Strategy*, 2: 165–71; Horne, *To Lose a Battle*, 277–8, 340, 462–3, 627.

47 Churchill, *The Second World War*, 1949, 2: 76–9; Horne, *To Lose a Battle*, 618–33, 651–2.

48 Horne, *To Lose a Battle*, 621.

49 Germany's pre-war GDP per capita was $3,648 while Austria-Hungary's was $1,986. Max-Stephan Schultze, 'Austria-Hungary's Economy in World War I,' in *The Economics of World War I*, ed. Stephen Broadberry and Mark Harrison (Cambridge: Cambridge University Press, 2005), 79.

50 German military expenditures were $23.2 billion in 1914 and $64.9 billion in 1915; its military numbered 862,000 in 1914 and 3.8 million in 1915. Singer and Small, *The Wages of War, 1816–1965*, 197; Schultze, 'Austria-Hungary's Economy in World War I,' 79.

51 Austria-Hungary's military expenditures were $13.6 billion in 1914 and $26 billion in 1915; its military grew from 839,000 people to 3.8 million. Singer and Small, *The Wages of War, 1816–1965*; Schultze, 'Austria-Hungary's Economy in World War I,' 79.

52 Fritz Fischer, *Germany's Aim in the First World War* (New York: Norton, 1967), 98.

53 Fischer, 101–13; Jack S. Levy, 'Preferences, Constraints, and Choices in July 1914,' *International Security* 15, no. 3 (Winter, 1990–1991): 151–86, https://doi.org/10.2307/2538910.

54 Herwig, *The First World War*, 47.

55 Roger Chickering, *Imperial Germany and the Great War, 1914–1918* (Cambridge: Cambridge University Press, 2014), 32–5.

56 Fischer, *Germany's Aim in the First World War*, 98.

57 Marvin B. Fried, *Austro-Hungarian Aims in the Balkans during World War I* (Basingstoke: Palgrave Macmillan, 2014), 8; Richard F. Hamilton and Holger H. Herwig, *Decisions for War, 1914–1917* (Cambridge: Cambridge University Press, 2004), 47.

58 Hamilton and Herwig, *Decisions for War, 1914–1917*, 40–1.

59 Bernadotte E. Schmitt, *The Coming of War 1914* (New York: Scribner, 1930), 368.

60 Samuel R. Williamson, *Austria-Hungary and the Origins of the First World War* (New York: St. Martin's, 1991), 16.

61 Williamson, 22.

62 Herwig, *The First World War*, 51–3.

63 Erich von Falkenhayn, *The German General Staff and Its Decisions, 1914–1916* (New York: Dodd, Mead, and Company, 1920), 20.
64 Herwig, *The First World War*, 52.
65 Falkenhayn, *The German General Staff and Its Decisions, 1914–1916*, 17.
66 Richard Lein, 'A Train Ride to Disaster: The Austro-Hungarian Eastern Front in 1914,' in *1914: Austria-Hungary, the Origins, and the First Year of World War I*, ed. Gunter Bischof, Ferdinand Karlhofer, and Samuel R. Williamson Jr. (New Orleans, LA: New Orleans University Press, 2014), 100–8; Graydon A. Tunstall, *Written in Blood: The Battles for Fortress Przemyl in WWI* (Bloomington, IN: Indiana University Press, 2016), 223–4.
67 Hamilton and Herwig, *Decisions for War, 1914–1917*, 136.
68 Falkenhayn, *The German General Staff and Its Decisions, 1914–1916*, 333; Lein, 'A Train Ride to Disaster,' 108, 112.
69 Falkenhayn, *The German General Staff and Its Decisions, 1914–1916*, 44.
70 Chickering, *Imperial Germany and the Great War, 1914–1918*, 53–4; Richard L. DiNardo, *Breakthrough: The Gorlice-Tarnów Campaign, 1915* (Santa Barbara, CA: Praeger, 2010), 21–48; Falkenhayn, *The German General Staff and Its Decisions, 1914–1916*, 24, 37.
71 Holger H. Herwig, 'Disjointed Allies: Coalition Warfare in Berlin and Vienna, 1914,' *Journal of Military History* 54, no. 3 (July 1990): 266.
72 The US's pre-war GDP per capita was $6,134; the UK's was $5,983 and China's was $778. Harrison, 'The Economics of World War II,' 3, 7.
73 The British figure excludes dominions and colonies and the Chinese figure excludes Japanese-controlled Manchuria. Harrison, 3, 7.
74 British military expenditures were $108 billion in 1942 and $123 billion in 1945; its military grew from approximately 4 million to more than 5 million during the same time. US military expenditures were $208.5 billion in 1942 and $653.5 billion in 1945. The American military grew from just under 4 million to almost 11.5 million during the Campaign. Singer and Small, *The Wages of War, 1816–1965*; Harrison, 'The Economics of World War II,' 3, 7, 14.
75 In 1942, China's defence expenditures were $495 million and its military numbered just over 3 million. In 1945, China invested $1.6 billion in its military, which by then numbered roughly 4.8 million. Singer and Small, *The Wages of War, 1816–1965*; Schultze, 'Austria-Hungary's Economy in World War I,' 7.
76 Winston Churchill, *The Second World War*, vol. 4: The Hinge of Fate (London: Cassell, 1951), 666; Albert C. Wedemeyer, *Wedemeyer Reports!* (New York: Henry Holt, 1958), 255.
77 Winston Churchill, *The Second World War*, vol. 5: Closing the Ring (London: Cassell, 1952), 493–96; Rothwell, *War Aims in the Second World War*, 206–8.
78 Rothwell, *War Aims in the Second World War*, 210–12.
79 Slim, *Defeat into Victory*, 441; Wedemeyer, *Wedemeyer Reports!*, 302–20.
80 Charles F. Romanus and Riley Sunderland, *United States Army in World War II: China-Burma-India Theater: Stilwell's Mission to China* (Washington, DC: United States Army Center of Military History, 1987), 172.
81 Charles F. Romanus and Riley Sunderland, *United States Army in World War II: China-Burma-India Theater: Stilwell's Command Problems* (Washington, DC: United States Army Center of Military History, 1987), 57–8.
82 Wedemeyer, *Wedemeyer Reports!*, 252.
83 Slim, *Defeat into Victory*, 236–7.
84 Churchill, *The Second World War*, 1952, 5: 502; Slim, *Defeat into Victory*, 507–8.
85 Slim, *Defeat into Victory*, 256–83.
86 Slim, 284–321.

87 Cappella Zielinski and Grauer, 'A Century of Coalitions in Battle,' Chapter 2, this volume.

Bibliography

Becker, Jordan, and Edmund Malesky. 'The Continent or the "Grand Large"? Strategic Culture and Operational Burden-Sharing in NATO.' *International Studies Quarterly* 61, no. 1 (March 2017): 163–80. 10.1093/isq/sqw039

Bennett, Andrew, Joseph Lepgold, and Danny Unger. 'Burden-Sharing in the Persian Gulf War.' *International Organization* 48, no. 1 (Winter 1994): 39–75. 10.1017/S0020818300000813

Bensahel, Nora. 'International Alliances and Military Effectiveness: Fighting Alongside Allies and Partners.' In *Creating Military Power: The Sources of Military Effectiveness*, edited by Risa A. Brooks and Elizabeth A. Stanley, 186–206. Stanford, CA: Stanford University Press, 2007.

Brooks, Risa. 'An Autocracy at War: Explaining Egypt's Military Effectiveness, 1967 and 1973.' *Security Studies* 15, no. 3 (September 2006): 396–430. 10.1080/09636410601028321

Bueno de Mesquita, Bruce, Alastair Smith, Randolph M. Siverson, and James D. Morrow. *The Logic of Political Survival*. Cambridge, MA: MIT Press, 2003.

Butler, J.R.M. *Grand Strategy*. Vol. 2: September 1939–June 1941. London: Her Majesty's Stationery Office, 1957.

Cappella Zielinski, Rosella, and Ryan Grauer. 'Organizing for Performance: Coalition Effectiveness on the Battlefield.' *European Journal of International Relations* 26, no. 4 (December 2020): 953–78. 10.1177/1354066120903369

Chickering, Roger. *Imperial Germany and the Great War, 1914–1918*. Cambridge: Cambridge University Press, 2014.

Choi, Ajin. 'Fighting to the Finish: Democracy and Commitment in Coalition War.' *Security Studies* 21, no. 4 (October 2012): 624–53. 10.1080/09636412.2012.734232

Choi, Ajin. 'The Power of Democratic Cooperation.' *International Security* 28, no. 1 (Summer 2003): 142–53. https://www.jstor.org/stable/4137578.

Churchill, Winston. *The Second World War*. Vol. 2: Their Finest Hour. London: Cassell, 1949.

Churchill, Winston. *The Second World War*. Vol. 4: The Hinge of Fate. London: Cassell, 1951.

Churchill, Winston. *The Second World War*. Vol. 5: Closing the Ring. London: Cassell, 1952.

Clodfelter, Michael. *Warfare and Armed Conflicts*. 3rd ed. Jefferson, NC: McFarland & Company, 2007.

Conybeare, John A. C., and Todd Sandler. 'The Triple Entente and the Triple Alliance 1880–1914: A Collective Goods Approach.' *The American Political Science Review* 84, no. 4 (December 1990): 1197–206. 10.2307/1963259

Craig, Gordon A. 'The World War I Alliance of the Central Powers in Retrospect: The Military Cohesion of the Alliance.' *The Journal of Modern History* 37, no. 3 (September 1965): 336–44. https://www.jstor.org/stable/1875406.

Desch, Michael C. *Power and Military Effectiveness: The Fallacy of Democratic Triumphalism*. Baltimore: Johns Hopkins University Press, 2008.

DiNardo, Richard L. *Breakthrough: The Gorlice-Tarnów Campaign, 1915*. Santa Barbara, CA: Praeger, 2010.

Falkenhayn, Erich von. *The German General Staff and Its Decisions, 1914–1916*. New York: Dodd, Mead, and Company, 1920.

Fang, Songying, and Kristopher W. Ramsay. 'Outside Options and Burden Sharing in Nonbinding Alliances.' *Political Research Quarterly* 63, no. 1 (March 2010): 188–202. 10.1177/1065912908327528

Fischer, Fritz. *Germany's Aim in the First World War*. New York: Norton, 1967.

Fried, Marvin B. *Austro-Hungarian Aims in the Balkans during World War I*. Basingstoke: Palgrave Macmillan, 2014.

Gartzke, Erik, and Kristian Skrede Gleditsch. 'Why Democracies May Actually Be Less Reliable Allies.' *American Journal of Political Science* 48, no. 4 (October 2004): 775–95. 10.1111/j.0092-5853.2004.00101.x

Gibbs, N.H. *Grand Strategy*. Vol. 1: Rearmament Policy. London: Her Majesty's Stationery Office, 1976.

Goemans, Hein, and William Spaniel. 'Multimethod Research: A Case for Formal Theory.' *Security Studies* 25, no. 1 (January 2016): 25–33. 10.1080/09636412.2016.1134176

Graham, Benjamin A. T., Erik Gartzke, and Christopher J. Fariss. 'The Bar Fight Theory of International Conflict: Regime Type, Coalition Size, and Victory.' *Political Science Research and Methods* 5, no. 4 (October 2017): 613–39. 10.1017/psrm.2015.52

Grauer, Ryan. *Commanding Military Power: Organizing for Victory and Defeat on the Battlefield*. New York: Cambridge University Press, 2016.

Greenhalgh, Elizabeth. *The French Army and the First World War*. Cambridge: Cambridge University Press, 2014.

Hamilton, Richard F., and Holger H. Herwig. *Decisions for War, 1914–1917*. Cambridge: Cambridge University Press, 2004.

Harrison, Mark. 'The Economics of World War II: An Overview.' In *The Economics of World War II: Six Great Powers in International Comparison*, edited by Mark Harrison, 1–42. Cambridge: Cambridge University Press, 1998.

Herwig, Holger H. 'Disjointed Allies: Coalition Warfare in Berlin and Vienna, 1914.' *Journal of Military History* 54, no. 3 (July 1990): 265–80.

Herwig, Holger H. *The First World War: Germany and Austria-Hungary 1914–1918*. New York: Bloomsbury, 2014.

Horne, Alistair. *To Lose a Battle: France 1940*. New York: Penguin, 2007.

Kreps, Sarah E. *Coalitions of Convenience: United States Military Interventions after the Cold War*. New York: Oxford University Press, 2011.

Lake, David A. 'Powerful Pacifists: Democratic States and War.' *American Political Science Review* 86, no. 1 (March 1992): 24–37. 10.2307/1964013

Leeds, Brett Ashley. 'Alliance Reliability in Times of War: Explaining State Decisions to Violate Treaties.' *International Organization* 57, no. 4 (September 2003): 801–27. 10.1017/S0020818303574057

Lein, Richard. 'A Train Ride to Disaster: The Austro-Hungarian Eastern Front in 1914.' In *1914: Austria-Hungary, the Origins, and the First Year of World War I*, edited by Gunter Bischof, Ferdinand Karlhofer, and Samuel R. Williamson Jr., 95–126. New Orleans, LA: New Orleans University Press, 2014.

Levy, Jack S. 'Preferences, Constraints, and Choices in July 1914.' *International Security* 15, no. 3 (Winter, 1990–1991): 151–86. 10.2307/2538910

Long, Stephen B. 'A Winning Proposition? States' Military Effectiveness and the Reliability of Their Allies.' *International Politics* 52, no. 3 (2015): 335–48. 10.1057/ip.2015.8

Lorentzen, Peter, M Taylor Fravel, and Jack Paine. 'Qualitative Investigation of Theoretical Models: The Value of Process Tracing.' *Journal of Theoretical Politics* 29, no. 3 (July 2017): 467–91. 10.1177/0951629816664420

Maurice, Sir Frederick. *Lessons Of Allied Co-Operation; Naval, Military, And Air, 1914–1918*. London: Oxford University Press, 1942.

May, Ernest R. *Strange Victory: Hitler's Conquest of France*. New York: Hill and Wang, 2000.

Mello, Patrick A. 'National Restrictions in Multinational Military Operations: A Conceptual Framework.' *Contemporary Security Policy* 40, no. 1 (January 2019): 38–55. 10.1080/13523260.2018.1503438

Mello, Patrick A. 'Paths Towards Coalition Defection: Democracies and Withdrawal from the Iraq War.' *European Journal of International Security* 5, no. 1 (February 2020): 45–76. 10.1017/eis.2019.10

Moller, Sara. 'Fighting Friends: Institutional Cooperation and Military Effectiveness in Multinational War.' PhD thesis, Columbia University, 2016.

Morrow, James D., Bruce Bueno de Mesquita, Randolph M. Siverson, and Alastair Smith. 'Selection Institutions and War Aims.' *Economics of Governance* 7, no. 1 (January 2006): 31–52. 10.1007/s10101-005-0108-z

Olson, Mancur. *The Logic of Collective Action*. Cambridge, MA: Harvard University Press, 1971.

Olson, Mancur, and Richard Zeckhauser. 'An Economic Theory of Alliances.' *The Review of Economics and Statistics* 48, no. 3 (1966): 266–79. 10.2307/1927082

Olson, Mancur, and Richard Zeckhauser. 'Collective Goods, Comparative Advantage, and Alliance Efficiency.' In *Issues of Defense Economics*, edited by Roland N. McKean, 25–63. New York: NBER, 1967.

Prete, Roy A. *Strategy and Command: The Anglo-French Coalition on the Western Front, 1914*. Montréal Québec: McGill-Queen's University Press, 2009.

Ramsay, Kristopher W. 'Settling It on the Field: Battlefield Events and War Termination.' *Journal of Conflict Resolution* 52, no. 6 (December 1, 2008): 850–79. 10.1177/0022002708324593

Reiter, Dan, and Allan C. Stam. *Democracies at War*. Princeton: Princeton University Press, 2002.

Resnick, Evan N. 'Hang Together or Hang Separately? Evaluating Rival Theories of Wartime Alliance Cohesion.' *Security Studies* 22, no. 4 (October 2013): 672–706. 10.1080/09636412.2013.844520

Riker, William H. *The Theory of Political Coalitions*. New Haven, CT: Yale University Press, 1962.

Romanus, Charles F., and Riley Sunderland. *United States Army in World War II: China-Burma-India Theater: Stilwell's Command Problems*. Washington, DC: United States Army Center of Military History, 1987.

Romanus, Charles F. , and Riley, Sunderland. *United States Army in World War II: China-Burma-India Theater: Stilwell's Mission to China*. Washington, DC: United States Army Center of Military History, 1987.

Rommel, Erwin, and Sir Basil H. Liddell Hart. *The Rommel Papers*. New York: Da Capo Press, 1953.

Rothwell, Victor. *War Aims in the Second World War: The War Aims of the Major Belligerents, 1939–1945*. Edinburgh: Edinburgh University Press, 2005.

Schmitt, Bernadotte E. *The Coming of War 1914*. New York: Scribner, 1930.

Schultze, Max-Stephan. 'Austria-Hungary's Economy in World War I.' In *The Economics of World War I*, edited by Stephen Broadberry and Mark Harrison, 77–111. Cambridge: Cambridge University Press, 2005.

Shimizu, Hirofumi, and Todd Sandler. 'Peacekeeping and Burden-Sharing, 1994–2000.' *Journal of Peace Research* 39, no. 6 (November 2002): 651–68. 10.1177/0022343302039006001

Singer, J. David, and Melvin Small. *The Wages of War, 1816–1965: A Statistical Handbook*. New York: John Wiley & Sons, Inc., 1972.

Slim, William. *Defeat into Victory*. London: Pan Books, 2009.

Tago, Atsushi. 'When Are Democratic Friends Unreliable? The Unilateral Withdrawal of Troops from the "Coalition of the Willing".' *Journal of Peace Research* 46, no. 2 (March 2009): 219–34. 10.1177/0022343308100716

Talmadge, Caitlin. *The Dictator's Army: Battlefield Effectiveness in Authoritarian Regimes*. Ithaca, NY: Cornell University Press, 2015.

Tunstall, Graydon A. *Written in Blood: The Battles for Fortress Przemyl in WWI*. Bloomington, IN: Indiana University Press, 2016.

Wedemeyer, Albert C. *Wedemeyer Reports!* New York: Henry Holt, 1958.

Weeks, Jessica L. P. *Dictators at War and Peace*. Ithaca, NY: Cornell University Press, 2014.

Weisiger, Alex. 'Exiting the Coalition: When Do States Abandon Coalition Partners during War?' *International Studies Quarterly* 60, no. 4 (December 2016): 753–65. 10.1093/isq/sqw029

Weitsman, Patricia. *Dangerous Alliances: Proponents of Peace, Weapons of War*. Stanford, CA: Stanford University Press, 2003.

Weitsman, Patricia. *Waging War: Alliances, Coalitions, and Institutions of Interstate Violence*. Stanford, CA: Stanford University Press, 2013.

Wilkins, Thomas Stow. 'Analysing Coalition Warfare from an Intra-Alliance Politics Perspective: The Normandy Campaign 1944.' *Journal of Strategic Studies* 29, no. 6 (December 2006): 1121–50. 10.1080/01402390601016592

Williamson, Samuel R. *Austria-Hungary and the Origins of the First World War*. New York: St. Martin's, 1991.

Wolford, Scott. *The Politics of Military Coalitions*. New York: Cambridge University Press, 2015.

9 Why Rebels Rely on Terrorists

The Persistence of the Taliban-al-Qaeda Battlefield Coalition in Afghanistan

Barbara Elias

Collaborations among rebels and terrorists have become increasingly common.[1] Branches of al-Qaeda have teamed up with insurgent rebels in Yemen, Afghanistan, Somalia, and Mali; the IRA joined military operations with Basque separatists (ETA) and FARC rebels in Colombia; and Somali Islamic Courts Union (ICU) units have fought alongside Hezbollah.[2] Some of these collaborations are persistent, with rebels and terrorists continuing to form battlefield coalitions over time. For rebels, partnering with terrorists can have significant advantages in terms of military and political benefits. Such partnerships are also potentially costly, however, as such couplings can come with reputational risks and are infused with uncertainty stemming from the unpredictability of terrorists.

This chapter investigates the question of why some rebels and terrorist groups persist in forming battlefield coalitions over time while others do not. Following the first chapter of this volume, 'Understanding Battlefield Coalitions' by Rosella Cappella Zielinski and Ryan Grauer, I define battlefield coalitions as collaborations among distinct armed groups that jointly wage combat in a single battlespace. Rebel groups are organisations that, as Bruce Hoffman explains, seek to 'seize and hold territory (even if only ephemerally during daylight hours), while also exercising some form of sovereignty or control over a defined geographical area and its population.'[3] Terrorists, by contrast, are politically driven organised groups that use violence to promote their political agendas but generally eschew territorial control.[4] In short, rebels seek to conquer and govern a local area, whereas terrorists have alternative, typically more radical – and often globalised – political objectives.

Battlefield coalitions formed by rebels and terrorists are often more temporary and fluid than those formed by state-based forces. Some sets of these actors, however, persist in forming battlefield coalition after battlefield coalition during a conflict. I argue that such persistence is most likely under three conditions: 1) the rebels and terrorists have non-rival, distinct overall political agendas; 2) the terrorist group is well-established and proficient, contributing military expertise, organisational capacity, and

DOI: 10.4324/9781003399896-13

political assistance in complex conflicts; and 3) the terrorist group's participation is locally institutionalised or maintained through the threat of violence to be directed at the rebels in the event the partnership disbands. These factors, collectively, ameliorate important costs and accentuate particular benefits for rebels and thus make continuing relationships more attractive.

I explore my theory of the persistence of rebel–terrorist battlefield coalitions in the context of the Taliban's partnership with al-Qaeda in Afghanistan, a pivotal case that looms large over debates about such collectives. US policymakers relentlessly and unsuccessfully pressured the Taliban to sever their ties with al-Qaeda for decades, finally withdrawing American troops in 2021 without concrete assurances that the two would cease working together in Afghanistan.[5] Relying on previously classified or captured primary source documents from rebel, terrorist, and state government officials that address rebel–terrorist coalition dynamics, my analysis demonstrates that a factor US policymakers often thought to be a boon in driving a wedge between the Taliban and al-Qaeda – their differing political objectives – actually worked to bind the two groups together. In addition, al-Qaeda provided the Taliban with military, organisational, and political advantages it would have otherwise lacked, and the relationship was reinforced by both the early institutionalisation of cooperation and al-Qaeda's latent threat of violence to be directed against the Taliban if it were abandoned. The case offers strong support for my theory.

The argument and findings presented in this chapter represent an important advance in our effort to understand how and why different types of non-state actors fight together on the battlefield. First, to date, few have focused directly on these questions. While there are important studies of 1) state sponsors of terrorism,[6] 2) battlefield coalitions among rebel groups,[7] and 3) coalitions among terrorist groups,[8] there are almost no studies examining rebel–terrorist battlefield coalitions as a unique form of non-state actor coalition.[9] This is a curious gap in our understanding since these types of partnerships have distinctive political and military dynamics, and their structure and coherence have profound consequences for counterterrorism. Second, while there are many different types of support provided between non-state actors, by analysing the motivations for some groups to repeatedly engage in joint operations in a shared battlespace, this analysis focuses on a limited set of tough cases in counterterrorism, where established terrorist organisations like al-Qaeda and Hezbollah become institutionalised in locations such as Afghanistan, Somalia, Lebanon, and Syria. Understanding the logic of those entrenched rebel–terrorist battlefield coalitions offers potential insights about how well-known terrorist groups maintain their relevance and embed themselves in local conflicts.

9.1 'I Don't Give a Damn About My Bad Reputation' – The Costs and Benefits of Collaborating with Terrorists

Terrorists are not often regarded as potential partners for actors seeking to advance their objectives in international politics. Some states and non-state actors work with terrorists, however, when the costs – which are often substantial – are outweighed by the benefits. Delineating the costs and benefits of partnership with terrorists thus begins to shed light on why battlefield coalitions including terrorists not only form, but sometimes persist.

State governments face substantial costs for maintaining coalitions with terrorist groups. As US Secretary of State Hillary Clinton commented about Pakistan's support for terrorist groups, 'You can't keep snakes in your backyard and expect them only to bite your neighbors. You know, eventually those snakes are going to turn on whoever has them in the backyard.'[10] Terrorist clients are unpredictable and known for using threats against their enemies, allies, and patrons alike. States face international condemnation and economic sanctions for maintaining ties to terrorists, raising the costs of keeping terrorists as partners. Yet, despite the significant drawbacks and risks, states such as Iran, Sudan, Syria, Libya, and Pakistan have nevertheless persistently formed battlefield coalitions with recognised terrorist organisations.[11]

Scholars have identified two key reasons why certain states are driven towards terrorists as coalitional partners, despite the costs. The first is that forming such partnerships provides a new avenue along which relatively weak states can apply pressure to powerful adversaries. Direct confrontation risks conventional warfare against a powerful adversary.[12] Terrorist coalition partners can help weak states challenge the regional status quo that keeps weak states at a disadvantage while minimising the risk to their state forces.[13] Relatively weak states may opt to employ terrorists as proxies in order to balance or challenge other states, such as Pakistan's support for Kashmiri terrorist groups as a way to pressure India. According to Daniel Byman, 'terrorists offer another means for states to influence their neighbors, topple a hostile regime, counter US hegemony, or achieve other aims ...'[14] As such, states like Iran or Pakistan calculate that foreign terrorist groups are relatively inexpensive to sponsor, with the benefits outweighing the risks.

Second, and in combination with the first benefit, relatively weak states gain from the degree of deniability that comes with shifting the blame to terrorist partners, which helps mitigate reputational costs and the likelihood of retaliation and escalating conflict.[15] In discussing Iran's reliance on Hezbollah, Afshon Ostovar argues that, despite the economic hardships from international sanctions and persistent insecurity within its own borders invited by partnering with terrorists, 'Iran successfully developed a network of clients that fought to undermine US influence ... [and] has been able to advance its core strategic objectives primarily through its cultivation of militant clients in the Middle East.'[16] In the case of Iran, coalitions with terrorist partners can endure international pressure and serve as a tool for resource-deficient actors to hedge their bets against an uncertain future and powerful adversaries.

In contrast to state sponsorship of terrorist groups, there is little scholarship analysing the motivations for non-state actors to engage with terrorists.[17] Relying on the insights of scholarship on state-terrorist battlefield coalitions, however, yields some clues. First, the costs for non-state actors relying on terrorists as coalition partners are potentially even more significant than the risks for state governments in terms of losing international support, funding, and local legitimacy. Regarding rebel reliance on al-Qaeda, officials in Mogadishu, for example, told US diplomats in 2009 'that widespread reporting in Somalia on the prominent role of foreign fighters in al-Shabaab's offensive had prompted a backlash in public opinion against al-Shabaab.'[18] Rebels risk losing popular support, as well as creating institutionalised dependencies on terrorist partners, and being pushed into extremism by committing to terrorist agendas.

Second, and also similar to states, non-state actors can opt to form coalitions with foreign terrorist groups as a way to counter powerful adversaries, such as Hezbollah's military alliance with the Islamic Courts Union against Israel.[19] Non-state groups may form battlefield coalitions with terrorist organisations to challenge other local state and non-state groups, expand their capacity for violence, and signal resolve.[20] Relying on foreign terrorist organisations is a strategy adopted by weak actors when they have few other options, whether they are states or rebels.[21] Maintaining a coalition with a foreign militant group is a calculated 'weapon of the weak,' a strategic decision for relatively resource-poor actors. Partnering with terrorists may not lead to definitive military victory, but embracing terrorists raises the costs of opposing the coalition's cause.

Considering the costs and benefits of non-state actors partnering with terrorists, it is possible to understand why, sometimes, such groups might form. What is not obvious, however, is what conditions would cause the benefits of forming such battlefield coalitions to consistently outweigh the costs of doing so over time. It is to that task that I now turn.

9.2 Desperate Times, Desperate Measures: Why Some Rebel–Terrorist Coalitions Persist

I argue that local rebels are likely to view the benefits of forming battlefield coalitions with foreign terrorists and persist in creating such collectives over time if: 1) they are non-rival, defined as maintaining distinct overall political agendas; 2) the terrorist group is well-established and proficient; and 3) the terrorist group is institutionally embedded in a variety of local contexts and is a potential threat to their rebel partners if they are not allied, which raises the cost of rebel defection from the coalition.

9.2.1 Agreeing to Disagree: Non-Rival Agendas and Enduring Rebel–Terrorist Coalitions

While there is extensive scholarship on the coherence of non-state actor battlefield coalitions in civil wars, most work focuses on fragmentation and coherence among various competing local rebel groups, as opposed to local rebels and foreign terrorists.[22] Fotini Christia, for example, argues that warring non-state rebel groups will choose to switch sides in forming and reforming coalitions, aiming 'to side with the winner, so long as they can have a credible guarantee that the winner will not strip them of power once victory is accomplished.'[23] This scholarship focuses on the strategic behaviour of rival groups vying for local control in highly uncertain environments, prioritising survival and autonomy. The logic for battlefield coalitions between local rebel and foreign terrorist groups differs from these dynamics along one key dimension: unlike local insurgents, foreign terrorist groups are not typically interested in local governance and state building.[24] Because foreign extremists are not competing to solidify their position as the local governing authority, they avoid the balancing behaviours embedded in rebel coalitions. Terrorist and rebel coalitions are also distinct from many terrorist alliances, which are often rival and short-lived.[25] As Bakke, Gallagher Cunningham, and Seymour detail about non-state battlefield coalitions, 'there is great diversity in the ways movements can be internally divided and in the implications this variation has on how conflict unfolds.'[26]

Competition among local rebel groups can be fierce in civil wars. Accordingly, battlefield coalitions comprised solely of rebels or similarly rival groups are often fragile. As Tricia Bacon explains, coalitions consisting of rival organisations feature partners that 'compete in the same political market,' and their partnerships are formed 'usually cautiously and temporarily.'[27] Side-switching is common, and the resultant relative fluidity between rival groups is evident in Yemen, Sudan, Bosnia, and Afghanistan.[28] Reporting on the US war there, journalist Dexter Filkins emphasises this point, explaining that 'people fought in Afghanistan, and people died, but not always in obvious ways. They had been fighting for so long, twenty-three years then, that by the time the Americans arrived the Afghans had developed an elaborate set of rules designed to spare as many fighters as they could. So the war could go on forever. Men fought, men switched sides, men lined up and fought again.'[29] A failure to maintain such fluidity was high, increasing the likelihood of casualties and groups being defeated and eliminated. Foreign groups, on the other hand, typically do not vie for local political control. Accordingly, when they join battlefield coalitions, they are not subject to the same pressures to switch allegiances or be subsumed. Foreign terrorists are thus often attractive as potential partners: so long as their long-term political goals differ from those of the local actors, they can offer support to rebels without rivalry.

Distinct and non-rival overarching political goals empower coalition partners to look for ways to collectively achieve shared strategic interests without concerns that such actions will compromise their autonomy or jeopardise their primary interests. While there are reputational risks for maintaining a coalition with terrorists, including international condemnation and potential liability for civilian casualties, working with terrorists can also create political opportunities to bolster rebel reputation and credibility. Rebels and states alike can view terrorists as groups that can be used to further their strategic interests and boost their credibility without competing as peers or rivals in their state-building efforts. As Daniel Byman explains regarding the political benefits of state-sponsored terrorism for Syria, 'Hafez al-Asad also helped a range of Palestinian groups in order to demonstrate his Arab nationalist bona fides.'[30] Gaining credibility as a steadfast supporter of a cause by partnering with motivated extremists can offer reputational benefits to both terrorists and rebels without the complications of competing with one another for the same political ends.

9.2.2 *Bring Your Own Bombs: Terrorist Proficiency and Enduring Rebel–Terrorist Coalitions*

Established terrorist organisations can bring military, organisational, and political benefits to coalitions. Practised terrorists are often more skilled and better equipped than the local rebels they are embedded with. In Somalia in 2008, for example, al-Shabab rebels were able to take control of the Middle Shabelle capital of Jowhar in part due to embedded 'well trained and equipped' al-Qaeda operatives.[31] Foreign terrorists can augment the military effectiveness and lethality of operations, in particular established, global, and networked groups like al-Qaeda.[32] Terrorist groups also often bring specialised military skills, such as bomb-making, communications, organisational skills, and tactical knowledge, such as suicide bombing.[33] Since most insurgencies fail, maintaining a battlefield coalition with recognised and experienced foreign terrorists can create opportunities to learn new skills for survival. As the US Department of State reported in 2009 regarding al-Qaeda's alliance with al-Shabab, 'suicide bombings, once unheard of in Somalia, have become increasingly routine as exemplified by the October 29, 2008, multiple simultaneous attacks (an al-Qaeda trademark) ... [rebel training facilities] have come to look increasingly like those run by al-Qaeda. They have been using foreign instructors, making use of similar training aids, physical training formations/activities, explosives and special tactics training undergone by al-Qaeda trainees elsewhere. The regimen ... differs markedly from that traditionally used in Somali camps, and seems to produce better results.'[34] Embedded terrorist commanders can train rebels and make local forces more militarily effective and resilient, actions that enable more effective collective operations.

Terrorist groups can also provide vital organisational capacity and political heft to coalitions. Because widely recognised terrorist groups like al-Qaeda are networked, they are able to connect local rebels to a web of potential fighters that can augment rebel military operations and offer new sources of external support. Organising, training, addressing collective action problems, and handling logistics significantly help coalitional performance on shared battlefields. Moreover, partnering with an established terrorist group provides leverage in negotiations with state and non-state adversaries, because it augments a rebel group's military reputation, raising the spectre of future violence.

These military, organisational, and political benefits provided by capable terrorists can augment and support successes in other areas. For example, military successes secured by the coalition can bolster rebel and terrorist group reputations, which can lead to increased defections from rival groups and serve as a potent recruiting tool, advertising battlefield glories and comradery that feed into future military gains.[35] As Harleen Gambhir writes regarding the interaction between military victory, ideological reputation, legitimacy, and recruitment for the Islamic State, 'the legitimacy of the [ISIS] Caliphate hangs on military victory and consolidation success, as proof of God's approval,' which ISIS can point to in its recruitment efforts that work to expand their military ranks further.[36] Battlefield victory reinforces a reputation for competence and likely success, which pulls the coalition together, minimising the costs and augmenting the strategic benefits for maintaining the partnership.

9.2.3 Local Institutionalised Accommodations: Implicit Threats of Terrorist Retaliation and Enduring Rebel–Terrorist Coalitions

Dependencies can institutionalise over time as rebel groups rely on foreign terrorists in collective warfighting. Because of the complex and competitive battlespaces of civil wars, there is significant variation in the ways rebel and terrorist coalition members accommodate each other across geography and time within a war. This suggests that, while dividing coalition members in one region or at one time may be feasible, dividing the coalition in another location or at another time may be more complicated. In her work on civil wars, Ana Arojna cautions that localised accommodations are chronically overlooked, yet macro-level processes, like conflict duration and outcome, are shaped by localised arrangements that inevitably emerge in long and complex wars.[37] 'Despite common depictions of war as chaotic and anarchic, order often emerges locally. Institutions vary greatly over time and space ...'[38] The structure of battlefield coalitions between rebels and terrorist groups varies widely within wars and can become institutionalised in diverse ways. This heterogeneity in accommodations strengthens these battlefield coalitions by making them dynamic and institutionalised. Members of terrorist groups embedded with rebels can leverage diverse localised arrangements by making it harder for rebel commanders to issue sweeping commands regarding coalition arrangements. Context-specific forms of local accommodation create a web of diverse relationships that are difficult to untangle.

This dynamic also suggests that local rebels may opt to maintain a coalition with international terrorist groups in part because they are partially captured by that terrorist organisation, which affects the cost-benefit analysis for maintaining the coalition. The costs of defection, including the possibility that terrorist partners could turn into terrorist adversaries, provide incentives to maintain existing coalitions. According to Steve Coll, Pakistani intelligence officials were pushed into maintaining ties to terrorists and supporting ever-increasing extremist activities in order 'to prove to its own restive clients that it was not going soft and that it should not be considered the enemy.'[39] There is a motivation for groups to preserve existing battlefield coalitions because the risks of blowback and retaliation for ending such coalitions can be higher than the costs of continued collaboration. Terrorists can also potentially threaten to switch sides and support alternative rebels, effectively further motivating rebels to preserve the coalition and avoid augmenting their adversaries.

While there are significant costs for maintaining ties, I argue that local rebels and foreign terrorists are more likely to maintain battlefield coalitions if: 1) they are non-rival; 2) the terrorist group is proficient, offering military, organisational, and political benefits; and 3) accommodations within the coalition are institutionalised locally and empower terrorists to implicitly threaten local rebel partners should the coalition dissolve.

9.3 The al-Qaeda–Taliban Coalition in Afghanistan

Despite enormous international pressure, including a $2.261 trillion investment by the United States in countering their combined efforts, the Taliban and al-Qaeda have persisted in forming battlefield coalitions over time.[40] In this section, I analyse my theory of rebel–terrorist persistence in such activities in Afghanistan from 1989 to 2021, relying on primary source documents from the Taliban, al-Qaeda, and the US to evaluate the coalition. While this case has received a lot of focus from academics and policymakers alike, the record on this coalition is far from complete.[41] Similar to Strick van Linschoten and Kuehn, my analysis advances our understanding of these groups and their partnership, with the caveat that, 'while the names "Taliban" and "al-Qaeda" are useful as general organizing principles, the networks and individuals associated with the two blocs are more significant as a means of understanding the precise interactions between the two, and, more often than not, evade clear categorization.'[42] As Stathis Kalyvas notes regarding mapping distinct groups in civil wars, 'positing coherent, identifiable political groups with clear preferences fails to match the vast complexity, fluidity, and ambiguity one encounters on the ground.'[43] As such,

I draw cautious conclusions from the available information and make note of potential weaknesses in the record. Afghanistan is a pivotal example of conflict driven by a persistent terrorist–rebel coalition and provides crucial data on coalition dynamics.

9.3.1 The Taliban and al-Qaeda's Non-Rival Agendas

At times during the five decades of civil war in Afghanistan, the marketplace for rebel groups was remarkably crowded. From 1978 to 1998, rivals such as Hizb-i-Islami, Ittihad-i-Islami, Jamiat-i-Islami, Junbish-i-Milli, Hizb-i-Wahdat, Shura-yi-Nezar, the Northern Alliance, the Taliban, and others all competed for power and control over the country. These groups entered and exited battlefield coalitions with one another perennially in an effort to survive and gain influence, engaging in complex balancing arrangements.[44] Also involved, at least on the periphery, were foreign fighters who were attracted to Afghanistan to gain combat experience and connect to extremist networks.

Unlike Afghan rebel groups, foreign fighters like al-Qaeda were not driven to control territory and govern Afghanistan, instead seeking to challenge 'apostate' dictatorships in the Middle East and North Africa and to challenge the United States for supporting those regimes.[45] In contrast, the Taliban as a local rebel group remained an avowed Afghan nationalist movement.[46] These political agendas have not changed significantly since the late 1980s. When asked about the Taliban–al-Qaeda coalition in 2007, for example, Afghan Taliban leader Mullah Mohammad Omar stated, 'We have never felt the need for a permanent relationship in the present circumstances. However, they [al-Qaida] have set jihad as their goal while we have set the expulsion of American troops from Afghanistan as our target. This is the common goal of all Muslims.'[47] These distinct, non-rival, yet complementary agendas allow al-Qaeda and the Taliban to view one another as collaborators as opposed to competitors. As Anne Stenersen explains regarding the Taliban's 1996 decision to permit al-Qaeda to operate in Afghanistan, 'Bin Laden's men were not involved in the civil war and therefore posed no immediate threat to the Taliban's conquest for power within Afghanistan.'[48] While tensions exist, a cooperative, dynamic relationship binds the coalition together on the battlefield. As US Central Command assessed in 2010, while there are possible fissures, 'the relationship between al-Qaida (AQ) and Taliban (TB) remains relatively strong ... . The two organizations maintain a mutually beneficial relationship characterized by tactical-level cooperation between AQ operatives and TB commanders in Afghanistan ... . The TB views itself as a legitimate insurgency trying to displace what they view as a corrupt and western led Afghan government with a principled Islamic one. AQ has a more self-centered goal desiring the Taliban to regain power in order to once again enjoy Afghanistan as a sanctuary. Below the senior level, both groups continue to operate together on training and tactical missions in

Afghanistan.'[49] The Taliban and al-Qaeda's divergent overarching political priorities allow the coalition to avoid the costs associated with rivalry while maintaining the benefits of working towards mutually beneficial arrangements.

While the Afghan Taliban and al-Qaeda have distinct overarching political goals, they share several operational priorities, including minimising American influence in the region and managing Pakistan as a patron. While not ascribing to identical ideologies (al-Qaeda's intellectual traditions stem from Saudi Arabia and Wahhabi Sunni Islam, while the Taliban are Indian-originated Deobandi Sunni Islamic), they nonetheless have a similar brand of uncompromising political Salafism and regularly describe their coalition as bound by ideological brotherhood and jihad. As bin Laden wrote in a 1999 letter captured by Americans, 'The position of the leader of the faithful [Mullah Omar] is too hard to comprehend by many who tailor their international relations on earthly interests. Otherwise, what motivates a poor country like Afghanistan to confront America at this age in which the politicians and leaders of the world rush to get close to it and seek its friendship? The reason why those (*sic*) find it hard to understand this stance is because they do not understand the secret of the influence of faith on the hearts of the faithful … we would like to reiterate our gratitude and our support to Mullah Omar and his [political/religious] positions which have revived faith in our hearts, strengthened our minds, and restored hope to the [Islamic] nation which was on the verge of collapse.'[50] While al-Qaeda benefits from its longstanding coalition with the Taliban by having access to frontlines and safe havens, expanding its credibility and experience, the Taliban's reputation is also bolstered by its coalition with al-Qaeda by demonstrating a willingness to withstand pressure from the US in an act of solidarity with others operating under a banner of political Islam. As Strick Van Linschoten and Kuehn argue, despite being pressured by international actors to surrender bin Laden, the Taliban nonetheless remained firm, 'they sought above all else to preserve the reputation that they believed they had in the Islamic world.'[51] Compromising on the coalition would pose political risks, potentially undermining the Taliban's political and military reputation.[52] As non-rival yet ideologically linked organisations, the Taliban and al-Qaeda are able to reinforce one another militarily and politically without competing.

9.3.2 Al-Qaeda: Proficient Terrorists and Enduring Battlefield Coalition Partners

Al-Qaeda offered the Taliban military, organisational, and political benefits, which made the coalition valuable for the latter, despite the costs. Al-Qaeda's logistical capacity and military expertise were particularly attractive. Al-Qaeda units did not serve as independently potent operational military forces, yet al-Qaeda forces were instrumental as reliable support for Taliban units on the frontlines, as well as recruiting, training, and logistics.[53]

Nigel Inkster, former Director for Operations and Intelligence for British Secret Intelligence Service, explained that al-Qaeda does not have the forces to 'do real operations in Afghanistan ... the most it can hope to be is a kind of force multiplier for the other entities that are already there. In that context it has been ready to provide some support and assistance to Afghan Taliban units--weapons training, some material assistance. Also, some of the foreign fighters are experienced fighters and know quite a lot about military tactics. So there has been, at the tactical level, a certain of amount of cooperation' with the Afghan Taliban.[54]

In recognition of their reputation as strong fighters, the Taliban incorporated al-Qaeda members on the frontlines at key moments to augment their forces.[55] Stenersen argues that this frontline participation was important for al-Qaeda, who saw Afghan battlefields as important proving grounds to harden, vet, and test members.[56] Yet, even though 'al-Qaida sent some of their most experienced commanders to the Taliban's frontline, it was still a drop in the ocean' of the civil war dominated by Afghan fighters.[57] An intelligence assessment from the US State Department further details that the '55th or 055, Brigade is made up-predominately of expatriate mujahidin forces loyal to bin Laden. Members of the brigade generally are employed in small groups where they are most useful on the battlefield rather than as a single large unit. The strength of the 55th probably ranges between 500 and 2,500. Al-Qaida probably rotates personnel through this brigade to gain combat experience prior to their dispatch abroad. The unit has a reputation for tenacity in battle and seems less susceptible to the defection and bribery that plague Taleban ranks ... . Brigade members are frequently employed as "shock troops," or to stiffen the line against the Northern Alliance.'[58] Al-Qaeda thus added to the military capabilities of the Taliban, while avoiding competing with the Taliban's political cause in Afghanistan.

Additionally, al-Qaeda elements introduced specialised military skills to Afghanistan, including tactical knowledge about using heavy artillery, rocket launchers, and mortars, reconnaissance, intelligence, and mapping.[59] Al-Qaeda's military specialisations and global network were important as the Taliban challenged American forces in the 2000s, expanding the latter's repertoire to include suicide attacks, assassinations of Afghan government officials, IEDs, and increased media engagement. As a US Central Command assessment explained, al-Qaeda's (AQ)'s independent activity is 'overshadowed; [and] integrated with Taliban ... . Seamless and extensive extremist networks will allow AQ to leverage and augment the insurgency as desired. Media and propaganda efforts remain crucial in 2010 to draw in recruits and maintain AQ's relevance via the battlefield.'[60] The Taliban have often faced scarcity, sanctions, and isolation, and al-Qaeda's willingness to provide funding, expertise, and independently funded fighters proved helpful.

Al-Qaeda also contributed organisational and logistical competence. In late 2001, for example, the US State Department explained the usefulness of the Taliban's access to al-Qaeda's global network. 'Al-Qaida's reach and scope is impressive, extending across several continents,' empowering the terrorist network to help the coalition with supplies, funding, and recruits.[61] Al-Qaeda solved various battlefield and logistical collective action problems in coordination with local Taliban. One of the most significant contributions al-Qaeda made in the coalition with the Taliban prior to the American invasion included running a host of training camps and coordinating a diverse group of foreign fighters, including Arabs seeking experience who would later head to locations such as Chechnya, as well as Pakistanis sympathetic to extremist positions regarding Kashmir. Stenersen explains that 'the lack of a central authority in war-torn Afghanistan and the border areas in Pakistan created a competitive environment where anyone with access to recruits and money could operate their own training camps and frontline sections ... . But it was al-Qaida that emerged as the most dominant actor in this environment – probably owing to its long experience in the region and Bin Laden's personal charisma, administrative skills and access to independent sources of funding.'[62] Organising, staffing, and establishing training routines and curricula was a bureaucratic feat that al-Qaeda managed well and augmented the coalition's ability for collective warfighting. Al-Qaeda's willingness to take on significant administrative tasks like running the camps was useful to the Taliban, who could welcome foreign fighters to the battlefield without the burden of recruiting, transporting, feeding, or training them.[63] As captured al-Qaeda documents illustrate, the terrorist group's approach to running the training camps was detailed and impressive. For example, one captured 19-page document clarifies budgetary procedures, staff compensation, and sick and vacation policies. Specifically, al-Qaeda specifies that:

> Every committee has a fixed budget approved by the Emir after it is discussed between the related committee chairman and the Regulation Committee. The expenditure must never exceed the approved budget regardless of the circumstances ... . Anything else outside the approved annual plan is covered from the account of the Supreme Emirate ... . For those who work in Peshawar, they are entitled for Fridays, the two holy feasts ... and a one month annual leave to be enjoyed at the end of the eleventh month of work, as well sick leave not to exceed 15 days annually. Those working in camps and in the frontlines: Married: Enjoys a 7 day monthly vacation. Single: Enjoys a 5 day monthly vacation. This in addition to the annual and sick leaves mentioned in the above paragraph.[64]

Al-Qaeda's work standardising training and managing logistics was a significant contribution to strengthening the force the Afghan Taliban could bring to bear against its adversaries. Furthermore, these actions simplified a complex battlefield, so foreigners and adventure-seekers drawn to

Afghanistan in the 1990s were not in the way of the Taliban or used by rival rebel groups. The bureaucratic specificity in al-Qaeda's training camps is a testament to their professionalism and organisational proficiency.[65] Standardising policies, training, and process brought some degree of order to a highly complex and varied battleground full of diverse actors and interests.

Al-Qaeda has also been a stalwart political ally of the Taliban, which has contributed to their longstanding battlefield coalition. Following the US invasion in 2001, for example, the Taliban was under siege and isolated. Even Pakistan paused its support as it recalibrated ways to respond to US demands that Islamabad abandon its longstanding support for the Taliban. In October 2001, the US Department of State summarised that 'with the decline in military aid by Pakistan over the last year, the Taleban have increasingly relied on Usama bin Laden and the al-Qaida organization for support. Bin Laden provides money, recruits, materiel, and training for the fight against the Northern Alliance.'[66] One of the motivations for the Taliban maintaining a coalition with al-Qaeda, even under enormous pressure from powerful international and local actors to break ties directly following 11 September 2001, is that not doing so risked becoming even more isolated and enfeebled. Maintaining a coalition with a capable terrorist organisation, even one that US counterterrorism agencies were fully mobilised to destroy, was a strategy adopted by the Taliban to maintain lasting sources of support, a key factor for success in rebellions.

9.3.3 An Institutionalised Partnership and the Potential Costs of Defecting from a Coalition with al-Qaeda

Many al-Qaeda elements in Afghanistan and Pakistan have carved out localised accommodations with Taliban elements, and it is unclear if those institutionalised webs can be untangled by a decree from Taliban commanders. Dismantling institutionalised coalition arrangements poses risks for high-ranking Taliban authorities due to the structure of the Taliban itself, effectively raising the costs of defecting from the established coalition. On its own, the Taliban is a coalition, a federation of different groups in Afghanistan and Pakistan that have diverse arrangements locally, with other Taliban components, and with members of al-Qaeda. In 2005, for example, a US Department of State internal memo noted how 'the Afghan insurgency is not monolithic. The Taliban, HIG, al-Qaida, Haqqani network, Jaish-i-Muslimeen, and other extremists have varying agendas, and lack internal cohesion. For example, one of our veteran international contacts continues to hear frictions between Kandahari and non-Kandahari elements of the Taliban. While these extremist elements may be able to agree on grand strategy, this generally does not translate into operational coordination.'[67] Attempting to extract al-Qaeda from Afghanistan is a potential threat to the Taliban as it could splinter key factions from the Taliban and create new local rivalries, a costly prospect for the rebels.

In addition to the risks to the Taliban's internal cohesion and reputation of being in solidarity with global Islamic fundamentalists that a potential split with al-Qaeda might foster, al-Qaeda could also turn its military expertise, capacity for violence, organisational abilities, and international network against the Taliban should they break ties with al-Qaeda. The alternative to an al-Qaeda ally for the Taliban is likely an al-Qaeda enemy. Prior to 9/11, for example, Taliban officials told American diplomats that expelling bin Laden (UBL) was not a straightforward task because of the risks of internal divisions within Taliban coalitions and retaliation from al-Qaeda. More specifically, 'when the [U.S.] Ambassador pushed [Taliban Deputy Foreign Minister Abdul] Jalil about whether he was saying that UBL was too strong for the Taliban to evict, Jalil admitted that it would not be "as easy" as banning poppy cultivation or ridding the country of illegal arms. Plus, said Jalil, if UBL can create problems for rich countries like Saudi Arabia and the United States, think of what he could do to a country like Afghanistan. The Taliban also fear, said Jalil, that other mullahs might issue fatwas against them should they expel UBL.'[68] In support of the third factor examined here, the potential costs of breaking a coalition between terrorists and rebels and avoiding potential retaliation from al-Qaeda have also motivated the Taliban to continue forming battlefield coalitions over time.

9.4 Conclusion: 'Taliban and AQ are unlikely to completely sever ties in the near future despite tensions'[69]

Why do some rebels persist in forming battlefield coalitions with terrorists, especially in the face of significant pressure for them to not do so? I argue that rebels and terrorists are more likely to maintain battlefield coalitions if: 1) they are non-rival, defined as maintaining distinct overall political agendas that overlap locally but do not compete in terms of grand political objectives; 2) the terrorist group is well-established and proficient, contributing military expertise, solving organisational problems in complex conflicts, and providing political benefits; and 3) the terrorist group is locally institutionalised and remains a potential threat to the local rebel partners if they are not allied, which raises the cost for rebel defection. These factors collectively make up for the costs and highlight particular benefits for rebels that make continuing such relationships more attractive.

Examining the al-Qaeda–Taliban coalition in Afghanistan, I find there are substantial structural factors holding this coalition together. First, as non-rival groups operating in a highly uncertain environment, al-Qaeda and the Taliban have context-specific accommodations that have enabled them to challenge rival organisations such as the Northern Alliance and state-based adversaries like the US Al-Qaeda has helped the Taliban solve organisational problems, such as developing standard military training protocols. While foreign fighters in Afghanistan have not typically been essential to Taliban military operations, they are nonetheless often found on the frontlines, offering specific military skills as well as bureaucratic and media know-how. Al-Qaeda and the Taliban have not been known to operationally coordinate independent units in battle, but instead tend to swap fighters in a general pool of talent, skills, and resources. This fusion and fluidity have made the coalition dynamic, adaptable, and durable.

Interestingly, while top US policymakers such as Zalmay Khalilzad, US Special Representative tasked with negotiating US withdrawal, have expressed confidence that the Taliban would turn against al-Qaeda in exchange for US withdrawal, declassified US documents demonstrate that US analysts have long been sceptical of such plans. As one US Central Command official explained in a classified 2008 assessment:

> the strategy of separating the Taliban from Al Qaida is a pretty farfetched concept since the majority of low level fighters for these organizations are known to be used by both. The probability of there being a split between Taliban and Al Qaida is functionally unlikely as these organizations are not operationally fused. Al Qaida plays a coordination and strategic role between several syndicate organizations, enabling global support while simultaneously ensuring the harmonization of these groups. These actions assist with command the control of the Taliban's main objective of forcing western forces out of Afghanistan and regaining control of the national government. Since Al Qaida's role is a strategic one which deals with multiple independent organizations the Taliban's ability to separate itself from Al Qaida would only be feasible at a propaganda level.[70]

Throughout the US war (2001–2021), the Taliban persistently maintained its coalition with al-Qaeda. In February 2021, the UN reported 'that the top leadership of Al-Qaeda is still under Taliban protection,' with the groups maintaining their close ties but trying to minimise the profile of their cooperation in order to avoid jeopardising the Taliban's position.[71] US Secretary of State Anthony Blinken claimed the Taliban's successful takeover of Afghanistan in August 2021 would inspire the rebels to reign in the terrorists out of 'self-interest … [since] they know what happened the last time they harbored a terrorist group that attacked the United States.'[72] While the Taliban may seek to diminish the likelihood of counterterrorism operations in Afghanistan, as analysed in this chapter, the Taliban–al-Qaeda nexus is nevertheless likely to persist as the Taliban transition away from insurgency towards political consolidation. The groups' goals are still non-rival, al-Qaeda can still provide key organisational and bureaucratic support to the Taliban as it reasserts itself across Afghanistan, and al-Qaeda still maintains a variety of institutionalised local arrangements and can retaliate against the Taliban, if necessary.

Overall, this study suggests that policies advocating for dividing rebels and terrorist coalitions by leveraging their divergent political priorities should be reconsidered. While some policymakers contend that preference divergence reliably offers opportunities to divide groups, I argue that divergent interests between rebels and terrorists make these groups non-rival and more likely to bind together. Because coalition members have different agendas, they can often work together without competing, effectively pooling scarce military, reputational, and political resources. For weak actors like rebel groups, the military, political, and organisational benefits of maintaining battlefield coalitions with notorious terrorists can outweigh the costs, leading to persistent partnerships.

Notes

1 Ryan Grauer and Dominic Tierney, 'The Arsenal of Insurrection: Explaining Rising Support for Rebels,' *Security Studies* 27, no. 2 (April 3, 2018): 269–70, 282–84, https://doi.org/10.1080/09636412.2017.1386936; Seth G. Jones, *Waging Insurgent Warfare: Lessons from the Vietcong to the Islamic State* (Oxford University Press, 2017), 148.

2 Matthew Levitt, *Hezbollah: The Global Footprint of Lebanon's Party of God* (Georgetown University Press, 2015), 269; Nicolas Desgrais, Yvan Guichaoua, and Andrew Lebovich, 'Unity Is the Exception. Alliance Formation and De-Formation among Armed Actors in Northern Mali,' *Small Wars & Insurgencies* 29, no. 4 (July 4, 2018): 654–79, https://doi.org/10.1080/09592318.2018.1488403; Maggie Michael, Trish Wilson, and Lee Keath, 'AP Investigation: US Allies, Al-Qaida Battle Rebels in Yemen,' *Associated Press*, August 6, 2018, https://apnews.com/article/saudi-arabia-united-states-ap-top-news-middle-east-international-news-f38788a5 61d74ca78c77cb43612d50da; Ely Karmon, 'Hamas in Dire Straits,' *Perspectives on Terrorism* 7, no. 5 (2013): 111–26; U.S. Embassy, Nairobi, 'Somalia – President Sharif Describes Current Fighting, Requests Immediate Aid' (U.S. Department of State, May 14, 2009), 09NAIROBI970; Cindy C. Combs and Martin W. Slann, *Encyclopedia of Terrorism, Revised Edition* (Infobase Publishing, 2009), 147.

3 Bruce Hoffman, *Inside Terrorism* (Columbia University Press, 1998), 41.

4 Hoffman, *Inside Terrorism*.

5 Kathy Gilsinan, 'The U.S. Once Wanted Peace in Afghanistan: Now It's Setting Its Sights Much Lower,' *The Atlantic*, February 29, 2020, https://www.theatlantic.com/politics/archive/2020/02/united-states-taliban-afghanistan-peace-deal/607234/.

6 Daniel Byman, *Deadly Connections: States That Sponsor Terrorism* (Cambridge; New York: Cambridge University Press, 2007); Daniel Byman and Sarah E. Kreps, 'Agents of Destruction? Applying Principal-Agent Analysis to State-Sponsored Terrorism,' *International Studies Perspectives* 11, no. 1 (February 1, 2010): 1–18, https://doi.org/10.1111/j.1528-3585.2009.00389.x; S. Paul Kapur and Sumit Ganguly, 'The Jihad Paradox: Pakistan and Islamist Militancy in South Asia,' *International Security* 37, no. 1 (2012): 111–41, https://doi.org/10.1162/ISEC_a_00090; Afshon Ostovar, 'The Grand Strategy of Militant Clients: Iran's Way of War,' *Security Studies* 28, no. 1 (January 1, 2019): 159–88, https://doi.org/10.1080/09636412.2018.1508862.

7 Fotini Christia, *Alliance Formation in Civil Wars* (New York, NY: Cambridge University Press, 2012); Evan N. Resnick, 'Hang Together or Hang Separately? Evaluating Rival Theories of Wartime Alliance Cohesion,' *Security Studies* 22, no. 4 (October 1, 2013): 672–706, https://doi.org/10.1080/09636412.2013.844520; Paul Staniland, *Networks of Rebellion: Explaining Insurgent Cohesion and Collapse* (Cornell University Press, 2014).

8 Michael C. Horowitz and Philip B. K. Potter, 'Allying to Kill: Terrorist Intergroup Cooperation and the Consequences for Lethality,' *Journal of Conflict Resolution* 58, no. 2 (March 1, 2014): 199–225, https://doi.org/10.1177/002200271246872 6; Brian J. Phillips, 'Enemies with Benefits? Violent Rivalry and Terrorist Group Longevity,' *Journal of Peace Research* 52, no. 1 (January 1, 2015): 62–75, https://doi.org/10.1177/0022343314550538; Tricia Bacon, *Why Terrorist Groups Form International Alliances* (Philadelphia: University of Pennsylvania Press, 2018).
 9 One exception is Assaf Moghadam and Michel Wyss, 'The Political Power of Proxies: Why Nonstate Actors Use Local Surrogates,' *International Security* 44, no. 4 (April 1, 2020): 119–57, https://doi.org/10.1162/isec_a_00377.
10 C. Christine Fair, 'Pakistan Has All the Leverage Over Trump: Why Islamabad Isn't Worried about Threats to Cut off U.S. Aid,' *Foreign Policy*, January 3, 2018, https://foreignpolicy.com/2018/01/03/pakistan-has-all-the-leverage-over-trump/.
11 Byman, *Deadly Connections*; Byman and Kreps, 'Agents of Destruction?'; David B. Carter, 'A Blessing or a Curse? State Support for Terrorist Groups,' *International Organization* 66, no. 1 (2012): 129–51, https://doi.org/10.1017/S0020818311000312; Ostovar, 'The Grand Strategy of Militant Clients.'
12 Kapur and Ganguly, 'The Jihad Paradox'; Fair, 'Pakistan Has All the Leverage Over Trump: Why Islamabad Isn't Worried about Threats to Cut off U.S. Aid.'
13 Tyrone L. Groh, *Proxy War: The Least Bad Option*, 1st edition (Stanford, California: Stanford University Press, 2019).
14 Byman, *Deadly Connections*, 4.
15 Ostovar, 'The Grand Strategy of Militant Clients,' 164; Kapur and Ganguly, 'The Jihad Paradox.'
16 Ostovar, 'The Grand Strategy of Militant Clients,' 162.
17 Assaf Moghadam and Michel Wyss, 'The Political Power of Proxies: Why Nonstate Actors Use Local Surrogates,' *International Security* 44, no. 4 (April 1, 2020): 119–57, https://doi.org/10.1162/isec_a_00377.
18 U.S. Ambassador to the UN, 'Somali FM Makes Urgent Plea for Aid to Amb. Rice' (U.S. Department of State, May 14, 2009), 09USUNNEWYORK502.
19 'U.N. Report Ties Somali Islamists to Hezbollah,' NPR.org, accessed June 8, 2021, https://www.npr.org/templates/story/story.php?storyId=6493083.
20 Kapur and Ganguly, 'The Jihad Paradox.'
21 Byman, *Deadly Connections*, 6.
22 Christia, *Alliance Formation in Civil Wars*; Lee J. M. Seymour, Kristin M. Bakke, and Kathleen Gallagher Cunningham, 'E Pluribus Unum, Ex Uno Plures: Competition, Violence, and Fragmentation in Ethnopolitical Movements,' *Journal of Peace Research* 53, no. 1 (January 1, 2016): 3–18, https://doi.org/1 0.1177/0022343315605571; Kristin M. Bakke, Kathleen Gallagher Cunningham, and Lee J. M. Seymour, 'A Plague of Initials: Fragmentation, Cohesion, and Infighting in Civil Wars,' *Perspectives on Politics* 10, no. 2 (2012): 265–83, https://doi.org/10.1017/S1537592712000667; Ben Oppenheim et al., 'True Believers, Deserters, and Traitors: Who Leaves Insurgent Groups and Why,' *Journal of Conflict Resolution* 59, no. 5 (August 1, 2015): 794–823, https://doi.org/10.1177/0022002715576750; Mohammed M. Hafez, 'Fratricidal Rebels: Ideological Extremity and Warring Factionalism in Civil Wars,' *Terrorism and Political Violence* 32, no. 3 (April 2, 2020): 604–29, https://doi.org/10.1080/09546553.2017.1389726; Eric S. Mosinger, 'Brothers or Others in Arms? Civilian Constituencies and Rebel Fragmentation in Civil War,' *Journal of Peace Research* 55, no. 1 (January 1, 2018): 62–77, https://doi.org/10.1177/0022343316675907; Emily Kalah Gade et al., 'Networks of Cooperation: Rebel Alliances in Fragmented Civil Wars,' *Journal of Conflict Resolution* 63, no. 9 (October 1, 2019): 2071–97, https://doi.org/10.1177/0022002719826234;

Emily Kalah Gade, Mohammed M Hafez, and Michael Gabbay, 'Fratricide in Rebel Movements: A Network Analysis of Syrian Militant Infighting,' *Journal of Peace Research* 56, no. 3 (May 1, 2019): 321–35, https://doi.org/10.1177/0022343318806940; Barbara F. Walter, 'Explaining the Number of Rebel Groups in Civil Wars,' *International Interactions* 45, no. 1 (January 2, 2019): 1–27, https://doi.org/10.1080/03050629.2019.1554573; Allard Duursma and Feike Fliervoet, 'Fueling Factionalism? The Impact of Peace Processes on Rebel Group Fragmentation in Civil Wars,' *Journal of Conflict Resolution* 65, no. 4 (April 1, 2021): 788–812, https://doi.org/10.1177/0022002720958062; Staniland, *Networks of Rebellion.*
23 Christia, *Alliance Formation in Civil Wars*, 3.
24 There are exceptions. The Islamic State (ISIS), for example, is a global terrorist group that has also pursued local state-building goals. As such, the theory presented here would predict different coalitional dynamics for local rebel groups allied with ISIS, as this competition to govern would lead to rival dynamics and balancing behaviours more closely resembling alliances among rebel groups.
25 Phillips, 'Enemies with Benefits?'; Horowitz and Potter, 'Allying to Kill.'
26 Bakke, Cunningham, and Seymour, 'A Plague of Initials,' 265.
27 Bacon, *Why Terrorist Groups Form International Alliances*, 9, 8; See also Phillips, 'Enemies with Benefits?'
28 Christia, *Alliance Formation in Civil Wars*; Lee J. M. Seymour, 'Why Factions Switch Sides in Civil Wars: Rivalry, Patronage, and Realignment in Sudan,' *International Security* 39, no. 2 (October 1, 2014): 92–131, https://doi.org/10.1162/ISEC_a_00179; Desgrais, Guichaoua, and Lebovich, 'Unity Is the Exception. Alliance Formation and De-Formation among Armed Actors in Northern Mali'; Sabine Otto, 'The Grass Is Always Greener? Armed Group Side Switching in Civil Wars,' *Journal of Conflict Resolution* 62, no. 7 (August 1, 2018): 1459–88, https://doi.org/10.1177/0022002717693047; Gade et al., 'Networks of Cooperation.'
29 Dexter Filkins, *The Forever War*, Reprint (Vintage, 2009), 50–1.
30 Byman, *Deadly Connections*, 5.
31 U.S. Embassy, Nairobi, 'Somalia – Al-Shabaab Takes Key Middle Shabelle Town: Possible Next Steps' (U.S. Department of State, May 18, 2009), 09NAIROBI1010.
32 Horowitz and Potter, 'Allying to Kill.'
33 Michael C. Horowitz, 'Nonstate Actors and the Diffusion of Innovations: The Case of Suicide Terrorism,' *International Organization* 64, no. 1 (2010): 33–64, https://doi.org/10.1017/S0020818309990233.
34 U.S. Embassy, Nairobi, 'Somalia – The TFG, Al-Shabaab, and Al-Qaeda' (U.S. Department of State, July 6, 2009), 09NAIROBI1395.
35 Mia Bloom, 'Constructing Expertise: Terrorist Recruitment and 'Talent Spotting' in the PIRA, Al Qaeda, and ISIS,' *Studies in Conflict & Terrorism* 40, no. 7 (July 3, 2017): 603–23, https://doi.org/10.1080/1057610X.2016.1237219.
36 Harleen K. Gambhir, 'Dabiq: The Strategic Messaging of the Islamic State' (Institute for the Study of War, August 15, 2014), 7, http://www.understandingwar.org/sites/default/files/Dabiq%20Backgrounder_Harleen%20Final_0.pdf.
37 Ana Arjona, 'Wartime Institutions: A Research Agenda,' *Journal of Conflict Resolution* 58, no. 8 (December 1, 2014): 1360, https://doi.org/10.1177/0022002714547904.
38 Arjona, 1360.
39 Steve Coll, *Directorate S: The C. I. A. and America's Secret Wars in Afghanistan and Pakistan* (New York: Penguin Press, 2019), 346.

40 'Human and Budgetary Costs to Date of the U.S. War in Afghanistan, 2001–2021, Figures.'
41 Alex Strick van Linschoten and Felix Kuehn, *An Enemy We Created: The Myth of the Taliban-Al Qaeda Merger in Afghanistan* (Oxford University Press, 2012), 4.
42 Linschoten and Kuehn, *An Enemy We Created*, 7.
43 Stathis N. Kalyvas, *The Logic of Violence in Civil War* (Cambridge: Cambridge University Press, 2006), 10.
44 Christia, *Alliance Formation in Civil Wars*.
45 Doran, 'The Pragmatic Fanaticism of al Qaeda'; Byman, 'Al-Qaeda as an Adversary.'
46 'The Al-Qaeda-Taliban Nexus' (Council on Foreign Relations, November 24, 2009), https://www.cfr.org/expert-roundup/al-qaeda-taliban-nexus; Vahid Brown, 'The Facade of Allegiance: Bin Ladin's Dubious Pledge to Mullah Omar,' *CTC Sentinel* 3, no. 1 (January 1, 2010), https://apps.dtic.mil/sti/citations/ADA516602.
47 Defense Intelligence Agency, 'DIA [Excised] Analytical Product - (U) Afghanistan: Taliban Claim Attempted Assassination of U.S. Vice President,' 3.
48 Anne Stenersen, *Al-Qaida in Afghanistan* (Cambridge University Press, 2017), 62.
49 U.S. Central Command, 'Exploitable Fissures between AQ Leadership and TB Leadership,' 1–2.
50 bin Laden, 'Portion of a Letter from Osama Bin Laden Regarding Motivation for Afghanistan to Confront America.'
51 Linschoten and Kuehn, *An Enemy We Created*, 234.
52 Barbara Elias, 'Know Thine Enemy,' *Foreign Affairs*, November 2, 2009, https://www.foreignaffairs.com/articles/pakistan/2009-11-02/know-thine-enemy.
53 Mustafa Hamid and Leah Farrall, *The Arabs at War in Afghanistan* (Oxford University Press, 2015); Stenersen, *Al-Qaida in Afghanistan*.
54 'The Al-Qaeda-Taliban Nexus.'
55 Stenersen, *Al-Qaida in Afghanistan*, 23–31 and 128–48; Hamid and Farrall, *The Arabs at War in Afghanistan*, 227.
56 Stenersen, *Al-Qaida in Afghanistan*, 133.
57 Stenersen, 132.
58 U.S. Department of State, Bureau of Intelligence and Research, 'Afghanistan: Al-Qaida Military Contribution to the Taleban,' October 1, 2001.
59 Stenersen, *Al-Qaida in Afghanistan*, 23, 29–30.
60 U.S. Central Command, 'Al-Qaida 101 – Afghanistan/Pakistan Intelligence Center of Excellence,' 9.
61 U.S. Department of State, Bureau of Intelligence and Research, 'Afghanistan: Al-Qaida Military Contribution to the Taleban,' October 1, 2001.
62 Stenersen, *Al-Qaida in Afghanistan*.
63 Stenersen, Chapter 5; 'Al Qaeda and Associated Movements Collection' (Conflict Records Research Center, n.d.), https://conflictrecords.wordpress.com/collections/aqam/.
64 'Advanced Draft of Al-Qaeda Bylaws, Including Fundamentals, Organizational Structure, Leadership, and Objectives.'
65 Stenersen, *Al-Qaida in Afghanistan*, 26.
66 U.S. Department of State, Bureau of Intelligence and Research, 'Afghanistan: Al-Qaida Military Contribution to the Taleban.'
67 U.S. Department of State, 'Counterterrorism Activities (Neo-Taliban),' December 10, 2009, 4; 'Kandahari Taliban' refers to the core members of the Taliban that are considered more traditional such as Mullah Omar. 'Non-Kandaharis' are affiliated Taliban from other areas such as the Haqqanis. See Thomas Ruttig,

'Have the Taliban Changed?,' *CTC Sentinel* 14, no. 3 (March 22, 2021), https://ctc.usma.edu/have-the-taliban-changed/.
68 U.S. Department of State, U.S. Embassy Islamabad, 'Taliban's Mullah Jalil's July 2 Meeting with the Ambassador, 01ISLAMA3702.'
69 Pakistan Intelligence Center for Excellence, U.S. Central Command, 'J2 Special Report: INS Network Leadership Relations, Afghanistan.'
70 International Security Assistance Force Afghanistan (ISAF), 'The Landing Zone: RC East OSINT Summary,' 4.
71 Dan De Luce, Ken Dilanian, and Mushtaq Yusufzai, 'Taliban Keep Close Ties with Al Qaeda despite Promise to U.S.,' *NBC News*, February 17, 2021, https://www.nbcnews.com/politics/national-security/taliban-keep-close-ties-al-qaeda-despite-promise-u-s-n1258033; 'Letter Dated 20 May 2021 from the Chair of the Security Council Committee Established Pursuant to Resolution 1988 (2011) Addressed to the President of the Security Council,' June 1, 2021, 13, https://www.undocs.org/S/2021/486.
72 Lara Jakes and Michael Crowley, 'Taliban Takeover Could Extinguish U.S. Influence in Kabul,' *The New York Times*, August 15, 2021, sec. U.S., https://www.nytimes.com/2021/08/15/us/politics/biden-taliban-afghanistan.html.

Bibliography

'Advanced Draft of Al-Qaeda Bylaws, Including Fundamentals, Organizational Structure, Leadership, and Objectives.' Undated (prior to 2002. Identifier: AQSH-PDAQ-SHPD-D-000-201_TF.pdf. al-Qaeda Collection, Conflict Records Research Center, Washington, DC.

'Al Qaeda and Associated Movements Collection.' Conflict Records Research Center, n.d. https://conflictrecords.wordpress.com/collections/aqam/

Arjona, Ana. 'Wartime Institutions: A Research Agenda.' *Journal of Conflict Resolution* 58, no. 8 (1 December 2014): 1360–89. 10.1177/0022002714547904

Bacon, Tricia. *Why Terrorist Groups Form International Alliances*. Philadelphia, PA: University of Pennsylvania Press, 2018.

Bakke, Kristin M., Kathleen Gallagher Cunningham, and Lee J. M. Seymour. 'A Plague of Initials: Fragmentation, Cohesion, and Infighting in Civil Wars.' *Perspectives on Politics* 10, no. 2 (2012): 265–83. 10.1017/S1537592712000667

Bloom, Mia. 'Constructing Expertise: Terrorist Recruitment and 'Talent Spotting' in the PIRA, Al Qaeda, and ISIS.' *Studies in Conflict & Terrorism* 40, no. 7 (3 July 2017): 603–23. 10.1080/1057610X.2016.1237219

Brown, Vahid. 'The Facade of Allegiance: Bin Ladin's Dubious Pledge to Mullah Omar.' *CTC Sentinel* 3, no. 1 (1 January 2010). https://apps.dtic.mil/sti/citations/ADA516602

Byman, Daniel. *Deadly Connections: States That Sponsor Terrorism*. Cambridge; New York, NY: Cambridge University Press, 2007.

Byman, Daniel L. 'Review: Al-Qaeda as an Adversary: Do We Understand Our Enemy?' *World Politics* 56, no. 1 (2003): 139–63.

Byman, Daniel, and Sarah E. Kreps. 'Agents of Destruction? Applying Principal-Agent Analysis to State-Sponsored Terrorism.' *International Studies Perspectives* 11, no. 1 (1 February 2010): 1–18. 10.1111/j.1528-3585.2009.00389.x

Carter, David B. 'A Blessing or a Curse? State Support for Terrorist Groups.' *International Organization* 66, no. 1 (2012): 129–51. 10.1017/S0020818311000312

Christia, Fotini. *Alliance Formation in Civil Wars*. New York, NY: Cambridge University Press, 2012.

Coll, Steve. *Directorate S: The C. I. A. and America's Secret Wars in Afghanistan and Pakistan*. New York, NY: Penguin Press, 2019.

Combs, Cindy C., and Martin W. Slann. *Encyclopedia of Terrorism, Revised Edition*. Infobase Publishing, 2009.

De Luce, Dan, Ken Dilanian, and Mushtaq Yusufzai. 'Taliban Keep Close Ties with Al Qaeda despite Promise to U.S.' *NBC News*, 17 February 2021. https://www.nbcnews.com/politics/national-security/taliban-keep-close-ties-al-qaeda-despite-promise-u-s-n1258033

Defense Intelligence Agency. 'DIA [Excised] Analytical Product - (U) Afghanistan: Taliban Claim Attempted Assassination of U.S. Vice President.' 27 February 2007.

Desgrais, Nicolas, Yvan Guichaoua, and Andrew Lebovich. 'Unity Is the Exception. Alliance Formation and De-Formation among Armed Actors in Northern Mali.' *Small Wars & Insurgencies* 29, no. 4 (4 July 2018): 654–79. 10.1080/09592318.2018.1488403

Doran, Michael. 'The Pragmatic Fanaticism of Al Qaeda: An Anatomy of Extremism in Middle Eastern Politics.' *Political Science Quarterly* 117, no. 2 (2002): 177–90. 10.2307/798179

Duursma, Allard, and Feike Fliervoet. 'Fueling Factionalism? The Impact of Peace Processes on Rebel Group Fragmentation in Civil Wars.' *Journal of Conflict Resolution* 65, no. 4 (1 April 2021): 788–812. 10.1177/0022002720958062

Elias, Barbara. 'Know Thine Enemy.' *Foreign Affairs*, 2 November 2009. https://www.foreignaffairs.com/articles/pakistan/2009-11-02/know-thine-enemy

Fair, C. Christine. 'Pakistan Has All the Leverage Over Trump: Why Islamabad Isn't Worried about Threats to Cut off U.S. Aid.' *Foreign Policy*, 3 January 2018. https://foreignpolicy.com/2018/01/03/pakistan-has-all-the-leverage-over-trump/

Filkins, Dexter. *The Forever War*. Reprint. Vintage, 2009.

Gade, Emily Kalah, Michael Gabbay, Mohammed M. Hafez, and Zane Kelly. 'Networks of Cooperation: Rebel Alliances in Fragmented Civil Wars.' *Journal of Conflict Resolution* 63, no. 9 (1 October 2019): 2071–97. 10.1177/0022002719826234

Gade, Emily Kalah, Mohammed M. Hafez, and Michael Gabbay. 'Fratricide in Rebel Movements: A Network Analysis of Syrian Militant Infighting.' *Journal of Peace Research* 56, no. 3 (1 May 2019): 321–35. 10.1177/0022343318806940

Gambhir, Harleen K. 'Dabiq: The Strategic Messaging of the Islamic State.' Institute for the Study of War, 15 August 2014. http://www.understandingwar.org/sites/default/files/Dabiq%20Backgrounder_Harleen%20Final_0.pdf

Gilsinan, Kathy. 'The U.S. Once Wanted Peace in Afghanistan: Now It's Setting Its Sights Much Lower.' *The Atlantic*, 29 February 2020. https://www.theatlantic.com/politics/archive/2020/02/united-states-taliban-afghanistan-peace-deal/607234/

Grauer, Ryan, and Dominic Tierney. 'The Arsenal of Insurrection: Explaining Rising Support for Rebels.' *Security Studies* 27, no. 2 (3 April 2018): 263–95. 10.1080/09636412.2017.1386936

Groh, Tyrone L. *Proxy War: The Least Bad Option*. 1st ed. Stanford, CA: Stanford University Press, 2019.

Hafez, Mohammed M. 'Fratricidal Rebels: Ideological Extremity and Warring Factionalism in Civil Wars.' *Terrorism and Political Violence* 32, no. 3 (2 April 2020): 604–29. 10.1080/09546553.2017.1389726

Hamid, Mustafa, and Leah Farrall. *The Arabs at War in Afghanistan*. Oxford University Press, 2015.

Hoffman, Bruce. *Inside Terrorism*. Columbia University Press, 1998.

Horowitz, Michael C. 'Nonstate Actors and the Diffusion of Innovations: The Case of Suicide Terrorism.' *International Organization* 64, no. 1 (2010): 33–64. 10.1017/S0020818309990233

Horowitz, Michael C., and Philip B. K. Potter. 'Allying to Kill: Terrorist Intergroup Cooperation and the Consequences for Lethality.' *Journal of Conflict Resolution* 58, no. 2 (1 March 2014): 199–225. 10.1177/0022002712468726

International Security Assistance Force Afghanistan (ISAF). 'The Landing Zone: RC East OSINT Summary.' 18 December 2008. Declassified, FOIA Case 10-0139.

Jakes, Lara, and Michael Crowley. 'Taliban Takeover Could Extinguish U.S. Influence in Kabul.' *The New York Times*, 15 August 2021, sec. U.S. https://www.nytimes.com/2021/08/15/us/politics/biden-taliban-afghanistan.html

Jones, Seth G. *Waging Insurgent Warfare: Lessons from the Vietcong to the Islamic State*. Oxford University Press, 2017.

Kalyvas, Stathis N. *The Logic of Violence in Civil War*. Cambridge: Cambridge University Press, 2006.

Kapur, S. Paul, and Sumit Ganguly. 'The Jihad Paradox: Pakistan and Islamist Militancy in South Asia.' *International Security* 37, no. 1 (2012): 111–41. 10.1162/ISEC_a_00090

Karmon, Ely. 'Hamas in Dire Straits.' *Perspectives on Terrorism* 7, no. 5 (2013): 111–26.

Laden, Osama bin. 'Portion of a Letter from Osama Bin Laden Regarding Motivation for Afghanistan to Confront America.' 18 July 1999. Identifier: AQSHPDAQ-SHPD-D-000-005_TF.pdf. al-Qaeda Collection, Conflict Records Research Center, Washington, DC.

'Letter Dated 20 May 2021 from the Chair of the Security Council Committee Established Pursuant to Resolution 1988 (2011) Addressed to the President of the Security Council.' 1 June 2021. https://www.undocs.org/S/2021/486

Levitt, Matthew. *Hezbollah: The Global Footprint of Lebanon's Party of God*. Georgetown University Press, 2015.

Linschoten, Alex Strick van, and Felix Kuehn. *An Enemy We Created: The Myth of the Taliban-Al Qaeda Merger in Afghanistan*. Oxford University Press, 2012.

Michael, Maggie, Trish Wilson, and Lee Keath. 'AP Investigation: US Allies, al-Qaida Battle Rebels in Yemen.' *Associated Press*, 6 August 2018. https://apnews.com/article/saudi-arabia-united-states-ap-top-news-middle-east-international-news-f38788a561d74ca78c77cb43612d50da

Moghadam, Assaf, and Michel Wyss. 'The Political Power of Proxies: Why Nonstate Actors Use Local Surrogates.' *International Security* 44, no. 4 (1 April 2020): 119–57. 10.1162/isec_a_00377

Mosinger, Eric S. 'Brothers or Others in Arms? Civilian Constituencies and Rebel Fragmentation in Civil War.' *Journal of Peace Research* 55, no. 1 (1 January 2018): 62–77. 10.1177/0022343316675907

NPR.org. 'U.N. Report Ties Somali Islamists to Hezbollah.' Accessed 8 June 2021. https://www.npr.org/templates/story/story.php?storyId=6493083

Oppenheim, Ben, Abbey Steele, Juan F. Vargas, and Michael Weintraub. 'True Believers, Deserters, and Traitors: Who Leaves Insurgent Groups and Why.'

Journal of Conflict Resolution 59, no. 5 (1 August 2015): 794–823. 10.1177/0022 002715576750

Ostovar, Afshon. 'The Grand Strategy of Militant Clients: Iran's Way of War.' *Security Studies* 28, no. 1 (1 January 2019): 159–88. 10.1080/09636412.2018.1508862

Otto, Sabine. 'The Grass Is Always Greener? Armed Group Side Switching in Civil Wars.' *Journal of Conflict Resolution* 62, no. 7 (1 August 2018): 1459–88. 10.1177/0022002717693047

Pakistan Intelligence Center for Excellence, U.S. Central Command. 'J2 Special Report: INS Network Leadership Relations, Afghanistan.' 13 December 2010.

Phillips, Brian J. 'Enemies with Benefits? Violent Rivalry and Terrorist Group Longevity.' *Journal of Peace Research* 52, no. 1 (1 January 2015): 62–75. 10.1177/0022343314550538

Resnick, Evan N. 'Hang Together or Hang Separately? Evaluating Rival Theories of Wartime Alliance Cohesion.' *Security Studies* 22, no. 4 (1 October 2013): 672–706. 10.1080/09636412.2013.844520

Ruttig, Thomas. 'Have the Taliban Changed?' *CTC Sentinel* 14, no. 3 (22 March 2021). https://ctc.usma.edu/have-the-taliban-changed/

Seymour, Lee J. M. 'Why Factions Switch Sides in Civil Wars: Rivalry, Patronage, and Realignment in Sudan.' *International Security* 39, no. 2 (1 October 2014): 92–131. 10.1162/ISEC_a_00179

Seymour, Lee J.M., Kristin M. Bakke, and Kathleen Gallagher Cunningham. 'E Pluribus Unum, Ex Uno Plures: Competition, Violence, and Fragmentation in Ethnopolitical Movements.' *Journal of Peace Research* 53, no. 1 (1 January 2016): 3–18. 10.1177/0022343315605571

Staniland, Paul. *Networks of Rebellion: Explaining Insurgent Cohesion and Collapse*. Cornell University Press, 2014.

Stenersen, Anne. *Al-Qaida in Afghanistan*. Cambridge University Press, 2017.

'The Al-Qaeda-Taliban Nexus.' Council on Foreign Relations, 24 November 2009. https://www.cfr.org/expert-roundup/al-qaeda-taliban-nexus

The Costs of War, Brown University. 'Human and Budgetary Costs to Date of the U.S. War in Afghanistan, 2001–2021, Figures.' April 2021. https://watson.brown.edu/costsofwar/figures/2021/human-and-budgetary-costs-date-us-war-afghanistan-2001-2021

U.S. Ambassador to the UN. 'Somali FM Makes Urgent Plea for Aid to Amb. Rice.' U.S. Department of State, 14 May 2009. 09USUNNEWYORK502.

U.S. Central Command. 'Al-Qaida 101 - Afghanistan/Pakistan Intelligence Center of Excellence.' 8 June 2010.

U.S. Central Command. 'Exploitable Fissures between AQ Leadership and TB Leadership.' Afghanistan-Pakistan Center of Excellence United States Central Command, 25 February 2010. J2 Intelligence Note.

U.S. Department of State. 'Counterterrorism Activities (Neo-Taliban).' 10 December 2009.

U.S. Department of State, Bureau of Intelligence and Research. 'Afghanistan: Al-Qaida Military Contribution to the Taleban.' 1 October 2001.

U.S. Department of State, U.S. Embassy Islamabad. 'Taliban's Mullah Jalil's July 2 Meeting with the Ambassador, 01ISLAMA3702.' 3 July 2001.

U.S. Embassy, Nairobi. 'Somalia - Al-Shabaab Takes Key Middle Shabelle Town: Possible Next Steps.' U.S. Department of State, 18 May 2009. 09NAIROBI1010.

U.S. Embassy, Nairobi. 'Somalia - President Sharif Describes Current Fighting, Requests Immediate Aid.' U.S. Department of State, 14 May 2009. 09NAIROBI970.

U.S. Embassy, Nairobi. 'Somalia - The TFG, Al-Shabaab, and Al-Qaeda.' U.S. Department of State, 6 July 2009. 09NAIROBI1395.

Walter, Barbara F. 'Explaining the Number of Rebel Groups in Civil Wars.' *International Interactions* 45, no. 1 (2 January 2019): 1–27. 10.1080/03050629. 2019.1554573

10 Coalitions and Wartime Diplomacy[1]

Speaking with One Voice

Eric Min

On 1 January 1942, representatives of the Allied 'Big Four' – President Franklin Roosevelt of the United States, Prime Minister Winston Churchill of the United Kingdom, Maxim Litvinov of the Soviet Union, and Soong Tse-vung of the Republic of China – signed what would be called the United Nations Declaration. Twenty-two additional nations signed the document on the following day, taking the first steps toward formalising a coalition of states seeking to 'defend life, liberty independence and religious freedom … in a common struggle against savage and brutal forces.' This 234-word document featured two actual declarations. The first was to commit all resources to defeat the Axis Powers. The second was a pledge that the signatory governments would pledge 'not to make a separate armistice or peace with the enemies.'

The United Nations Declaration was not unique. On 11 December 1941, as Germany and Italy declared war against the United States and four days after the Japanese attack on Pearl Harbor, representatives from Germany, Italy, and Japan signed an agreement to collectively fight the war 'which has been imposed on them by the United States of America and England.' The second of four articles explicitly disallowed any of the three powers from negotiating separately with the United States or Britain.

Histories of conflict extending beyond World War II demonstrate that states fighting in coalitions harbour significant concerns about disunity. A group of actors, ostensibly banded together in a common cause, may worry that a member may unilaterally embrace a new posture that leaves other parties to haphazardly adjust. While some recent scholarship has analysed the dynamics of states abandoning their coalitions during war,[2] wholesale exit from conflict is a dramatic outcome that elides analysis of the intermediate steps that actors take in order to navigate hostilities. Diplomatic negotiations, being a necessary step to amend one's participation in conflict short of military victory or defeat, are essential in revealing this intervening stage where states may reveal discord within their ranks.

Despite these clear opportunities, scholarly literature has made little progress in investigating how wartime coalitions negotiate. Two lacunae account for this. First, research has only recently begun to analyse

DOI: 10.4324/9781003399896-14

negotiations as a significant dimension to understanding war, but this line of enquiry defines belligerents using features such as total material strength or domestic political institutions, but not by the composition of states fighting together as a coalition. Second, while meaningful progress has been made in understanding the impact of coalitions on overall war outcomes and duration, much less has been said about the intra-war activity that fundamentally sculpts when and how a conflict comes to an end.

This piece casts new light on these puzzles. I argue that a wartime coalition's propensity to negotiate is a function of two primary factors: the imbalance of military contributions by its members – which I will refer to as a coalition's asymmetry – and the coalition's overall battlefield fortunes. These dimensions concurrently affect each coalition member's calculus regarding its desire to obtain public benefits from overall victory versus private benefits from pursuing its specific policy aims. I contend and find evidence that asymmetric coalitions in which a single state provides the majority of military force will be more likely to stick together and refuse to negotiate when enjoying success but will be prone to seek negotiations when fighting trends unfavourably. Conversely, symmetric coalitions in which multiple states supply equitable forces will be more likely to engage in rushed and more frequent negotiations when victory seems likely but will forgo negotiations in the face of mounting losses.

Two sets of evidence support these claims. I first perform a brief quantitative analysis of coalition wars across two centuries of interstate conflict. The results support my theoretical expectations. I then analyse how Great Britain and France navigated developments on the battlefield and the negotiating table as they fought Russia in the 1853–56 Crimean War. The case offers an instructive example of how more symmetric coalitions may have fragmented diplomatic strategies that hasten the cessation of hostilities when victory feels within reach.

At least one-quarter of all wars between states over the last two centuries have involved coalitions,[3] and approximately one-half of the United States' uses of force between 1948 and 1998 have relied on multilateral coalitions.[4] If negotiations represent a necessary but strategically treacherous step towards terminating one's participation in conflict, then understanding what features influence how coalitions engage in negotiations could reveal a great deal about the strategic impulses that underlie states' decisions to form, remain in, or splinter these increasingly prominent wartime groups.

10.1 What We Know about Wartime Negotiations

Seminal work on international conflict over the last three decades views war as a violent form of bargaining that occurs when opposing parties cannot settle their disagreements through words alone. Actors in a dispute over something of value – whether land, policy, regime type, or the like – have strategic incentives to overstate their military capabilities or resolve in order to extract a more favourable set of concessions and overall negotiated settlement.[5] Status quo powers that see a rising power which could challenge their supremacy also have incentives to prevent further growth, while those rising powers would try to reassure all parties that they have no revisionist goals.[6] In both scenarios, the actors' threats or assurances may not be credible; a weak actor would want to exaggerate its strength, while an aggressive rising actor would want to downplay any hostile intentions. War becomes a costly but honest way to reveal each side's actual abilities, resilience, and motives in a way that words cannot.

A consequence of this perspective is that diplomacy during war tends to be deemed irrelevant or a mere reflection of the battlefield. If diplomacy failed to prevent conflict, it seems logical that diplomacy would have minimal value during the conflict itself. Yet there are multiple reasons this view is overly reductive. Political and military leaders certainly act as if they care deeply about the choice to communicate with the enemy during hostilities. Two-thirds of all interstate wars over the last two centuries have been concluded through negotiated settlements short of complete military victory or defeat,[7] and patterns of negotiation behaviour vary widely across wars in a manner that suggests strategic motivations.[8] The act of negotiating represents a concrete but precarious step forward in the conflict resolution process.[9]

Recent work by Mastro emphasises that the choice to negotiate with the enemy during war is highly strategic and dictated by whether belligerents believe an offer to talk will be construed as a sign of weakness.[10] At the outset of conflict, states refuse to negotiate with the enemy and to adopt what Mastro calls a closed diplomatic posture. Over the course of hostilities, states take great care to decide when to consider talks and to transition to an open diplomatic posture.

10.2 The Dynamics of Coalition Diplomacy

The aforementioned works represent significant advances in our understanding of wartime negotiations. Nonetheless, much of this scholarship assumes that war takes place between two monolithic actors that each wholly define their war aims, capabilities, resolve, and diplomatic posture. The situation is more complicated when multiple states fight together in a coalition. On one hand, all members share an overarching and immediate goal of defeating a common enemy and benefit from working together to accomplish this aim. These efforts often pay off, as coalitions generally enjoy higher chances of winning wars than individual states.[11] Yet on the other hand, each actor has their own private interests that may not be shared by other members, as well as a desire to minimise their own costs from fighting. Individual states may pursue these personal concerns by seeking negotiations, fracturing the coalition's diplomatic posture and ability to realise greater success.

This strategic dilemma is captured in broad strokes using a classic game theory model called the stag hunt. In a standard formulation, n identical actors independently choose to either hunt a large stag together or a much smaller rabbit on their own. The only way to hunt down a stag is for at least $M \leq n$ individuals to participate. Any single actor who chooses to hunt a rabbit will easily do so, but at the cost of getting far less sustenance and also jeopardising the success of any actors that chose to pursue the stag. The stag hunt game highlights a trade-off between a risky but efficient outcome and a safer but inefficient outcome.

Instead of choosing to hunt either a stag or a rabbit, wartime coalition members choose either to fight without negotiating (a closed diplomatic posture) or to negotiate with the enemy and possibly exit the conflict (an open diplomatic posture). All n states in a wartime coalition begin with a closed diplomatic posture. The coalition's prospects of realising their shared goals are highest when all members eschew diplomacy and continue hostilities, and they decrease as individual members splinter off and engage in negotiations that prioritise their private goals of minimising personal costs or obtaining their own policy aims.

The classic stag hunt game offers a useful lens to describe the underlying coordination dilemma. That said, the game features multiple Nash equilibria or stable outcomes, so the question of what outcome ultimately transpires is a matter of psychology and additional context beyond the confines of the game. Two informal elaborations of the standard stag hunt game prove useful in addressing when and why states in wartime coalitions may fracture and eventually choose to negotiate with the enemy over the course of hostilities: (1) the relative contributions that each group member can potentially make and (2) the group's apparent prospects for success.

First, certain individuals may have smaller or larger marginal impacts on the group's ability to realise its common goal. A group in which one hunter possesses a powerful rifle while all others have sticks is fundamentally different from a group in which all hunters wield sticks. This analogy has a straightforward parallel in coalition warfare, where I will use the term *coalition asymmetry* to refer to the imbalance of military contributions made by each member.[12] A coalition is asymmetric when a single core coalition member provides the vast majority of military force while the remaining auxiliary states contribute far less. Smaller auxiliary states have greater leeway to reshape their level of participation in conflict without significantly affecting the group's ability to accomplish its common policy goals, while core states bear most responsibility for the group's outcomes. The most extreme of asymmetric coalitions thus resemble solo belligerents that dictate their own fate and fortunes.

Each state, whether a core or auxiliary member, considers the group's apparent prospects for victory when deciding whether to negotiate with the enemy.[13] Warring parties consider the appeal of fighting versus seeking settlement by observing outcomes on the battlefield, which provide information about the future trajectory of the conflict.[14] I will refer to recent trends in fighting as *battlefield momentum*. If an asymmetric coalition enjoys favourable momentum on the battlefield, auxiliary states will have greater incentives to maintain silence and continue fighting. Doing so not only leads to the provision of the coalition's common goal of security but is the most viable path to obtain one's desired private benefits in a final settlement with a defeated adversary. The existence of a powerful coalition member augments the likelihood that the terms of agreement can be credibly enforced among the coalition members themselves,[15] bolstering the appeal of remaining in the conflict and opting not to prematurely negotiate with the enemy. Feeling confident in victory, core states will also continue fighting to realise both public and private benefits.

The 1864–70 War of the Triple Alliance, or the Lopez War, is a distinct example of an asymmetric coalition experiencing success on the battlefield. One of the most devastating wars in the Americas, the conflict pitted Francisco Solano López of Paraguay against Brazil, Argentina, and Uruguay. While López's army of 80,000 managed to claim some early successes in late 1864, the Triple Alliance sustained a steady stream of victories thereafter. This coalition was highly asymmetric, with Brazil providing the lion's share of resources and suffering the highest losses. Uruguay only had 3,000 soldiers, Argentina peaked at 16,000 during the war, and Brazil eventually amassed an Imperial Army of over 111,000.[16] Brazil came to dictate the alliance's diplomatic strategy. All three states sought López's complete defeat, but Brazil maintained the most publicly hardline stance throughout the war and became the state that regional neighbours often contacted as the alliance's representative.[17]

However, the calculus changes when an asymmetric coalition experiences negative battlefield momentum that suggests possible defeat. An auxiliary state which believes that its material contribution to the coalition has small marginal effect, or that it has limited resources to defend itself from harm, may seek to minimise its own private costs via negotiations. This may help to explain why smaller states consider exiting their coalition if victory seems out of reach.[18] Concurrently, a beleaguered core state will act and feel like a state fighting alone. Extant research, which presumes monolithic belligerents, has demonstrated that negotiated settlements are most likely when battlefield momentum distinctly favours one side: The disadvantaged party becomes more willing to make concessions and the advantaged side eventually becomes willing to accept an increasingly large concession.[19] Both auxiliary and core states will thus seek to negotiate their way out of suffering further losses.

This was exhibited in the 1879–83 War of the Pacific, which pitted Chile against a heavily asymmetric coalition of Peru and Bolivia. Bolivia was the clear auxiliary state, providing about 8,000 poorly equipped troops and no naval forces compared to Peru's 40,000 troops and four ironclad ships.[20] Bolivia was also the state most eager to engage in diplomacy, especially as Chilean forces gained the upper hand in late 1879 and 1880. The government in La Paz proactively sought mediation by the United States in 1879, and it informally accepted an American mediation offer in September 1880 after Chilean forces won a decisive battle in Tacna (a region of Peru) that effectively neutralised the Bolivian military and left Peru to fight alone. Consequently, Peru also accepted this American offer two weeks later.[21] The resulting conference in Lackawanna in late October 1880 failed to forge peace. It was only in 1883, once Chile essentially created and recognised a separate Peruvian government, that the two belligerents forged peace.[22]

Transitioning away from asymmetric coalitions, now consider a coalition that is symmetric in that each member supplies comparable levels of military force, rendering the terms 'core' and 'auxiliary' irrelevant. Each member may not add as much force in absolute terms, but each member also has a larger marginal impact on a coalition's ability to avoid defeat and achieve their overlapping interests. Any single member's exit may put the remaining coalition below the threshold of force necessary to realise either public or private benefits. This suggests that when a symmetric coalition worries about potential defeat, all members – even as they suffer significant private costs – would recognise that their continued participation is vital to preventing an overall loss. In the face of unfavourable momentum, states in a symmetric coalition should be more willing to abstain from negotiations and continue fighting.

The United Nations Command (UNC), a coalition of multinational forces militarily led by the United States, implicitly demonstrated this logic during the Korean War.[23] As the war literally went south following China's intervention in late 1950, the Truman administration contemplated whether the UNC should seek negotiations with North Korea and China. The idea was eventually shot down over fears that such a gesture, particularly as the Communist states made significant progress, would indicate weakness and undercut any bargaining leverage.[24] It was only once lines of control drifted back to a more equitable position and became stagnant that negotiations began in July 1951.

If each state's contributions are essential to realising the group's over-arching aims, this also implies that the coalition's members were more likely to have set aside or overlooked divergences in their specific or long-term policy objectives because they required one another to achieve a common goal in the shorter term. A logical consequence of this situation is that when a symmetric coalition experiences favourable battlefield momentum and surmises that the commonly shared benefit of victory is likely, each member becomes more concerned with securing and codifying its particular private interests.[25] Each state may harbour concerns about whether the terms of a comprehensive negotiated agreement would be upheld by their fellow members,[26] who are of relatively equal strength. Doubts about the stability of a peace among all members may motivate a state to seize agenda-setting power to seek a more favourable peace on its own terms.

The Seven Weeks War of 1866 presents an instructive example. Squaring off against Austria, Prussia and Italy constituted a somewhat symmetric

		Battlefield Momentum	
		Unfavourable	*Favourable*
Coalition Asymmetry	*Asymmetric*	High	Low
	Symmetric	Low	High

Figure 10.1 Expected Likelihood of Negotiations Taking Place during War, Conditional on Battlefield Momentum and a Coalition's Asymmetry.

coalition. Prussia fielded approximately two-thirds of all active troops, while Italy supplied one-third.[27] As soon as Bismarck believed he had accomplished his goal of breaking apart the German Confederation, he promptly sought negotiations with Austria, leaving Italy out of these machinations. The preliminary peace signed in Nikolsburg only involved Prussia and Austria, forcing Italy to hurriedly strike its own armistice with Austria two weeks later. The final terms of peace that involved all belligerents eventually gave Italy its key prize of Venetia from the Austrians but did so in an indirect manner,[28] forcing Italy to pay a cash indemnity and creating national humiliation.[29]

This discussion leads to four principles and observable implications regarding the propensity for wartime negotiations, summarised in Figure 10.1. Members of asymmetric coalitions stick together to maximise the common and private terms of victory but splinter to minimise the private terms of defeat, while members of symmetric coalitions splinter to maximise the private terms of victory but stick together to minimise common and private terms of defeat.

One may be concerned that differences in negotiating behaviour between asymmetric and symmetric coalitions are driven by purely functional considerations. Asymmetric coalitions feature a leader with apparent decision-making power, while symmetric coalitions feature leaders with a multiplicity of goals and much more ambiguous decision-making procedures, which facilitates divisions and undercuts the coalition's ability to speak with one voice. These factors are certainly relevant, yet they exist irrespective of battlefield activity. Any evidence that different coalition structures respond differently to identical battlefield conditions would offer compelling evidence that coalitions' negotiation behaviour is not merely dictated by logistical issues, but also a deeper calculus concerning each coalition member's balance of public and private costs and benefits.

Importantly, my argument does not invalidate the potential role of non-rational factors in shaping attitudes toward diplomacy. For example, the desire to restore one's honour may undermine willingness to negotiate even when battlefield realities are dire.[30] Nonetheless, a purely non-rational argument may not produce sufficient predictions on its own. Confirmation

of my argument would point to a broad, reliable baseline of rational negotiation behaviour, which can be modulated by psychological and emotional
considerations.

We now turn to empirical analysis to assess these claims.

10.3 Analysis

I adopt a mixed-methods approach that brings to bear two forms of evidence, each with its strengths. A brief statistical analysis of interstate wars
since 1816 reveals the broad validity of the aforementioned implications
across multiple conflicts. Thereafter, a qualitative review of the Allies in the
Crimean War permits a much closer view of the strategic dynamics that
shape each member's incentives to negotiate in the midst of armed conflict.

10.3.1 Quantitative Evidence from Two Centuries of War

We begin with a brief statistical study of multiple conflicts. The initial list of
candidate conflicts included in the analysis comes from the Correlates of War
(COW) dataset, which enumerates 95 violent interstate clashes involving at
least 1,000 casualties between 1816 and the present day.

Previous work by Morey identifies COW conflicts where states actively
coordinated their political and/or military activities during conflict as coalition wars. A total of 26 coalition wars are included in this analysis.[31] These
are distinguished from 10 other multilateral conflicts, which Morey calls
'wars in parallel,' where multiple belligerents fought against a common
enemy but did not engage in meaningful coordination.[32]

Lacking data on intra-war activity, most quantitative studies of war have
used the entire conflict as a single unit of analysis, precluding an intra-
conflict study of the relationship between the battlefield and the bargaining
table. New data from addresses this situation by recording negotiations and
battles that took place across these wars.[33]

For each day of an Interstate War, the negotiation data records whether
formal representatives from at least one actor from each side of the conflict
engaged in direct or mediated communication with the ostensible aim of
reaching a mutually acceptable agreement. These discussions can either be
publicly known or secret at the time, and they need not result in any
meaningful progress. Approximately 23% of days in the data feature
negotiations that satisfy these criteria. The data do not identify which specific states of any relevant coalition participated, but note that my argument
does not hinge upon the identity of specific participants. In asymmetric
coalitions, both core and auxiliary members have their own reasons to seek
negotiations when battlefield momentum is unfavourable and the odds of
defeat increase. In symmetric coalitions, all members have incentives to
negotiate a deal that prioritises their private benefits when battlefield
momentum is favourable and the odds of victory increase.

These diplomatic data are paired with corresponding information on battlefield hostilities. Battles are a distinct unit of analysis that leaders use to understand and plan military situations.[34] The Interstate War Battle Dataset, or IWBD, features information on over 1,700 battles across international wars from 1816 to the present.[35] A battle is defined here as a clash at a specific time and location between organised state-level forces over a contested strategic objective.[36] The IWBD greatly extends and improves upon the U.S. Army's Concepts Analysis Agency Database of Battles Version 1990, or CDB-90.[37] which is an extant resource that contains information on approximately 500 battles across the same period.

For each battle, the IWBD contains information on the date(s) of the clash; which side first attacked the other; and the ultimate outcome. The IWBD does not identify which specific states participated in battles during coalition wars. However, this does not unduly impact our ability to test my theoretical argument, which implies that victories or losses experienced by any subset of coalition members will still affect strategic calculations regarding diplomacy for all coalition members.

Each battle may either end with one of the two sides claiming victory (either by seizing a strategic objective central to the battle from the enemy or defending an objective from being taken by the enemy) or with an inconclusive outcome. In the present analysis, I assign a score of +1 when one side wins a battle, −1 when it loses a battle, and 0 if inconclusive. I then gauge battlefield momentum by calculating the running sum of all battle outcomes for one side over the previous 60 days of fighting.

Finally, I create a measure of coalition asymmetry using a variant of the Herfindahl-Hirschman index (HHI), which is typically used to gauge inter-firm competition but is also adapted to track ethnic fractionalisation.[38] Suppose one side of a conflict consists of n states and that s_i represents the share of military resources possessed by coalition member i. For each side of a conflict, the following formula defines a coalition's asymmetry:

$$\sum_{i=1}^{n} s_i^2 = s_1^2 + s_2^2 + \ldots + s_n^2$$

This measure of resource concentration can theoretically range between $1/n$ (completely even contributions by all coalition states) and 1 (all contributions by a single state). I gauge states' contributions in terms of their overall military personnel per year as reported in the National Military Capabilities dataset.[39] It bears noting that this is an admittedly blunt measure of asymmetry that is based solely on states' total military personnel. It does not reflect the likely lower numbers of troops that were deployed in a war, the effectiveness of each state's military forces, or states' other material and economic contributions during the conflict. Nonetheless, this coalition asymmetry measure represents the most consistent metric that can be produced across all relevant wars, and it does reflect broad power differentials between coalition members that likely affect the functional asymmetry between them.

The unit of analysis throughout the quantitative study is the side-war-day, where each observation represents a single day of war for one of the two sides of the conflict. A 'side' may be a single state or a coalition, but every war included in this analysis has at least one side that is a coalition. A total of 31,626 side-war-days, representing 15,813 days of hostilities across 26 coalition wars, constitute the core data. The main outcome of interest is the occurrence of negotiations on each side-war-day. I thus utilise logistic regressions, which are appropriate for studying binary outcomes. The main explanatory variables are battlefield momentum, coalition asymmetry, and most importantly, the interaction of these two dimensions.

The statistical analysis accounts for several factors that may confound the relationship between the coalition's composition, performance, and negotiation behaviour. These include the primary issues at hand in the war, which may affect belligerents' motivation to fight; the presence of democratic states, who may be more effective but impatient belligerents; the participation of major or nuclear powers, who may seek to keep conflicts limited; the entry and exit of coalition members over the previous 60 days, which may influence current belligerents' interests in negotiating; the opponent's level of asymmetry; and the number of states in the coalition. I also account for the relative military capabilities of each side. Doing so is crucial, as it permits the analysis to focus on whether the relative contributions of multiple actors affect bargaining behaviour, independent of the overall strength that the coalition possesses. Finally, we may expect that coalitions with more members will face greater risks of collective action problems from member states defecting, free-riding, and pursuing their own goals. Larger coalitions may also have lower asymmetry measures due to how the metric is calculated, as the lowest value possible is $1/n$. I therefore include a variable accounting for the number of states fighting on the relevant side.

Full regression results and more technical details are available in the Appendix.[40] Figure 10.2 succinctly captures the key finding in a visual manner. This diagram shows the predicted probability of a belligerent engaging in negotiations on a given war-day, conditional on their asymmetry and battlefield momentum.

The diagram mirrors the expectations summarised in Figure 10.1 and illustrates a clear link between the state of the battlefield and the character of a coalition. The right half of the image represents situations where recent fighting favours the side in question, while the top half represents conditions in which the side is highly asymmetric. As such, the right-hand portion of the figure indicates that when coalitions are fighting effectively, asymmetric belligerents are less likely to engage in negotiations that show any interest in slowing down or settling hostilities, while symmetric parties will be much likelier to engage in talks with the embattled adversary. The left-hand side

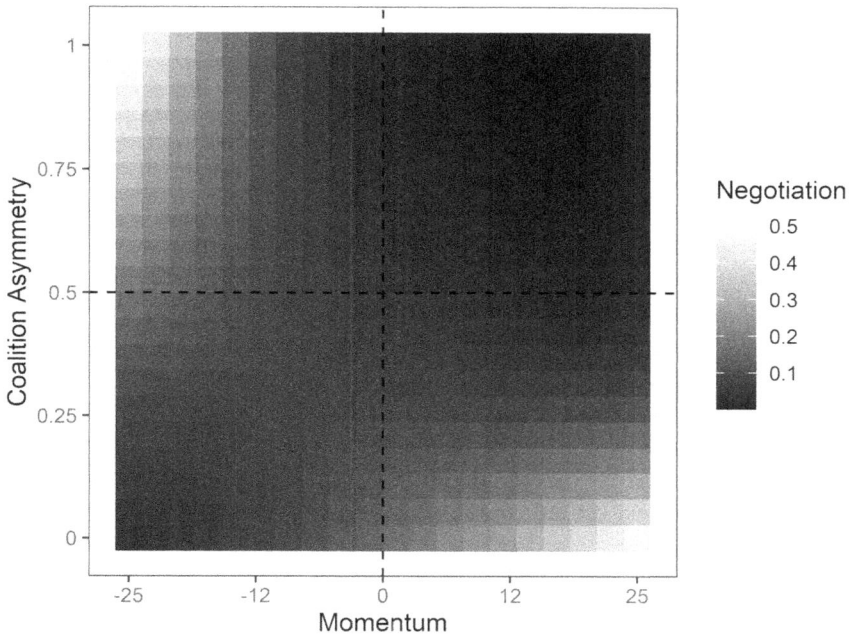

Figure 10.2 Predicted Probability of Negotiation on a War-Day, Conditional on Battlefield Momentum and Belligerent's Asymmetry.

shows that when coalitions' recent battles are trending poorly, members of asymmetric coalitions are more likely to engage in negotiations than states from symmetric groups. The highest likelihood for negotiations overall exists when a highly asymmetric coalition suffers significant losses in recent battlefield hostilities (the top left corner). These effects are not only substantively meaningful but also statistically significant, indicating that these results are not a consequence of random chance. Such findings precisely overlap with the theoretical implications previously established and emphasise how states' diplomatic approaches are strategically affected by their contribution to the coalition.[41]

Additional statistical analysis in the Appendix indicates that negotiations which take place alongside particularly high levels of battlefield momentum are far more likely to result in early exits by at least one individual coalition member. Meanwhile, negotiations are associated with a higher likelihood of war termination, regardless of the state of the battlefield.[42] These results buttress the claim that coalition members are most liable to obtain their preferred private interests through diplomacy, whether that involves minimising costs or codifying a particular policy outcome, when the conflict appears to be trending strongly in one side's favour.

These quantitative findings, which involve numerous conflicts spanning time and space, broadly support my argument regarding when coalition members are likely to engage in negotiations during war. However, this large-scale study uses a highly coarse measure of coalition asymmetry and cannot reveal whether the underlying reasons why coalition members talk to the enemy are consistent with my theoretical claims. To address these issues, I now turn to a qualitative analysis of the negotiation behaviour and strategies of the Allied states in the Crimean War of 1853–56. This war is not only historically consequential but is substantively relevant, as the Allied side features major powers – a situation mirrored in recent coalitions featuring the United States.

10.3.2 *Qualitative Evidence from the Crimean War*

Over the course of the early nineteenth century, European states became increasingly aware of the Ottoman Empire's decline and internal fragmentation. The Turks' most recent war defeat to Russia in 1828–29 highlighted the fact that the empire's still vast territorial claims could be up for grabs.[43] In 1850, a small dispute regarding the rights of Christian minorities in the Holy Land in Palestine, which was part of the Ottoman Empire, became wrapped up in this larger picture. France pressed the Turkish government to support the Catholics, while Russia sought to empower the Orthodox Christians. The major powers of Britain, France, Austria, and Russia engaged in numerous talks to find a diplomatic solution. Yet for reasons not necessary to review here, by late June 1853, Tsar Nicholas I pressed the issue and sent forces into the Danubian Principalities of Moldavia and Wallachia to take control by force. All subsequent diplomatic options appearing to fail, including up to eleven attempted peace proposals,[44] the aggrieved Ottomans formally declared war against Russia in October 1853 and believed that Britain and France would support the move. Figure 10.3 illustrates the trajectory of battlefield activity during the entire war, using measures of fighting created with the battle data described in the quantitative analysis. The left-hand side of the figure makes clear that in late 1853 and early 1854, the Russians enjoyed success in battles at Sinope and Bashgedikler as the Turks fought alone.

For several months, Britain and France – already concerned about Russia's expansionist goals – experienced deteriorating relations with Nicholas I, who refused to withdraw from the Principalities. The conflict became a coalitional war on 27 and 28 March 1854, when Great Britain and France respectively and formally declared war against Russia. The Crimean War from this point forward represents a telling example of how diplomatic activity is complicated by the involvement of multiple active and influential coalition members.

The Allies of the Crimean War constituted a relatively symmetric coalition. At its peak, Russia marshalled a total of 888,000 troops. The Allies

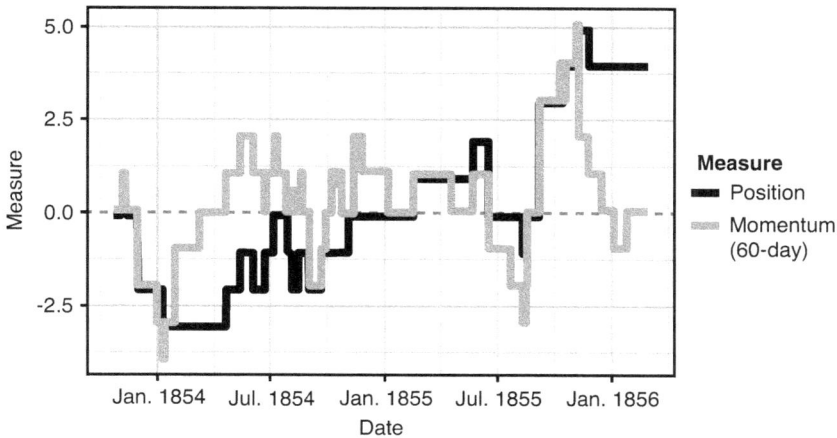

Figure 10.3 Fighting during the Crimean War. *Note: Positive values represent success for the Allies; negative values represent success for Russia. Position represents the sum of all battle outcomes; momentum is the sum of all battle outcomes in the last 60 days.*

eventually had a combined strength of less than 600,000. The Ottomans constituted 165,000 individuals, while Great Britain and France offered 98,000 and 309,000 soldiers respectively. Upon entering the war much later in January 1855, Sardinia provided an additional 21,000, though they proved largely unimportant to active hostilities.[45] If we were to calculate the coalition asymmetry measure using these statistics, we get 0.40 before Sardinia's entry and 0.38 afterwards. These numbers are similar to the coalition asymmetry measures of 0.38 and 0.39 that emerge from my quantitative data of overall military personnel.[46] This places the Allies in the top third of most symmetric coalitions across all coalitional wars.

Britain and France entered the war with common immediate motivations. Both wanted Russia to leave the Principalities and to cease further expansion in Europe, particularly as the Ottomans slipped into obvious decline. However, this shared short-term goal gave both Prime Minister Lord Palmerston and Napoleon III the space to discount the fact that they possessed different plans for how the war could forward their broader policy objectives. The two leaders generally knew of each other's grandiose visions about the transformation of Europe, and while their visions were not mutually exclusive, they did not exhibit the same set of priorities.

For his part, Palmerston sought to dismember Russia and to distribute each of its parts among the great powers. The Russians' taking of the Principalities was a vivid sign of the Ottoman Empire's vulnerability and the spectre of Russia rapidly expanding its territorial claims. Although Napoleon III also worried about Russian aggression, he primarily sought the collapse of the European order established in 1815, which had been explicitly designed to hobble France. This would involve breaking the Holy Alliance between Austria, Prussia, and Russia – the caretakers of the present order – and unleashing tides of nationalism and popular sovereignty across the continent and creating a 'United States of Europe.'[47] These steps together would help Napoleon III boost his shaky domestic and international standing.

The dissonance between these somewhat fantastical objectives was workable at the outset, as the Allies were focused on the immediate need to push back against Russian troops in the Principalities. Over the late spring and early summer of 1854, the Allies turned the tides of conflict, successfully repelling Russian warships at Odessa in April 1854 and withstanding a siege at Silistria between May and June 1854. Given that the Allies were attempting to mitigate previous losses to the Russians, no coalition member had an interest in adopting an open diplomatic posture that could plausibly lead to exit from the conflict. It is therefore unsurprising that no negotiations took place over this period.

The war entered a new and more complicated phase in the summer of 1854. Throughout the entire conflict, Austria had walked a fine line in remaining neutral. On one hand, both Austria and Russia were members of the Holy Alliance, and Russia had helped Austria quell a Hungarian revolt in 1849. On the other hand, Austria did not approve of Russia's provocations and was open to the Allies restoring the pre-war status quo. Austria swung sharply toward the Allies on 3 June, when it threatened to join the Allies on the battlefield if Russia did not evacuate the Principalities. Feeling betrayed, Nicholas I announced on 7 August that Russian forces would withdraw from the Principalities. This move shocked the Allies, as Russia had effectively ceded the main issue which had triggered the war.[48] The British and French governments were suddenly forced to contemplate their diplomatic posture and whether they wanted to negotiate an end to the war or continue pressing further on their own ambitions.

Interpreting Nicholas I's decision as a sign of vulnerability, the Allies and Austria ultimately chose to exploit the Allies' suddenly favourable position to realise broader goals that would further tilt the European balance of power. On 8 August 1854, one day after the withdrawal announcement, Great Britain, France, and Austria exchanged notes over what the policy goals of the war should be. These would become known as the 'Four Points' that summarised Allied conditions for a negotiated peace. The first point, and most immediately tied to the initiation of hostilities, demanded that Russia give up its claims to the Principalities and that they be administered collectively by the great powers. The second guaranteed unrestricted navigation of the Danube River. The third sought a revision of a key 1841 convention regarding the passage of warships in the Dardanelles, which allows passage to the Black Sea. The fourth declared that Russia end any claims to protecting the Orthodox populations in the Ottoman Empire. This was an attempt to redefine the war in broader terms to justify the extension of the war to match great power concerns.[49] Yet again, these more expansive goals did not draw many bright lines. A subsequent 'fifth' point asserted that the Allies could later add more points to the list. By August 16, the Allies won one of their first offensive battles at Bomarsund, shifting away from a mostly defensive stance. An undeterred Nicholas I angrily rejected the terms wholesale on 3 September.

In October, as the battlefield entered a phase of stalemate, Austria led Britain and France to sign an alliance document pledging not to strike a separate peace with Russia and to base any negotiated settlement on the Four Points.[50] Austria went one step further in December, agreeing to consult one another once again if peace was not reached by year's end.[51] This was a major development: France had succeeded in splitting Austria from Russia, tearing apart the Holy Alliance and fulfilling one of Napoleon III's key policy objectives. France did not yet consider exiting the conflict, as Napoleon III hoped that the Allies' efforts could now help fulfil his more fanciful goal of spreading nationalism. However, Napoleon III had realised a vital private policy objective and required fewer reasons to consider a diplomatic settlement.

Russia responded to each development by verbally agreeing to accept the Allies' Four Points as the basis of talks. Britain and France both worried that Russia's statements were a ploy to goad Austria into pushing them toward a premature peace agreement, even though Palmerston and Napoleon III sought further progress toward their broader aims.[52] They in fact underestimated the situation: Count Boul, the foreign minister of Austria, was instrumental in convincing the Russians to express interest in talks. Neither Britain nor France was interested in settlement but felt compelled to accept negotiations that were organised by their new non-belligerent ally.

These forces came to a head from 15 March to 4 June 1855, when the four parties met in Vienna to have the first formal negotiations of the war. Austria provided a compromise solution to push talks forward, but the effort was essentially dead on arrival. Even though Britain and France could not refuse to negotiate, they intentionally sank prospects for a diplomatic agreement by presenting unacceptable proposals. British and French officials used the 'fifth' point to redefine the third point, demanding that Russia severely and unilaterally draw back its naval presence in the Dardanelles. Austria privately promised Russia (now led by Tsar Alexander II following the death of Nicholas I in early March) that it would not support this proposal. The British and French aversion to peace became most apparent when their diplomats managed to formulate a complex deal on naval presence in the Dardanelles with Russia, only to be disavowed by their own governments.[53] The Vienna Conference laid bare the disunity of positions between Britain and France on one side and Austria on the other. The Four Points were exposed as an insincere set of conditions by parties with no genuine interest in peace. This public schism may have helped Russia avoid diplomatic defeat and resulted in chilled relations between Austria and its two allies that were actually on the battlefield. Two weeks after negotiations ended, Russia temporarily reclaimed some initiative. Russian forces defended positions in battles at Redan and Malakov in June, stalling the Allies and forcing them to continue fighting together without negotiations to avoid an impasse or defeat.

However, major Allied victories in early September 1855 marked a decisive inflection point that cast light on the widening gaps in policy objectives between Britain and France. Allied forces' victories at Redan and Malakov on 8 September, which broke through Russian defences that had held months before, were crucial to the subsequent fall of Sevastopol – the port city which hosted the Russian Black Sea Fleet and wielded control over the Mediterranean Sea – after an eleven-month siege. Figure 10.2 captures this dramatic upswing in the Allies' fortunes. The loss of Sevastopol severely weakened Russia's ability to prosecute the war; most battles from this point on were initiated and won by the Allies. Hearkening back to his overarching goal of dismantling Russia, Palmerston wanted to continue fighting in Russia to fully destroy its military might. Napoleon III was meanwhile focused on spreading nationalism across Europe and wanted to do so by restoring Poland, which was then part of the Russian empire. This would involve extending the war into Central Europe. Palmerston eventually made clear that he had no interest in shifting hostilities into Poland or pursuing popular sovereignty. Napoleon III knew that he could not liberate Poland using French strength alone, but he also knew that he had his other goal of breaking apart the 1815 order by riving the Holy Alliance. Realising that he could not achieve his wider aims, Napoleon III embraced his own open diplomatic posture and sought to end the war on his own terms without being pulled into Palmerston's plans of continuing the war and losing lives in Russia.

Between September and November of 1855, as the Allies claimed additional victories at Kinburn and Ingur, Austrian Foreign Minister Karl Ferdinand von Buol and French Ambassador Adolph de Bourqueney created an updated proposal for peace that revised the Four Points. Central to these revisions was the third point regarding naval presence in the Dardanelles: Instead of pushing Russia to eliminate its naval forces unilaterally, the entire Black Sea area would be neutralised. Austria would present this document to Russia as an ultimatum, threatening to enter the war if Russia refused. The two finalised the proposal on 14 November and only sent the terms belatedly to Britain on November 18.[54]

Palmerston's government was blindsided. British foreign minister Lord Clarendon even remarked that Austria was 'cramming her ultimatum down our throats rather than that of Russia.'[55] The cabinet discussed the terms on November 20 but from a position of impotence. The British had been locked out of the conversation and could not make serious changes to the drafted ultimatum. It was also apparent that Britain could not prosecute the war against Russia without France and that doing so would garner condemnation across Europe and even the United States. In subsequent discussions between Britain and France, British officials ended up dropping most of their already minor requests for adjustments.

The final Austrian-French ultimatum was given to Russia on 28 December 1855. The Russian government acceded to the ultimatum on 16 January 1856, and hostilities ended on 29 February. The Treaty of Paris formalised terms on 14 March and hostilities officially ended on 30 March. Russia was humiliated, lost control over the Black Sea, and ceded a small strip of land to the Ottomans to ensure the security of the Danubian Principalities,[56] but it was nowhere near as weakened as Britain had sought. Indeed, by 1870, Russia managed to reclaim much of what it lost during this conflict.

The Crimean War illustrates the promises and pitfalls of having a symmetric coalition in a conflict. Because of the high marginal contributions each state made as well as the enormity of their rival's forces, Britain and France knew that they relied on one another to have any success. This proved valuable in the early stages of the conflict when the Allies collectively sought to reverse Russia's gains against the Turks and occupation of the Principalities – common and proximate goals that would provide them public benefits in terms of security and regional stability. Unity also proved important in the summer of 1855 when Russian forces briefly managed to blunt the Allies' attacks. However, when the Allies had the upper hand and sensed the potential for victory, private goals became more pressing and undercut their ability to adopt a unified negotiating stance. Since France had accomplished its private goal of ending the post-1815 order and realised Britain would not support its nationalist agenda, Napoleon III adopted an open diplomatic posture to satisfy his own ends and dragged Britain into ending the conflict. He chose to stop the war because France could not liberate Poland without British troops. Palmerston reluctantly accepted peace, even though much of Russia's power was left intact because it could not fight without French support.

It is worth emphasising that the war would likely have gone far beyond the siege of Sevastopol if the symmetric coalition were hypothetically replaced by an asymmetric coalition where either Britain or France made overwhelming contributions in forces, or by one of the two countries fighting alone with just as many forces. Under British command, troops would have kept battering Russia with hopes of crippling the empire's reach. Under French command, troops would have pressed forward into Central Europe to bring nationalism to Poland.

10.4 Conclusion

Even though coalition members may have acute collective interests that tie them together, working together successfully is another matter. Many works planted in military practice emphasise the enormous logistical frictions and complications that arise from coalition warfighting, which may engender an overall effort that is somewhat less than the sum of its parts. A quotation from Napoleon Bonaparte represents this belief: 'If I must make war, I prefer it to be against a coalition.'[57] More than a century later, Churchill adds that '[t]he history of all coalitions is a tale of the reciprocal complaints of allies.'[58] If frictions exist in orchestrating battlefield activity, similar complications can logically exist when states attempt to shape their diplomatic postures with respect to the war. This chapter has outlined a framework to understand when and how those complications arise. The composition of the coalition and its recent fortunes on the battlefield play decisive roles in shaping the extent to which coalition members prioritise public and common goals over their own private and parochial ones.

All that said, one key fact must be reiterated: Coalitions do not appear randomly; they are created. My findings therefore highlight the importance of thoughtful planning when forming a coalition designed to engage in conflict. When a state can largely fight a war on its own, it may consider adding minor coalition members to bolster an image of legitimacy or international support. This may be fruitful if the coalition is sure to succeed, but it may backfire and prematurely trigger pressures to negotiate if hostilities trend in the wrong direction. Efforts should be made to maximise success or to only include states that offer important strategic benefits to help the core member accomplish its key objectives. Conversely, if a coalition must be symmetric out of material necessity, it is essential that members establish a unified and transparent strategy prior to engaging in combat. At least two options exist here. First, prior to conflict, states may agree upon a common and specific set of policy goals that dictate their negotiation strategy in order to ensure that no single party deviates to pursue a separate agenda that endangers the collective endeavour. Second, all members can establish a clear system of leadership by ceding operational control of military forces and political decision-making power to a single state. The Republic of Korea (ROK) supplied almost half of all forces in the Korean War, but the ROK Army granted all control to the United Nations Command. Indeed, the coalition's imbalance was most evident at the bargaining table: South Korean representatives did not speak a single time during the armistice negotiations between 1951 and 1953.

The future of combat is likely to become even more reliant on multilateral coalitions than we have seen over the last two centuries.[59] Thus, scholars and policymakers must heed the growing necessity of understanding how coalitions function in both battlefield effectiveness and diplomatic behaviour, both of which critically inform the overall nature of global conflict.

Notes

1 Funding details: The data used in this work was partially supported by the National Science Foundation's Graduate Research Fellowship #DGE-114747.
2 Ajin Choi, 'Fighting to the Finish: Democracy and Commitment in Coalition War', *Security Studies* 21, no. 4 (2012), 624–53, https://doi.org/10.1080/0963 6412.2012.734232; Alex Weisiger, 'Exiting the Coalition: When Do States

Abandon Coalition Partners during War?', *International Studies Quarterly* 60, no. 4 (2016), 753–65, https://doi.org/10.1080/09636412.2012.734232.

3 Daniel S. Morey, 'Military Coalitions and the Outcome of Interstate Wars', *Foreign Policy Analysis* 12/4 (2016), 533–51, https://doi.org/10.1111/fpa.12083.

4 Atsushi Tago, 'Determinants of Multilateralism in US Use of Force: State of Economy, Election Cycle, and Divided Government', *Journal of Peace Research* 42, no. 5 (2005), 146–57, https://doi.org/10.1177/0022343305056235.

5 James D. Fearon, 'Rationalist Explanations for War', *International Organization* 49, no. 3 (1995), 379–414, https://doi.org/10.1017/S0020818300033324.

6 Robert Powell, 'War as a Commitment Problem', *International Organization* 60, no. 1 (2006), 169–203, https://doi.org/10.1017/S0020818306060061.

7 Bahar Leventoğlu and Branislav L. Slantchev, 'The Armed Peace: A Punctuated Equilibrium Theory of War', *American Journal of Political Science* 51, no. 4 (2007), 755–71, https://doi.org/10.1111/j.1540-5907.2007.00279.x.

8 Eric Min, 'Talking While Fighting: Understanding the Role of Wartime Negotiations', *International Organization* 74, no. 3 (2020), 610–32, https://doi.org/10.1017/S0020818320000168.

9 Michael G. Findley, 'Bargaining and the Interdependent Stages of Civil War Resolution', *Journal of Conflict Resolution* 57/5 (2012), 905–32, https://doi.org/10.1177/0022002712453703; Jeffrey M. Kaplow, 'The Negotiation Calculus: Why Parties to Civil Conflict Refuse to Talk', *International Studies Quarterly* 60/1 (2016), 38–46, https://doi.org/10.1093/isq/sqv005.

10 Oriana S. Mastro, *The Costs of Conversation: Obstacles to Peace Talks in Wartime* (Ithaca, NY: Cornell University Press, 2019).

11 Morey, 'Military Coalitions'; Patricia A. Weitsman, *Waging War: Alliances, Coalitions, and Institutions of Interstate Violence* (Stanford, CA: Stanford University Press, 2014).

12 See also Rosella Capella Zielinski and Ryan Grauer, 'Organizing for performance: coalition effectiveness on the battlefield', *European Journal of International Relations* 26, no. 4 (2020), 953–78, https://doi.org/10.1177/1354066120903369. In the quantitative analysis, I define a measure of coalition asymmetry that can range between $1/n$ (identical contributions by all n members) and 1 (all contributions from a single member). The median level of asymmetry across all wartime coalitions is 0.66. This number is calculated after omitting all belligerents that fight alone. Coalitions with asymmetry measures lower than 0.66 can be considered more symmetric, while those higher than 0.66 are more asymmetric.

13 Returning briefly to the hunting metaphor, group members that manage to tightly surround a stag will be more likely to continue their pursuit, while members of a group repeatedly injured by an aggressive stag may consider seeking rabbits instead.

14 Kristopher W. Ramsay, 'Information, Uncertainty, and War', *Annual Review of Political Science* 20, no. 1 (2017), 505–27, https://doi.org/10.1146/annurev-polisci-051215-022729; Dan Reiter, 'Exploring the Bargaining Model of War', *Perspectives on Politics* 1, no. 1 (2003), 27–43, https://doi.org/10.1017/S1537592703000033.

15 Scott Wolford, 'The Problem of Shared Victory: War-Winning Coalitions and Postwar Peace', *Journal of Politics* 79, no. 2 (2017), 702–16, https://doi.org/10.1086/688700.

16 Michael Clodfelter, *Warfare and Armed Conflicts: A Statistical Encyclopedia of Casualty and Other Figures, 1494–2007* (Jefferson, NC: McFarland & Company, 2017), 317–18. Based on these totals, the coalition's asymmetry measure is 0.74, which is higher than the observed median of 0.66 (see footnote 11).

17 Chris Leuchars, *To the Bitter End: Paraguay and the War of the Triple Alliance* (Westport, CT: Greenwood Press, 2002).

18 Weisiger, 'Exiting the Coalition.'
19 Branislav L. Slantchev, 'The Principle of Convergence in Wartime Negotiations', *American Political Science Review* 97, no. 4 (2003), 621–32, https://doi.org/10.1 017/S0003055403000911; Suzanne Werner and Amy Yuen, 'Making and Keeping Peace', *International Organization* 59, no. 2 (2005), 261–92, https://doi.org/10.1017/S0020818305050095.
20 Clodfelter, *Warfare and Armed Conflicts*, 324. This coalition's asymmetry value using these strengths is 0.72, which is higher than the median of 0.66 (see footnote 11).
21 Jorge G. Granier, *United States and the Bolivian Seacoast* (La Paz, Bolivia: Ministerio de Relaciones Exteriores y Culto, 1988).
22 William J. Dennis, *Documentary History of the Tacna-Arica Dispute* (Iowa City, IA: University of Iowa, 1927).
23 Using the UNC's personnel figures from late 1950, the UNC's asymmetry measure is approximately 0.25, which is lower than the median of 0.66 for all coalitions in my data (see footnote 11).
24 George F. Kennan, *Memoirs, 1950–1963* (Boston, MA: Little, Brown, 1972).
25 At the risk of stretching the stag hunt analogy too far, hunters that have surrounded a stag may race one another to kill the stag and take as much of the best meat for themselves as possible before the other hunters do.
26 Wolford, 'The Problem of Shared Victory.'
27 This coalition's asymmetry measure using these numbers is 0.57, which is lower than the median value of 0.66 across all coalitions in my dataset (see footnote 11).
28 Geoffrey Wawro, *The Austro-Prussian War: Austria's War with Prussia and Italy in 1866* (Cambridge: Cambridge University Press, 1993).
29 Bolton King, *A History of Italian Unity: Being a Political History of Italy from 1814 to 1871, Volume II* (London: Nisbet & Co., 1899).
30 Alexander Lanoszka and Michael A. Hunzeker, 'Rage of Honor: Entente Indignation and the Lost Chance for Peace in the First World War', *Security Studies* 24, no. 4 (2015), 662–95, https://doi.org/10.1080/09636412.2015.1103135.
31 Section A of the Appendix features a complete list of coalition wars.
32 Morey, 'Military Coalitions.'
33 Min, 'Talking While Fighting'; Eric Min, 'Interstate War Battle dataset (1823–2003)', *Journal of Peace Research* 58, no. 2 (2021): 294–303, https://doi.org/10.1017/S0020818320000168.
34 Trevor N. Dupuy, *Understanding War: History and Theory of Combat* (New York: Paragon House, 1987).
35 Min, 'Interstate War Battle dataset (1823–2003).'
36 David Eggenberger, *An Encyclopedia of Battles: Accounts of Over 1,560 Battles from 1479 B.C. to the Present'* (New York: Dover Publications, 1985); Tony Jaques, *Dictionary of Battles and Sieges: A Guide to 8,500 Battles from Antiquity through the Twenty-First Century* (Westport, CT: Greenwood Press, 2007).
37 Robert L. Helmbold, 'Personnel Attrition Rates in Historical Land Combat Operations: A Catalog of Attrition and Casualty Data Bases on Diskettes Usable with Personal Computers', Defense Technical Information Center, ADA279069 (1993).
38 William Easterly and Ross Levine, 'Africa's Growth Tragedy: Policies and Ethnic Divisions', *Quarterly Journal of Economics* 112, no. 4 (1997), 1203–50, https://doi.org/10.1162/003355300555466.
39 J. David Singer, 'Reconstructing the Correlates of War Dataset on Material Capabilities of States, 1816–1995', *International Interactions* 14, no. 2 (1987), 115–32, https://doi.org/10.1080/03050628808434695.

40 The Appendix is available online at https://www.tandfonline.com/doi/suppl/10.1 080/01402390.2021.2011232.
41 Full results in Section C of the Appendix show that results are unaffected by removing side-war-days where the side is a single state. They are also not affected by using a 30-day or 90-day temporal window for battlefield momentum, nor are they changed by removing the Korean War (where the United States wielded full operational control of the United Nations Command).
42 Refer to Section D in the Appendix. See also Weisiger, 'Exiting the Coalition.'
43 Bernadotte E. Schmitt, 'The Diplomatic Preliminaries of the Crimean War', *American Historical Review* 25, no. 1 (1919), 36–67, https://doi.org/10.1086/ ahr/25.1.36.
44 Winfried Baumgart, *The Crimean War, 1853–1856* (London, Arnold, 1999), 14.
45 Clodfelter, *Warfare and Armed Conflicts*, 180.
46 Both figures are also lower than the median asymmetry value of 0.66 observed across all coalitions (see footnote 11).
47 David Wetzel, *The Crimean War: A Diplomatic History* (New York: Columbia University Press, 1985).
48 Henry M. Stephens, *Syllabus of Eighty-Seven Lectures on Modern European History* (New York: Macmillan, 1899).
49 Gavin B. Henderson, 'The Two Interpretations of the Four Points, December 1854', *English Historical Review* 52, no. 205 (1937), 48–66, https://doi.org/10.1 093/ehr/LII.CCV.48.
50 Clive Ponting, *The Crimean War: The Truth Behind the Myth* (London: Chatto & Windus, 2004), 216–17.
51 Andrew Lambert, *The Crimean War: British Grand Strategy against Russia, 1853–56* (Burlington, VT: Greenwood Press, 2002).
52 Paul W. Schroeder, *Austria, Great Britain, and the Crimean War* (Ithaca, NY: Cornell University Press, 1972).
53 Baumgart, *The Crimean War*.
54 Ponting, *The Crimean War*, 315–16.
55 J. B. Conacher, *Britain and the Crimea 1855–56: Problems of War and Peace* (New York: Palgrave Macmillan, 1988), 154.
56 Erik Goldstein, *Wars and Peace Treaties: 1816–1991* (New York: Routledge, 1992), 28.
57 Stuart E. Johnson, 'In This Issue', *Joint Force Quarterly* 3/1 (1993–4), 6.
58 Winston S. Churchill, *Marlborough: His Life and Times* (New York: Charles Scribner's Sons, 1937), 246.
59 Joint Chiefs of Staff, *Multinational Operations*, Joint Publication 3–16 (2019); Kathleen J. McInnis, 'Lessons in Coalition Warfare: Past, Present and Implications for the Future', *International Politics Reviews* 1, no. 1 (2013), 78–90, https://doi.org/10.1057/ipr.2013.8.

Bibliography

Baumgart, Winfried.*The Crimean War, 1853–1856*. London: Arnold, 1999.
Capella Zielinski, Rosella, and Ryan Grauer. 'Organizing for Performance: Coalition Effectiveness on the Battlefield.' *European Journal of International Relations* 26, no. 4 (2020): 953–78. 10.1177/1354066120903369.
Choi, Ajin. 'Fighting to the Finish: Democracy and Commitment in Coalition War.' *Security Studies* 21, no. 4 (2012): 624–53. 10.1080/09636412.2012.734232.
Churchill, Winston S. *Marlborough: His Life and Times*. New York: Charles Scribner's Sons, 1937.

Clodfelter, Michael. *Warfare and Armed Conflicts: A Statistical Encyclopedia of Casualty and Other Figures, 1494–2007*. Jefferson, NC: McFarland & Company, 2017.

Conacher, J. B. *Britain and the Crimea 1855–56: Problems of War and Peace*. New York: Palgrave Macmillan, 1988.

Dennis, William J. *Documentary History of the Tacna-Arica Dispute*. Iowa City, IA: University of Iowa, 1927.

Easterly, William, and Ross Levine. 'Africa's Growth Tragedy: Policies and Ethnic Divisions.' *Quarterly Journal of Economics* 112, no. 4 (1997): 1203–50. 10.1162/003355300555466.

Eggenberger, David. *An Encyclopedia of Battles: Accounts of Over 1,560 Battles from 1479 B.C. to the Present*. New York: Dover Publications, 1985.

Fearon, James D. 'Rationalist Explanations for War.' *International Organization* 49, no. 3 (1995): 379–414. 10.1017/S0020818300033324.

Findley, Michael G. 'Bargaining and the Interdependent Stages of Civil War Resolution.' *Journal of Conflict Resolution* 57, no. 5 (2012): 905–32. 10.1177/0022002712453703.

Goldstein, Erik. *Wars and Peace Treaties: 1816–1991*. New York: Routledge, 1992.

Granier, Jorge G.. *United States and the Bolivian Seacoast*. La Paz, Bolivia: Ministerio de Relaciones Exteriores y Culto, 1988.

Helmbold, Robert L. 'Personnel Attrition Rates in Historical Land Combat Operations: A Catalog of Attrition and Casualty Data Bases on Diskettes Usable with Personal Computers', Defense Technical Information Center, ADA279069 (1993).

Henderson, Gavin B. 'The Two Interpretations of the Four Points, December 1854.' *English Historical Review* 52, no. 205 (1937): 48–66. 10.1093/ehr/LII.CCV.48.

Jaques, Tony. *Dictionary of Battles and Sieges: A Guide to 8,500 Battles from Antiquity through the Twenty-First Century*. Westport, CT: Greenwood Press, 2007.

Johnson, Stuart E. 'In This Issue.' *Joint Force Quarterly* 3, no. 1 (1993–1994), 6.

Joint Chiefs of Staff. *Multinational Operations*, Joint Publication 3–16 (2019).

Kaplow, Jeffrey M. 'The Negotiation Calculus: Why Parties to Civil Conflict Refuse to Talk.' *International Studies Quarterly* 60, no. 1 (2016): 38–46. 10.1093/isq/sqv005.

Kennan, George F. *Memoirs, 1950–1963*. Boston, MA: Little, Brown, 1972.

King, Bolton. *A History of Italian Unity: Being a Political History of Italy from 1814 to 1871, Volume II*. London: Nisbet & Co. 1899.

Lambert, Andrew. *The Crimean War: British Grand Strategy against Russia, 1853–56*. Burlington, VT: Greenwood Press, 2002.

Lanoszka, Alexander and Michael A. Hunzeker. 'Rage of Honor: Entente Indignation and the Lost Change for Peace in the First World War.' *Security Studies* 24, no. 4 (2015): 662–95. 10.1080/09636412.2015.1103135.

Leuchars, Chris. *To the Bitter End: Paraguay and the War of the Triple Alliance*. Westport, CT: Greenwood Press, 2002.

Leventoğlu, Bahar and Branislav L. Slantchev. 'The Armed Peace: A Punctuated Equilibrium Theory of War.' *American Journal of Political Science* 51, no. 4 (2007): 755–71. 10.1111/j.1540-5907.2007.00279.x.

Mastro, Oriana S. *The Costs of Conversation: Obstacles to Peace Talks in Wartime*. Ithaca, NY: Cornell University Press, 2019.

McInnis, Kathleen J. 'Lessons in Coalition Warfare: Past, Present and Implications for the Future.' *International Politics Reviews* 1, no. 1 (2013): 78–90. 10.1057/ipr.2013.8.

Min, Eric. 'Interstate War Battle Dataset (1823–2003).' *Journal of Peace Research* 58, no. 2 (2021): 294–303. 10.1177/0022343320913305.

Min, Eric. 'Talking While Fighting: Understanding the Role of Wartime Negotiations.' *International Organization* 74, no. 3 (2020): 610–32. 10.1017/S0020818320000168.

Morey, Daniel S. 'Military Coalitions and the Outcome of Interstate Wars.' *Foreign Policy Analysis* 12, no. 4 (2016): 533–51. 10.1111/fpa.12083.

Ponting, Clive. *The Crimean War: The Truth Behind the Myth*, 216–217. London: Chatto & Windus, 2004.

Powell, Robert. 'War as a Commitment Problem.' *International Organization* 60, no. 1 (2006): 169–203. 10.1017/S0020818306060061.

Ramsay, Kristopher W. 'Information, Uncertainty, and War.' *Annual Review of Political Science* 20, no. 1 (2017): 505–27. 10.1146/annurev-polisci-051215-022729.

Reiter, Dan. 'Exploring the Bargaining Model of War.' *Perspectives on Politics* 1, no. 1 (2003): 27–43. 10.1017/S1537592703000033.

Schmitt, Bernadotte E. 'The Diplomatic Preliminaries of the Crimean War.' *American Historical Review* 25, no. 1 (1919). 10.1086/ahr/25.1.36.

Schroeder, Paul W. *Austria, Great Britain, and the Crimean War*. Ithaca, NY: Cornell University Press, 1972.

Singer, J. David. 'Reconstructing the Correlates of War Dataset on Material Capabilities of States, 1816–1995.' *International Interactions* 14, no. 2 (1987): 115–32. 10.1080/03050628808434695.

Slantchev, Branislav L. 'The Principle of Convergence in Wartime Negotiations.' *American Political Science Review* 97, no. 4 (2003): 621–32. 10.1017/S0003055403000911.

Stephens, Henry M. *Syllabus of Eighty-Seven Lectures on Modern European History*. New York: Macmillan, 1899.

Tago, Atsushi. 'Determinants of Multilateralism in US Use of Force: State of Economy, Election Cycle, and Divided Government.' *Journal of Peace Research* 42, no. 5 (2005): 146–57. 10.1177/0022343305056235.

Wawro, Geoffrey. *The Austro-Prussian War: Austria's War with Prussia and Italy in 1866*. Cambridge: Cambridge University Press, 1993.

Weisiger, Alex. 'Exiting the Coalition: When Do States Abandon Coalition Partners during War?.' *International Studies Quarterly* 60, no. 4 (2016): 753–65. 10.1093/isq/sqw029.

Weitsman, Patricia A. *Waging War: Alliances, Coalitions, and Institutions of Interstate Violence*. Stanford, CA: Stanford University Press, 2014.

Werner, Suzanne and Amy Yuen. 'Making and Keeping Peace.' *International Organization* 59, no. 2 (2005): 261–92. 10.1017/S0020818305050095.

Wetzel, David. *The Crimean War: A Diplomatic History*. New York: Columbia University Press, 1985.

Wolford, Scott. 'The Problem of Shared Victory: War-Winning Coalitions and Postwar Peace.' *Journal of Peace* 79, no. 2 (2017): 702–16. 10.1086/688700.

Part V
Looking Ahead

11 Next Steps in the Study of Battlefield Coalitions

Alex Weisiger

This volume provides the first in-depth examination of battlefield coalition in political science. Alliances have, of course, been an enduring focus of international relations research, dating back to Athenian success in converting a balancing alliance against the Persians into an empire. Yet the overwhelming majority of this scholarship has focused on the high politics of alliances – who allies with whom, when states do or do not uphold their alliance commitments, or when alliances end – rather than the details of how states actively coordinate their alliance relations, especially on the battlefield. The contributions in this volume thus substantially improve our understanding of battlefield coalitions.

This conclusion opens by briefly summarising several key observations that emerge from these studies. In particular, I highlight three key implications largely neglected by prior scholarship: that battlefield coalitions evolve over time, that non-state actors frequently are important participants in coalitions, and that battlefield coalitions influence outcomes beyond relations among coalition members and the outcome of war.

These contributions notwithstanding, as the field moves forward, future studies will need to contend with several significant challenges, both theoretical and empirical. These challenges include basic definitional questions about who constitutes an actor, theoretical questions about what differentiates battlefield coalitions from other forms of security cooperation, and, likely most fundamentally, empirical questions about how well findings will travel across wars that vary in the kinds of actors participating in the wars and the military strategies that they adopt. In the end, while some findings may apply equally to, for example, interstate coalitions in the World Wars and rebel coalitions in Afghanistan, others will not, and determining how general a given finding is will prove challenging. Fortunately, there are opportunities for addressing some of these challenges, and for better unpacking how we should expect arguments to travel.

While these challenges will prove daunting, there are important opportunities to further improve our understanding of the politics of battlefield coalitions. In the final section, I discuss potential avenues for improving our understanding of battlefield coalitions. The study of battlefield coalitions

DOI: 10.4324/9781003399896-16

would benefit in particular from more systematic theorising about the strategic options available to coalition members, and from more explicitly integrating military strategy and military technology into theories.

11.1 Contributions of This Volume

The chapters in this volume constitute the cutting edge in the study of battlefield coalitions. While each study makes individual contributions, this section highlights several general observations that each apply to multiple studies.

First, while previous studies have demonstrated the importance of coalition institutions, there has been comparatively little examination of how those institutions evolve over the course of conflict.[1] Thus, while prior work has concluded that integrated command produces better battlefield performance, Moller provides a useful theory of the emergence of integrated command in the face of strong political disincentives to place one's forces under another actor's control, highlighting the importance of experiences of failure.[2] Von Hlatky and Juneau demonstrate within the Libya coalition that the institutional and political relations among coalition members frequently drive coalition behaviour in ways that cannot be explained solely by examining either the preexisting interests of the most powerful states or prior institutionalised practises.[3] Mahoney unpacks the concept of coalition institutions and demonstrates through historical examples that they can vary substantially both across and within wars in substantively important ways.[4] Min moves beyond the relatively blunt measures of abandonment of coalitions used in prior literature to examine within-war negotiations, a key indicator of potential dissociation from allies, allowing for a much richer story of coalition diplomacy during war.[5]

Second, while there has been some prior scholarship that examines cooperation among similar non-state actors, whether rebel groups or terrorist organisations, several studies in this volume push this agenda forward, particularly in their focus on what Reiter refers to as 'hybrid' coalitions.[6] Cappella Zielinski and Grauer provide the first systematic coding of non-state actors as coalition members in interstate wars, and demonstrate several important trends related to their involvement in war.[7] Reiter extends existing theories of coalition effectiveness to hybrid coalitions, highlighting places where established predictions are unlikely to hold.[8] Elias extends this logic further by examining coalitions between rebels and terrorist groups, developing an argument for why such alliances might be unexpectedly durable when the terrorist group is powerful and has interests that do not directly compete with those of the rebels.[9]

Third, and somewhat more amorphously, several of these articles identify important implications for how battlefield coalitions influence factors other than relations among coalition members or the outcome of war. Cappella Zelinski, Grauer, and Smith demonstrate important differences between comparatively more and less democratic regimes in the way and extent to which they contribute to coalition operations.[10] Boehlefeld and Grieco discuss the implications of multilateral military exercises for international stability.[11] Von Hlatky and Juneau discuss the ways in which coalition structure influences the mandate of an ongoing mission.[12] Moving forward, there likely will be important opportunities for thinking about battlefield coalitions affect variables such as treatment of civilians during conflict, the responses of the international community to a conflict, and relations among coalition members after a war is over, among others.[13]

11.2 Challenges for Future Research

The studies in this volume also help to illustrate a number of the challenges that future research will need to confront. One immediate challenge is that, as we move to studying hybrid and rebel coalitions, identifying coalition members will prove more challenging. Should a rebel group composed of forces loyal to several different warlords be treated as a single actor or as a coalition? In practice, so long as the warlords continue to cooperate, existing studies overwhelmingly treat such rebel groups as single actors, at least until such time as they fracture.[14] Yet this approach conceals important political and military dynamics.

To take one important, if relatively little-known, example, the 1924 Second Zhili-Fengtian War, which precipitated the collapse of the Zhili Faction in China's warlord period, is consistently treated as a bilateral conflict between the government forces (Zhili) and Fengtian rebels, with Fengtian emerging victorious. This outcome, which Waldron constitutes a crucial turning point in Chinese history, cannot be understood without recognising that Zhili was in fact a battlefield coalition of multiple major warlords, who were cooperating under the nominal command of Zhili leader Wu Peifu, but who had distinct political interests and maintained independent control of their armies. As Zhili forces neared victory, Feng Yuxian, nominally a general in the Zhili army but in practice commander of his own independent force, withdrew his army from the battlefield and captured Beijing, leaving the remaining Zhili forces in an indefensible position and precipitating the Fengtian victory and Zhili's broader political collapse.[15]

Alternately, researchers will need to address questions about when coalitions unify to become single actors. These challenges arise in interstate wars: should cooperation in the World Wars between the Commonwealth countries – who recognised a common head of state – be considered to be somehow different than that between Britain and France? Had the French accepted Winston Churchill's proposal of Anglo-French unification in June 1940 would their forces no longer have been fighting in coalition?[16] Did the German coalition in the Franco-Prussian War dissolve when the constituent German states joined the new German Empire?

These challenges will undoubtedly be more significant when studying hybrid and rebel coalitions. When the Dowager Empress Cixi formally recognised the independent Black Flag Army as representing China in the Sino-French War of 1884-5, did the resulting force constitute a hybrid coalition, or did it (as for example, the Correlates of War project has coded it) simply constitute the army of China?[17] Did the Taliban's path to power in the early 1990s and in 2021 – in both cases with rival warlords defecting to the Taliban side as its victory seemed ever more likely – constitute the creation of an overwhelming coalition or the consolidation of power by a single actor? Most discussions treat the Taliban as a single actor, yet the quick collapse of Taliban rule in 2001 – facilitated by the defection of many of the same warlords who had aligned with the Taliban in the early 1990s – suggests that it could quite reasonably be considered a continued coalition.[18]

From one perspective, these observations identify an important opportunity: there are cases of effective battlefield coalitions that are not generally acknowledged as such.[19] Recognising that many rebel groups in fact constitute battlefield coalitions implies both that there are opportunities for theories and findings from the study of battlefield coalitions to speak to a broader range of cases than we otherwise would have expected and that there are additional cases on which to test theories, which could help ameliorate some of the challenges discussed immediately below. That said, once we treat a rebel group in which relatively independent units could easily splinter (but have not yet done so) as a battlefield coalition, we must confront the challenges of precisely delineating how much independence is enough and of systematically collecting data on factions within rebel groups. This combination of challenges suggests that research pursuing this avenue will almost certainly be qualitative, at least in the near future.

Whatever the challenges associated with defining coalition membership, however, a bigger challenge for any effort to identify broad generalisations about battlefield coalitions will be the incommensurability of many wars involving battlefield coalitions. Existing studies have generally accepted the standard division of wars into interstate and civil conflicts, yet that division may both obscure places where findings generalise from across types of conflicts and, more problematically, implicitly treat cases that differ in fundamental ways as commensurable. As Reiter discusses, for example, there is good reason to expect that findings about the benefits of unity of command will not translate directly to guerrilla as compared to conventional wars.[20] The existence of conventionally-fought civil wars suggests opportunities to apply existing arguments in some civil wars (though Reiter's observations about the dangers of unified command for rebel groups and the greater variation in authority relations in wars involving rebel groups both provide reason for caution).[21] More problematically, interstate conflicts in which one side wages an insurgency may well not be an appropriate basis for testing arguments about battlefield coalitions, yet quantitative studies routinely include guerrilla wars such as the Vietnam War.[22] To take a specific example, what should we draw from Moller's findings about the significance of battlefield losses for the emergence of unitary command in wars without set-piece battles?[23]

The distinction between conventional and irregular warfare is only one of a number of salient differences that may well limit generalisation across cases. Within conventional war, Biddle draws a distinction between wars fought using the modern system and those fought with more basic attritional tactics; Reiter however notes reasons to think that the requirement for independent action in the modern system of warfare will attenuate the benefits of unitary command relative to wars fought using pre-modern system tactics.[24]

Even setting aside military strategy, we must confront significant variation on a range of additional variables. In the dataset developed by Cappella Zielinski and Grauer, the World Wars constitute 3% of all observations, but account for 30% of the battles.[25] These wars are unrepresentative along many dimensions, including the political stakes and human costs of the war, the broad geographic range over which they were fought, and even the presence of a coalition on both sides of the conflict (a phenomenon otherwise observed, excepting hybrid coalitions, in only two other conflicts, the Korean War and the War over Angola). Best practise thus will entail at a minimum checking, as Cappella Zielinksi and Grauer do, whether results are dependent on inclusion of data from the World Wars.

These challenges grow as we consider the obstacles to testing arguments about the effects of institutional or strategic choices in battlefield coalitions, where selection effects are likely to be quite pronounced. Consider, for example, the finding in Cappella Zielinski and Grauer that hybrid coalitions fared particularly poorly in battle. Taken at face value, that finding would suggest a conclusion (which, to be explicit, the authors do not draw) that allying with a non-state actor will decrease your military effectiveness. The reality, of course, is that hybrid coalitions emerge in situations in which military victory is relatively difficult, whether because of the strength of one's adversary, the nature of the war, or some other factor.

Similarly, diversity in prior circumstances raises questions about whether Min's interesting conclusions about the sources of openness to negotiation in coalition wars would apply to the other half of the Crimean War coalition.[26] As a weak state, the Ottoman Empire had every incentive to push for the greatest possible military victory over Russia, the state that most directly threatened it, while Sardinia, like many participants in American-led coalitions in Afghanistan and Iraq, was more interested in maintaining good relations with its great power allies than in the core issues at stake in the war; the odds that military victory would lead either to open separate negotiations with Russia were thus low.

What can be done to address the obstacles to valid inference in this setting? For hypotheses that make predictions at a more micro-level, such as for example the willingness of individual soldiers or small units to fight or flee, or local levels of friendly fire incidents, there may be cases in which the tools of the causal inference revolution might be applied.[27] When, as in most cases, those tools cannot be credibly applied, researchers will need to fall back on more traditional solutions. Critical cases that hold specific variables in unusual combinations will be useful for isolating effects of specific variables. The Libya case analysed by Van Hlatky and Juneau, for example, provides an unusual opportunity to isolate the effects of imbalanced capabilities within a coalition from the effects of political leadership by an unusually strong state, given the reluctance of the Obama administration to commit to the intervention. In other settings, a comparative case approach might take advantage of variation within a single war. Most commonly, this would be intertemporal variation (as seen for example in Moller), but other options may exist. In some settings, a promising option consistent with Mahoney's discussion would be to examine cases in which a common set of countries fight together across different fronts in a single war, allowing the analyst to control for a wide range of factors that are common across the war and the actors while focusing on the effects of variables that vary across fronts.

While these research strategies offer an opportunity to increase confidence in findings in a specific conflict or small set of conflicts, they obviously do not provide a solution to the generalisation problem discussed above. Given the typical infeasibility of conducting in-depth case studies across a wide range of cases, quantitative studies will provide the only reasonable alternative, useful for testing hypotheses about variables that can be coded across cases. Less obviously, but evident for example in the studies in this volume by Cappella Zielinski and Grauer, Reiter, and Min, quantitative research, when done well, provides scholars with a broad knowledge of cases that can be useful for recognising the limits to the generalisability of findings.

11.3 Opportunities and Gaps in Our Knowledge of Battlefield Coalitions

Confronting these challenges will be important given the range of opportunities for further improving our understanding of battlefield coalitions. This section focuses on two such opportunities: fleshing out theories of battlefield coalitions, and better integrating traditional military variables.

One major research opportunity involves improving our theories of battlefield coalitions. There is obviously an extensive history of theories of alliance behaviour, originating in balance of power theory and extending to debates over logics of alliance formation, management, reliability, and termination.[28] In practise, however, this literature has focused largely on peacetime alliances or on the effects of alliances on war onset, and has largely neglected cooperation during war. Cappella Zielinski and Grauer note this dearth of scholarship, and several studies in this volume (especially Reiter's) provide guidance about how productive directions for the field to go, but we remain far from a general theory of battlefield coalitions.

Part of what is interesting about battlefield coalitions is the particular mix of strategic incentives participants confront, which motivate a wide range of different strategic behaviours. An informal survey of the past several centuries of battlefield coalitions would identify a bewildering array of different approaches that coalition members have taken, including wholehearted commitment to a shared cause, political commitment undermined by ineffective military cooperation, cooperation mixed with positioning for anticipated conflict with allies in the future, free riding on the pursuit of a shared objective, withdrawal from the war effort, and even defection to the opposing side.

Participants must weigh the desire to achieve political goals, which can be shared with other coalition members or particular, against the potential downsides of continued fighting, including most obviously the costs of war but also risks of defeat, political embarrassment for the leader, or even positioning a country poorly for the postwar political environment. Standard political science variables such as relative capabilities, regime type, alliance institutions, political ideology, and the extent of external threat then likely will influence policymakers' decisions about how to approach the fundamental tradeoffs between commitment to a common goal that may be achieved (at cost and with risks) only through cooperation and pursuit of parochial interests, with all the political costs that pursuit might generate.

Many of these factors do, of course, also exist for peacetime alliances, but the tradeoffs differ markedly. To take just one example, shirking on contributions to NATO defence is not the same as abandoning coalition partners during war. As Min discusses, the French decision to end their commitment to the Crimean War forced the British to dramatically scale back their war aims, producing a substantially different postwar environment than would have existed had the Allies pushed forward after the victory at Sevastopol. Shortfalls in peacetime defence spending do of course have the potential to swing war outcomes, but their consequences are far more probabilistic, depending both on whether a war is fought and how any war proceeds. At the same time, the costs and risks of continued fighting are far greater than the those of peacetime arming.[29]

Indeed, the focus of the literature on peacetime alliances in many ways has things backwards, insofar as peacetime alliance decisions are ultimately informed by beliefs about what (potential) partners can and will do during war. A general theory of battlefield coalitions, informed by but going substantially beyond existing theories of peacetime alliances, thus has the potential to generate important implications for cooperation both during war and during peace.

As good political scientists, the contributors to this volume focus on many of the standard variables emphasised in political science theories: relative power (Min's discussion of balanced vs. unbalanced coalitions, or Boehlefeld and Grieco's argument about the implications of high-integration, high-status multilateral military exercises for the local distribution of power), interests (Elias's discussion of competitive and non-competitive goals), institutions (unitary command in Moller and Reiter, or alliance depth in Mahoney), and regime type (the relative merits of democratic vs. non-democratic coalitions in Cappella Zielinski and Grauer, or the impact of regime type on battlefield cooperation in Cappella Zielinski, Grauer, and Smith). Extending the battlefield coalition literature will require addressing difficult conceptual challenges on some of these variables, however. For example, Cappella Zielinski and Grauer treat all non-state coalition members as nondemocratic, a reasonable approach but one that stands in contrast to studies in the literature that find a relevant distinction between comparatively more democratic and less democratic rebels.[30] Similarly, whatever the weaknesses of the standard measure of the standard CINC score measure of state capabilities, the field lacks anything comparable that could be used to capture the capabilities of rebel groups, either relative to states or relative to one another.[31]

While it will be essential to more precisely specify how to apply traditional variables in the comparatively unfamiliar environments of hybrid and rebel coalitions, bigger gains may be available from incorporating traditional military variables that existing literature largely neglects. The previous section noted questions about whether findings would generalise from conventional to unconventional wars, or across coalitions using traditional or modern system tactics. That said, the range of potentially relevant variables extends far beyond these.

An obvious candidate here is geography. While some studies of coalition warfare have considered the implications of fighting being spread across multiple fronts, the focus has been on isolating the effects of other variables rather than considering the direct implications of geography for coalition behaviour.[32] In practise, however, those implications are likely quite significant. Consider, for example, the relative positions of Germany's Eastern European partners in World War II. Hungary and Romania had the misfortune of lying along the inevitable invasion path between Germany and the Soviet Union, and hence were forced to couple themselves to Germany, even at the cost of exceptionally high military losses in pursuit of political goals that they did not necessarily share, until such time as the Soviets largely occupied their territory. By contrast, Finland's comparatively peripheral position meant that the Finns could limit their commitment to German military efforts, and were ultimately able to defect from the German coalition without being fully occupied by the Soviets, allowing them to establish a neutral position during the Cold War. Geography similarly played an important role in Arab anti-Israeli coalitions, with Israel's immediate neighbours – Syria, Jordan, and Egypt – forced to confront a more direct threat from Israeli forces, while more distant coalition members such as Iraq and Saudi Arabia were able to adopt more provocative diplomatic stances while limiting immediate military involvement.

While these examples focus on the implications of geography for exposure to direct military threats, geography also matters in other ways. Consistent with Morrow's observation that states within alliances produce different kinds of goods, with stronger states typically providing military capacity (security) while weaker states facilitate the effective deployment of that capacity (autonomy), geography influences the effectiveness and relative political power of members of battlefield coalitions in important ways.[33] In the Persian Gulf War, Saudi Arabia exercised greater political influence than would be expected from its military capacity because of the importance of access to Saudi territory for the liberation of Kuwait. In World War I, Romania's entry into the war – which despite Romania's comparative military weakness introduced a new threat on Austria-Hungary's flank at the same time that German forces were tied up in the Battle of the Somme – caused Kaiser Wilhelm to despair that the war was 'definitely lost.'[34]

Technology is also underrepresented in existing studies of battlefield coalitions. With the increased importance of advanced technology for military success, the technological capabilities of that armies bring to the battlefield play an outsized role. While the coalition in the Kosovo War in principle involved all members of NATO, five members were unable to contribute any air power to the war, while three countries – the United Kingdom, France, and Italy – contributed roughly 60% of the non-American air power to the campaign. The inability of many NATO members to contribute militarily in turn limited their influence, as can be seen from their exclusion from the critical May 27 meeting of foreign ministers at which the effective decision was made to prepare seriously for a ground invasion.[35] These kinds of limitations are likely to only grow in future coalition wars involving the United States.

One possible response to this situation is, as American policymakers have advocated, for allies to specialise in particular aspects of warfighting, such as refuelling, transport, or particular types of military operations. Greater attention to topics such as logistics and the interoperability of forces will permit future studies to move beyond general observations about the political obstacles to specialisation to consider the practical effectiveness of such types of cooperation.[36] These kinds of questions will be particularly salient for hybrid coalitions, in which different types of participants will bring both dramatically different military and political capabilities to a conflict, with non-state participants typically limited in military capacity but willing to engage in types of operations – such as ground combat and local counter-insurgency – that state actors with lower political investment in a conflict may shy away from.

11.4 Conclusion

Battlefield coalitions are a constant feature of contemporary warfare, from the American preference to fighting with allies to Russian reliance on separatist and mercenary forces in Ukraine to the involvement of Eritrean forces in the Tigray war in Ethiopia. The inherent mix of incentives for cooperation to achieve a shared goal and competition to prioritise one's own interests and minimise one's costs introduces interesting strategic dynamics, which play out in varying and interesting ways across the many different types of coalition warfare. The chapters in this volume improve our understanding of the internal dynamics of coalitions and the ways in which coalition politics influences conflict onset, outcome, and termination, demonstrating at the same time the important complications that emerge across different types of battlefield coalitions. They also point the way to further progress, highlighting important opportunities for greater theoretical clarity about the political logic of battlefield coalitions, especially within rebel and hybrid coalitions, and helping us to identify salient factors, especially in the realm of hard military, geographical, and technological variable, that existing work has comparatively neglected.

Notes

1 Prominent previous studies of coalition institutions include Patricia Weitsman, *Waging War: Alliances, Coalitions, and Institutions of Interstate Violence* (Stanford, CA: Stanford University Press, 2013); Rosella Cappella Zielinski and Ryan Grauer, 'Organizing for Performance: Coalition Effectiveness on the Battlefield,' *European Journal of International Relations* 26, no. 4 (December 2020), 953–78, https://doi.org/10.1177/1354066120903369; Daniel S. Morey, 'Centralized Command and Coalition Victory,' *Conflict Management and Peace Science* 37, no. 6 (November 2020), 716–34, https://doi.org/10.1177/073889422 0934884.

2 Sarah Bjerg Moller, 'Learning from Losing: How Defeat Shapes Coalition Command Arrangements in Wartime,' Chapter 7, this volume.

3 Stéfanie von Hlatsky and Thomas Juneau, 'When the Coalition Determines the Mission: NATO's Detour in Libya,' Chapter 4, this volume.

4 Casey Mahoney, 'International Institutions in Wartime and Battlefield Coalitions: A Conceptual Framework,' Chapter 5, this volume.

5 Eric Min, 'Speaking with One Voice: Coalitions and Wartime Diplomacy,' Chapter 10, this volume.

6 Prominent earlier studies that examine cooperation among violent nonstate actors include Fotini Christia, *Alliance Formation in Civil Wars* (New York: Cambridge University Press, 2012); Michael C. Horowitz and Philip BK Potter, 'Allying to Kill: Terrorist Intergroup Cooperation and the Consequences for Lethality,' *Journal of Conflict Resolution* 58, no. 2 (March 2014), 199–225, https://doi.org/0.1177/0022002712468726; Emily Kalah Gade, et al., 'Networks of Cooperation: Rebel Alliances in Fragmented Civil Wars,' *Journal of Conflict Resolution* 63, no. 9 (October 2019), 2071–97, https://doi.org/10.1177/0022002 719826234; Christopher W. Blair, et al, 'Honor Among Thieves: Understanding Rhetorical and Material Cooperation Among Violent Nonstate Actors,' *International Organization* 76, no. 1 (Winter 2022), 164–203, https://doi.org/ 10.1017/S0020818321000114. These studies, however, focus primarily on cooperation among like actors, rather than on hybrid coalitions.

7 Rosella Cappella Zielinski and Ryan Grauer, 'A Century of Coalitions in Battle: Incidence, Composition, and Performance, 1900–2003,' Chapter 2, this volume.

8 Dan Reiter, 'Command and Military Effectiveness in Rebel and Hybrid Battlefield Coalitions,' Chapter 6, this volume.

9 Barbara Elias, 'Why Rebels Rely on Terrorists: The Persistence of the Taliban-al-Qaeda Battlefield Coalition in Afghanistan,' Chapter 9, this volume.

10 Rosella Cappella Zelinski et al., 'Regime Type, War Aims, and Coalition Member Effort in Combat,' Chapter 8, this volume.

11 Kathryn M.G. Boehlefeld and Kelly A. Grieco, 'Exercising Escalation: Do Multinational Military Exercises Provoke Interstate Security Crises?' Chapter 3, this volume.

12 Von Hlatsky and Juneau, 'When the Coalition Determines the Mission,' Chapter 4, this volume.

13 Some studies addressing these points of course already exist. See for example Scott Wolford, *The Politics of Military Coalitions* (New York: Cambridge University Press, 2015).

14 For a useful discussion of the complications of coding fragmentation, see Kristin M. Bakke et al., 'A Plague of Initials: Fragmentation, Cohesion, and Infighting in Civil Wars,' *Perspectives on Politics* 10, no. 2 (June 2012), 265–83, https://doi.org/10.1017/S1537592712000667.

15 The seminal study of this conflict is Arthur Waldron, *From War to Nationalism: China's Turning Point, 1924–1925* (New York: Cambridge University Press, 1996).

16 For the history of the British proposal, see for example Avi Schlaim, 'Prelude to Downfall: The British Offer of Union to France, June 1940,' *Journal of Contemporary History* 9, no. 3 (July 1974), 27–63.

17 Hugh McAleavy, *Black Flags in Vietnam: The Story of a Chinese Intervention* (New York: The Macmillan Company, 1968).

18 For a useful perspective on the complicated internal politics of the Taliban, see Abdulkader H. Sinno, *Organizations at War in Afghanistan and Beyond* (Ithaca, NY: Cornell University Press, 2008).

19 A small number of studies do examine rebel groups as coalitions, albeit typically without engaging with core concepts in the study of battlefield coalitions such as unity of command or military effectiveness. See for example Bakke et al., 'A Plague of Initials,' 265–83.

20 Reiter, 'Command and Military Effectiveness in Rebel and Hybrid Battlefield Coalitions,' Chapter 6, this volume.

21 For the often-neglected point that a significant number of civil wars are fought conventionally, see Stathis N. Kalyvas and Laia Balcells, 'International System and Technologies of Rebellion: How the End of the Cold War Shaped Internal Conflict,' *American Political Science Review* 104, no. 3 (August 2010), 415–29, https://doi.org/10.1017/S0003055410000286. The opportunity to shift between irregular and conventional tactics, central to the Maoist theory of revolution, has important strategic implications for cooperation in civil conflicts. See for example Jonah Schulhofer-Wohl, *Quagmire in Civil War* (New York: Cambridge University Press, 2020).

22 This concern is somewhat ameliorated by the fact that most irregular interstate wars had only two state participants, and hence do not appear on lists of conflicts involving battlefield coalitions (though hybrid coalitions did exist in many cases, as with cooperation between the French and Mexican conservatives in the Franco-Mexican War of the 1860s). The Kosovo War introduces a separate but related challenge, insofar as NATO forces waged an exclusively air war (though again as part of a hybrid coalition with the Kosovo Liberation Army).

23 Moller, 'Learning from Losing,' Chapter 7, this volume.

24 Stephen Biddle, *Military Power: Explaining Victory and Defeat in Modern Battle* (Princeton, NJ: Princeton University Press, 2004); Reiter, 'Command and

Military Effectiveness in Rebel and Hybrid Battlefield Coalitions,' Chapter 6, this volume.

25 Cappella Zielinski and Grauer, 'A Century of Coalitions in Battle,' Chapter 2, this volume.

26 Min, 'Speaking with One Voice,' Chapter 10, this volume.

27 For a useful discussion of the causal inference revolution, including in international relations, see Cyrus Samii, 'Causal Empiricism in Quantitative Research,' *Journal of Politics* 78, no. 3 (July 2016), 941–55, https://doi.org/10.1086/686690.

28 For balance of power theory, see Inis L. Claude, Jr., *Power and International Relations* (New York: Random House, 1962), ch. 2–3; Daniel H. Nexon, 'The Balance of Power in the Balance,' *World Politics* 61, no. 2 (April 2009), 330–59. For other classic studies of alliance politics, see Stephen M. Walt, *The Origins of Alliances* (Ithaca, NY: Cornell University Press, 1987); Patricia Weitsman, *Dangerous Alliances: Proponents of Peace, Weapons of* War (Stanford, CA: Stanford University Press, 2004); Glenn Snyder, *Alliance Politics* (Ithaca, NY: Cornell University Press, 2007); Brett Ashley Leeds, 'Alliance Reliability in Times of War: Explaining State Decisions to Violate Treaties,' *International Organization* 57, no. 4 (Fall 2003), 801–27, https://doi.org/10.1017/S00208183 03574057; Brett Ashley Leeds and Burcu Savun, 'Terminating Alliances: When Do States Abrogate Agreements?' *Journal of Politics* 69, no. 4 (November 2007), 1118–32, https://doi.org/10.1111/j.1468-2508.2007.00612.x.

29 This point is not to deny the observation that, because money spent on the military cannot be spent on other socially desirable goods, military spending is socially inefficient in a way that is analogous to the inefficiency puzzle at the heart of the bargaining model of war. Andrew J. Coe and Jane Vaynman, 'Why Arms Control Is So Rare,' *American Political Science Review* 114, no. 2 (May 2020), 342–55, https://doi.org/10.1017/S000305541900073X. In practice, the costs of war are far more socially salient than the costs of peacetime military spending, as is evident in the observation that, while the frequency that countries are involved in interstate wars has declined substantially over time, Panama and Costa Rica are the only two countries that are neither islands nor microstates to have eliminated their standing army.

30 See for example Jessica Stanton, *Violence and Restraint in Civil War: Civilian Targeting in the Shadow of International Law* (New York: Cambridge University Press, 2016), e.g. 34–35.

31 For a discussion of weaknesses of CINC scores, see Kelly Kadera and Gerald Sorokin, 'Measuring National Power,' *International Interactions* 30, no. 3 (July 2004), 211–30, https://doi.org/10.1080/03050620490492097. For a well-regarded attempt to capture rebel group capabilities that nonetheless has significant limitations, see David E. Cunningham et al., 'It Takes Two: A Dyadic Analysis of Civil War Duration and Outcome,' *Journal of Conflict Resolution* 53, no. 4 (August 2009), 570–97, https://doi.org/10.1177/0022002 709336458.

32 See for example Alex Weisiger, 'Exiting the Coalition: When Do States Abandon Coalition Partners during War?' *International Studies Quarterly* 60, no. 4 (December 2016), 753–65, https://doi.org/10.1093/isq/sqw029.

33 James D. Morrow, 'Alliances and Asymmetry: An Alternative to the Capability Aggregation Model of Alliances,' *American Journal of Political Science* 35, no. 4 (November 1991), 904–33, https://doi.org/10.2307/2111499.

34 Glenn E. Torrey, *The Romanian Battlefront in World War I* (Lawrence, KS: University of Kansas Press, 2011), 32. Ultimately, the transfer of German forces to the Eastern Front allowed the Central Powers to stabilise the situation and ultimately to defeat Russia and Romania.

35 John E. Peters et al., *European Contributions to Operation Allied Force: Implications for Transatlantic Cooperation* (RAND: Santa Monica, CA, 2001), 19; Benjamin S. Lambeth, *NATO's Air War for Kosovo: A Strategic and Operational Assessment* (RAND, Santa Monica, CA: 2001), 47–48.
36 Political scientists have largely neglected logistics. For a historical discussion of logistics in war, see Martin Van Creveld, *Supplying War: Logistics from Wallenstein to Patton* (New York: Cambridge University Press, 2004).

Bibliography

Bakke, Kristin M., Kathleen Gallagher Cunningham, and Lee J. M. Seymour. 'A Plague of Initials: Fragmentation, Cohesion, and Infighting in Civil Wars.' *Perspectives on Politics* 10, no. 2 (June 2012): 265–83. 10.1017/S1537592712000667

Biddle, Stephen. *Military Power: Explaining Victory and Defeat in Modern Battle.* Princeton, NJ: Princeton University Press, 2004. 10.1515/9781400837823

Blair, Christopher W., Erica Chenoweth, Michael C. Horowitz, Evan Perkoski, and Philip B.K. Potter. 'Honor Among Thieves: Understanding Rhetorical and Material Cooperation Among Violent Nonstate Actors.' *International Organization* 76, no. 1 (Winter 2022): 164–203. 10.1017/S0020818321000114

Cappella Zielinski, Rosella, and Ryan Grauer. 'Organizing for Performance: Coalition Effectiveness on the Battlefield.' *European Journal of International Relations* 26, no. 4 (December 2020): 953–78. 10.1177/1354066120903369

Christia, Fotini. *Alliance Formation in Civil Wars.* New York: Cambridge University Press, 2012. 10.1017/CBO9781139149426

Coe, Andrew J., and Jane Vaynman. 'Why Arms Control Is So Rare.' *American Political Science Review* 114, no. 2 (May 2020): 342–55. 10.1017/S000305541 900073X

Claude, Jr., Inis L. *Power and International Relations.* New York: Random House, 1962.

Cunningham, David E., Kristian Skrede Gleditsch, and Idean Salehyan. 'It Takes Two: A Dyadic Analysis of Civil War Duration and Outcome.' *Journal of Conflict Resolution* 53, no. 4 (August 2009): 570–97. 10.1177/0022002709336458

Gade, Emily Kalah, Michael Gabbar, Mohammed M. Hafez, and Zane Kelly. 'Networks of Cooperation: Rebel Alliances in Fragmented Civil Wars.' *Journal of Conflict Resolution* 63, no. 9 (October 2019): 2071–97. 10.1177/0022002719826234

Horowitz, Michael C., and Philip B.K. Potter. 'Allying to Kill: Terrorist Intergroup Cooperation and the Consequences for Lethality.' *Journal of Conflict Resolution* 58, no. 2 (March 2014): 199–225. 10.1177/0022002712468726

Kalyvas, Stathis N., and Laia Balcells. 'International System and Technologies of Rebellion: How the End of the Cold War Shaped Internal Conflict.' *American Political Science Review* 104, no. 3 (August 2010): 415–29. 10.1017/S000305541 0000286

Kadera, Kelly, and Gerald Sorokin. 'Measuring National Power.' *International Interactions* 30, no. 3 (July 2004): 211–30. 10.1080/03050620490492097

Lambeth, Benjamin S. *NATO's Air War for Kosovo: A Strategic and Operational Assessment*, 47–48. Santa Monica, CA: RAND, 2001.

Leeds, Brett Ashley. 'Alliance Reliability in Times of War: Explaining State Decisions to Violate Treaties.' *International Organization* 57, no. 4 (Fall 2003): 801–27. 10.1017/S0020818303574057

Leeds, Brett Ashley, and Burcu Savun. 'Terminating Alliances: Why Do States Abrogate Agreements?' *Journal of Politics* 69, no. 4 (November 2007): 1118–32. 10.1111/j.1468-2508.2007.00612.x

McAleavy, Hugh. *Black Flags in Vietnam: The Story of a Chinese Intervention.* New York: The Macmillan Company, 1968.

Morey, Daniel S. 'Centralized Command and Coalition Victory.' *Conflict Management and Peace Science* 37, no. 6 (November 2020): 716–34. 10.1177/0738894220934884

Morrow, James D. 'Alliances and Asymmetry: An Alternative to the Capability Aggregation Model of Alliances.' *American Journal of Political Science* 35, no. 4 (November 1991): 904–33. 10.2307/2111499

Nexon, Daniel H. 'The Balance of Power in the Balance.' *World Politics* 61, no. 2 (April 2009): 330–59. 10.1017/S0043887109000124

Peters, John E., Stuart Johnson, Nora Bensahel, Timothy Liston, and Traci Williams. *European Contributions to Operation Allied Force: Implications for Transatlantic Cooperation.* Santa Monica, CA: RAND, 2001.

Samii, Cyrus. 'Causal Empiricism in Quantitative Research.' *Journal of Politics* 78, no. 3 (July 2016): 941–55. 10.1086/686690

Schlaim, Avi. 'Prelude to Downfall: The British Offer of Union to France, June 1940.' *Journal of Contemporary History* 9, no. 3 (July 1974): 27–63.

Schulhofer-Wohl, Jonah. *Quagmire in Civil War.* New York: Cambridge University Press, 2020. 10.1017/9781108762465

Sinno, Abdulkader H. *Organizations at War in Afghanistan and Beyond.* Ithaca, NY: Cornell University Press, 2008.

Snyder, Glenn. *Alliance Politics.* Ithaca, NY: Cornell University Press, 2007.

Stanton, Jessica. *Violence and Restraint in Civil War: Civilian Targeting in the Shadow of International Law.* New York: Cambridge University Press, 2016. 10.1017/9781107706477

Torrey, Glenn E. *The Romanian Battlefront in World War I.* Lawrence, KS: University of Kansas Press, 2011.

Van Creveld, Martin. *Supplying War: Logistics from Wallenstein to Patton.* New York: Cambridge University Press, 2004.

Waldron, Arthur. *From War to Nationalism: China's Turning Point, 1924–1925.* New York: Cambridge University Press, 1996.

Walt, Stephen M. *The Origins of Alliances.* Ithaca, NY: Cornell University Press, 1987.

Weisiger, Alex. 'Exiting the Coalition: When Do States Abandon Coalition Partners during War?' *International Studies Quarterly* 60, no. 4 (December 2016): 753–65. 10.1093/isq/sqw029

Weitsman, Patricia. *Dangerous Alliances: Proponents of Peace, Weapons of War.* Stanford, CA: Stanford University Press, 2004.

Weitsman, Patricia. *Waging War: Alliances, Coalitions, and Institutions of Interstate Violence.* Stanford, CA: Stanford University Press, 2013.

Wolford, Scott. *The Politics of Military Coalitions.* New York: Cambridge University Press, 2015. 10.1017/CBO9781316179154

Index

Note: *Italicized* and **bold** page numbers refer to figures and tables. Page numbers followed by "n" refer to notes.

For Product Safety Concerns and Information please contact our EU
representative GPSR@taylorandfrancis.com
Taylor & Francis Verlag GmbH, Kaufingerstraße 24, 80331 München, Germany